P9-APQ-056

NUMERICAL METHODS FOR SCIENTISTS AND ENGINEERS

INTERNATIONAL SERIES IN PURE AND APPLIED MATHEMATICS

William Ted Martin and E. H. Spanier
CONSULTING EDITORS

AHLFORS · Complex Analysis
BELLMAN · Stability Theory of Differential Equations
BUCK · Advanced Calculus
BUSACKER AND SAATY · Finite Graphs and Networks
CHENEY · Introduction to Approximation Theory
CODDINGTON AND LEVINSON · Theory of Ordinary Differential Equations
DETTMAN · Mathematical Methods in Physics and Engineering
EPSTEIN · Partial Differential Equations
GOLOMB AND SHANKS · Elements of Ordinary Differential Equations
GRAVES · The Theory of Functions of Real Variables
GREENSPAN · Introduction to Partial Differential Equations
GRIFFIN · Elementary Theory of Numbers
HAMMING · Numerical Methods for Scientists and Engineers
HILDEBRAND · Introduction to Numerical Analysis
HOUSEHOLDER · Principles of Numerical Analysis
JAMMER · The Conceptual Development of Quantum Mechanics
LASS · Elements of Pure and Applied Mathematics
LASS · Vector and Tensor Analysis
LEPAGE · Complex Variables and the Laplace Transform for Engineers
NEHARI · Conformal Mapping
NEWELL · Vector Analysis
RALSTON · A First Course in Numerical Analysis
ROSSER · Logic for Mathematicians
RUDIN · Principles of Mathematical Analysis
SAATY AND BRAM · Nonlinear Mathematics
SIMMONS · Introduction to Topology and Modern Analysis
SNEDDON · Elements of Partial Differential Equations
SNEDDON · Fourier Transforms
STOLL · Linear Algebra and Matrix Theory
STRUBLE · Nonlinear Differential Equations
WEINSTOCK · Calculus of Variations
WEISS · Algebraic Number Theory
ZEMANIAN · Distribution Theory and Transform Analysis

R. W. HAMMING

Bell Telephone Laboratories

NUMERICAL METHODS FOR SCIENTISTS AND ENGINEERS

McGraw-Hill Book Company

New York San Francisco Toronto London

1962

NUMERICAL METHODS FOR SCIENTISTS AND ENGINEERS

 Library of Congress Catalog Card Number 61-18727

8 9 - MP - 9

25886

THE PURPOSE OF COMPUTING IS INSIGHT, NOT NUMBERS

Preface

This book is written for scientists and engineers who expect to use modern large-scale digital computers as a tool for research and development and as a text for advanced undergraduate-level courses in numerical analysis. It is the author's belief that they do not need a book of formulas or a collection of isolated results but rather an organized presentation of the main ideas in the computing field so that they can understand what relation the numbers computed by a machine bear to a particular problem. As the motto of the book states, it is insight, not numbers, that is being sought.

This book differs from the currently available books in one or more of the following ways:

1. There are many excellent books written from the point of view of a person using a desk calculator; this book assumes that a large-scale digital computer is available and will be used. The difference is not merely that larger problems can be handled but it involves a completely different point of view.
2. There are available a number of very good books which are collections of unrelated chapters (often written by different authors) and which fail to show any unity in the whole field. One of the main purposes of this book is to indicate how the many different specific results may be fitted together in a common framework of ideas and techniques. Thus the user can hope to understand the relation between some of the many different formulas which purport to do the same thing. He can also create many new formulas to fit his particular needs.
3. There are a number of good books written by and for mathematicians; this book tries to present the material in a form suitable for those who are more interested in the use of this new, powerful tool of computing than in the beauty of the derivations of the formulas or in further research.
4. The use of polynomial approximation has dominated the presentation of numerical methods in most books. In the polynomial approach, the error term is usually expressed in terms of some high-order derivative

which, practically, is seldom available even for crude estimates. This book includes the band-limited-function approach which is well known to electrical engineers but which has not been widely used in computing circles. In this approach, the error is expressed in terms of the function itself. The band-limited model is also much easier to relate to many physical problems than is the polynomial model and hence is a powerful tool in planning and understanding a computation. The band-limited-function approach also relates computing with the sampling theorem of information theory and a number of other topics.

5. There is a tendency in the literature to emphasize the solution of simultaneous linear algebraic equations, matrix inversion, matrix eigenvalues, and finding the zeros of polynomials. As a matter of practice, the user should expect the computing center to supply not only routines for finding the special functions such as sin, cos, log, and exp but also routines for the above processes, and it is not important how these are programmed provided that he can impose criteria relevant to his problem and can understand the results obtained (and also that the process does not use too much machine time). Indeed, it is doubtful if the user wants to solve a system of ill-conditioned equations; rather he wants to know why they are ill-conditioned in order to find out what is wrong with the particular formulation of his problem. As a result, this book discusses these topics just enough for the reader to understand what is involved and what he can expect and assumes that, if trouble occurs, he will go to an expert rather than try to rescue himself.

6. The book does not contain numerous illustrative examples, computed to many decimal places. The author originally planned to include many machine listings but gradually came to realize two things: First, a simple low-accuracy illustrative example which the reader can follow with his eye, or at most with a slide rule, is far more educational than a 10-decimal computation at which the reader glances and then turns the page, since it so obviously will require a lot of work on his part to follow the computation. Secondly, as a matter of actual fact, most planning and programming of problems these days are done in the absence of any worked examples. It is sometimes argued that this is all the more reason for including many elaborate worked examples, but the author regards the argument invalid. If people are going to develop a plan and do the coding in the absence of any numerical examples, then the sooner we admit this and teach them to get along without examples the better. It is, of course, helpful if a person has had a year or two of experience computing many problems with a desk calculator, but it is also a great waste of a scientist's limited time. The scientist needs almost all his time to keep up with his own field, and it is wrong to waste it on inefficient desk calculations.

In any book designed for a year's course, it is necessary to omit a great deal of useful material. As a result, there are a number of things that this book is not.

1. This book does not tell how to code for a computing machine but assumes that the reader has been exposed to a short course in some synthetic language such as FORTRAN or ALGOL.

2. Functions of several variables have been excluded to keep the book within a reasonable size. As a result, the important topic of partial differential equations has not been included. This is a serious omission, and we can only refer the reader to other books such as G. E. Forsythe and W. R. Wasow, "Finite Difference Methods for Partial Differential Equations" [9].

3. As noted before, simultaneous linear algebraic equations, matrix inversion, eigenvalues of matrices, and the solution of polynomials are treated only far enough to give the reader a feeling for what can be done, what he can expect, and what he should look for.

4. The treatment of algorithms in general is inadequate, but a reasonably adequate theory is only now beginning to appear in the literature. In the present state of chaos, the topic does not seem suitable for a first course.

5. This book was written in one year while the author was a visiting professor at Stanford University and engaged in teaching such a course. Since most of the material was not copied from other sources but came from scattered notes and from memory accumulated over some 15 years of practice, where the interest was more in obtaining results than in giving historical credit, it follows that what is new and novel is mixed up with what is old and well known.

The organization of this book is indicated by the following outline. Part I covers the discrete finite difference calculus which is the basis of most numerical work. Thus the techniques and methods of numerical analysis can be examined without becoming involved in questions of limits, errors of approximation, etc. Part II assumes that between the sample points the function can be approximated by a polynomial of degree n. This part covers what may be considered the classical part of numerical methods. Part III mainly examines the results of assuming that between the (in this case) equally spaced sample points the function is represented by a band-limited function. This part connects many parts of computing with other known fields of knowledge, such as the sampling theorem of information theory and the filter design, transfer functions, etc., of electrical engineering. Part III also examines briefly the topic of approximation by exponentials, and the treatment of singularities. Part IV begins with algorithms and passes on to heuristics and random processes and ends with a chapter on the art of computing.

The problem of giving proper credit is a vexing one in this case. It is obvious that the author did not invent all that is new in this book, although he is clearly responsible for the presentation and any errors. All that he can do is to indicate what seem to him to be some of the main sources beyond the standard texts that have been available for some years.

The author believes that most credit should go to those at the Bell Telephone Laboratories who have made it a place where it is natural to do research and creative work—even when at times it has been 3,000 miles away! Secondly, a great deal of credit goes to Prof. J. W. Tukey, of Princeton University and the Bell Telephone Laboratories, for the education that he gave the author through his patient teaching of the elements of good computing and for setting an example of excellence in his own work. Acknowledgment is also due to Miss R. A. Weiss for many years of patient computation, useful suggestions, and encouragement. More general acknowledgment is due to the author's colleagues at the Bell Telephone Laboratories for bringing to him a wide variety of interesting and stimulating problems and many times suggesting a proper attack for the numerical solution. Thanks are also due to Stanford University, especially Profs. G. E. Forsythe, J. Herriot, R. V. Oakford, W. Linvill, E. Parzen, and N. Abramson. Finally we give an alphabetical list of some of the people who have helped (excluding the classes of students to whom much is due): M. Epstein, Milton Levy, Roger Pinkham, R. Rausch, Henry C. Thacher, and J. T. Ung. Thanks are also due to Mrs. Margaret Guthrie and to Miss Judy Pinsen for typing and to Mrs. Betty Jo Prine for mimeographing early versions of the book.

R. W. Hamming

Contents

PART III. NONPOLYNOMIAL APPROXIMATION

PART I

The Discrete Finite Difference Calculus

CHAPTER 1

The Difference Calculus

1.1 INTRODUCTION AND NOTATION

Part I is devoted to the study of functions of a single variable which are defined at a set of discrete, equally spaced points; for example,

$$f(a), \quad f(a+h), \quad f(a+2h), \quad \ldots, \quad f(a+(n-1)h)$$

are n values of a function $f(x)$ at n equally spaced points $x = a, a + h, a + 2h, \ldots, a + (n-1)h$. Another example of a function defined at a set of equally spaced values is the sequence of partial sums of a series

$$S(r) = \sum_{k=1}^{r} a_k \qquad (r = 1, 2, \ldots, n)$$

Throughout Part I, except in Chap. 4 where infinite series are discussed, the number of function values is finite, and there are no difficult questions of limits or existence proofs, since it is generally obvious when the arithmetic or algebra required can or cannot be performed. By limiting Part I to such simple situations, much of the mechanics of numerical analysis can be discussed without considering the more difficult question of the relationships between expressions in analysis and the corresponding expressions in the finite difference calculus.

Since computing machines can do only a finite number of operations, it is clear that the limiting processes of analysis cannot be done directly by a machine. Much of the art of numerical analysis consists in obtaining estimates of the results of limiting processes while using only finite processes.

It is worth remembering that the differential equations which occur in practice are often derived from the physical situation via a finite difference process. If the reverse problem of passing from the continuous situation to the discrete is not valid, then there are some grounds for doubting the physical significance of the differential equations.

Part II assumes that the function is represented (or approximated) between the given points by a polynomial or by a rational function;

Part III, by a band-limited function,[1] by a sum of exponential functions, or by special functional forms determined by the nature of a singularity. In all these cases, however, we can compute only with the values of the functions at the given set of discrete points.

Occasionally it is advantageous to use a set of unequally spaced points, but even then the ideas and formal manipulations are often closely related to those of equally spaced points. Thus, the methods used in the discrete equally spaced finite difference calculus are basic to much of numerical analysis.

Part I has been arranged in a form to exploit the parallels between the difference calculus and the differential calculus. The reader is assumed to be familiar with the differential calculus, and any difficulties that he may have in Chaps. 1, 3, and 5 are likely to be due to lack of understanding of the corresponding material in the differential calculus.

Most modern large-scale computers work in the "floating-point mode." For a decimal machine this means that numbers are written in the form

$$\begin{aligned} \pi &= 3.1415927 \cdot 10^0 \\ 10\pi &= 3.1415927 \cdot 10^1 \\ \frac{\pi}{10} &= 3.1415927 \cdot 10^{-1} \end{aligned}$$

etc. This is the usual scientific notation,[2] and we shall always assume that the computation is in the floating-point mode unless it is explicitly stated otherwise.

This book also assumes that a large-scale computer is available and that a satisfactory system of programming, a satisfactory monitor system, and a reasonable library of subroutines are supplied by the computation center. The only hand computation considered is that necessary to check that a particular program is functioning properly and the arithmetic necessary to estimate the solution time on a computer.

[1] "A band-limited function" is a technical expression meaning, loosely, that only frequencies in a given band are present in the function; thus, a function having frequencies only in the band 0 to 10 cycles per second describes a particular class of band-limited functions.

[2] Many machines place the decimal point *before* the first digit:

$$\pi = 0.31415927 \cdot 10^1$$

and some put an excess 50 in the exponent when representing number; thus, in

$$\pi = 3.1415927 \cdot 10^{50}$$

the written exponent is larger by 50 than it should be. The excess 50 notation allows the use of positive and negative exponents without using a sign digit for the exponent.

In the first example of this section a change of variables from the x_k variable

$$x_k = a + kh$$

to a new variable

$$t_k = \frac{x_k - a}{h} = k \qquad (k = 0, 1, \ldots, n-1) \tag{1.1-1}$$

reduces the situation to a standard, convenient form. Thus, in Part I, unless otherwise indicated, the expression $f(x)$ will have values for $x = 0, 1, \ldots, n - 1$. It is sometimes convenient to write

$$f(x) = f_x$$

when there can be no confusion, and we often write

$$\begin{aligned} f(0) &= f_0 \\ f(1) &= f_1 \\ &\cdots\cdots \end{aligned}$$

Whether we change a given problem to unit spacing or transform the difference calculus expression to the spacing of the problem is often a matter of indifference as far as the amount of actual computation is concerned. It is similar to the situation in analytic geometry of transforming the circle

$$(x - a)^2 + (y - b)^2 = r^2$$

by the change of variables

$$\begin{aligned} x - a &= x' \\ y - b &= y' \end{aligned}$$

to

$$x'^2 + y'^2 = r^2$$

We can regard the transformation as an "alias" or an "alibi"—an "alias" if we assume that the coordinate system moves, giving rise to new names for the same points of the circle; an "alibi" if we assume that the circle moves to a new location while the coordinate system stays fixed.

EXERCISES

1.1-1. If x takes the values 11, 9, 7, . . . , -11, find the transformation that reduces this to the standard form 0, 1, . . . , 11.

Ans. $t = (-x + 11)/2$

1.1-2. Reduce to standard form $x = 3, 3.5, 4.0, 4.5, \ldots, 10$.

Ans. $t = 2(x - 3)$

1.2 THE DIFFERENCE OPERATOR

The basic operation Δ in the finite difference calculus is *differencing*, which is defined by

$$\Delta f(x) = f(x + h) - f(x) \tag{1.2-1}$$

This operation is familiar to the student of calculus where it is used in the process of defining the derivative. The close relationship between the derivative and the difference provides the basis for much of the use of the finite difference calculus as a means of approximating expressions in the differential calculus.

We can imagine the operator Δ as being separate from, and acting upon, the function $f(x)$, just as we often regard $df(x)/dx$ as being the operator d/dx operating on $f(x)$ and as we often regard $\int f(x)\,dx$ as being the integration operator $\int \cdots dx$ acting on the function $f(x)$.

The difference operator is *linear*, as are the differentiation and integration operators, since if a and b are constants

$$\Delta[af(x) + bg(x)] = a\,\Delta f(x) + b\,\Delta g(x) \tag{1.2-2}$$

The linear property of an operator is very important, and it makes the Δ operator easy to use in many situations.

As an example of using the difference operator, let

$$y = ax^2 + bx + c$$

Then, using (1.2-2),

$$\begin{aligned} \Delta y &= a\,\Delta x^2 + b\,\Delta x + c\,\Delta 1 \\ &= a[(x+h)^2 - x^2] + b[(x+h) - x] + c[1-1] \\ &= 2ahx + ah^2 + bh \end{aligned} \tag{1.2-3}$$

The difference operator operating on a product gives

$$\begin{aligned} \Delta[f(x)g(x)] &= f(x+h)g(x+h) - f(x)g(x) \\ &= f(x+h)g(x+h) - f(x+h)g(x) + f(x+h)g(x) \\ &\qquad - f(x)g(x) \\ &= f(x+h)\,\Delta g(x) + g(x)\,\Delta f(x) \end{aligned} \tag{1.2-4}$$

Alternatively,

$$\Delta[f(x)g(x)] = f(x)\,\Delta g(x) + g(x+h)\,\Delta f(x)$$

Note the appearance of *one* $x + h$ argument. Except that the argument of one of the functions is $x + h$, the formula corresponds to the differential calculus formula

$$\frac{d}{dx}[f(x)g(x)] = f(x)g'(x) + g(x)f'(x)$$

Similarly for a quotient,

$$\begin{aligned} \Delta\frac{f(x)}{g(x)} &= \frac{f(x+h)}{g(x+h)} - \frac{f(x)}{g(x)} \\ &= \frac{f(x+h)g(x) - g(x+h)f(x) + [f(x)g(x) - f(x)g(x)]}{g(x)g(x+h)} \\ &= \frac{g(x)\,\Delta f(x) - f(x)\,\Delta g(x)}{g(x)g(x+h)} \end{aligned} \tag{1.2-5}$$

which again resembles the corresponding formula from the differential calculus, namely,

$$\frac{d}{dx}\frac{f(x)}{g(x)} = \frac{g(x)f'(x) - f(x)g'(x)}{g^2(x)}$$

Calculus books usually give a long list of formulas for derivatives, and a similar list is useful in the difference calculus. We give the following short list:

$$\Delta \sin (ax + b) = 2 \sin \frac{ah}{2} \cos \left[a\left(x + \frac{h}{2}\right) + b\right] \quad (1.2\text{-}6)$$

$$\Delta \cos (ax + b) = -2 \sin \frac{ah}{2} \sin \left[a\left(x + \frac{h}{2}\right) + b\right] \quad (1.2\text{-}7)$$

$$\Delta \tan (ax + b) = \sin ah \sec (ax + b) \sec [a(x + h) + b] \quad (1.2\text{-}8)$$

$$\Delta a^x = a^x(a^h - 1) \quad (1.2\text{-}9)$$

$$\Delta 2^x = 2^x(2^h - 1) \quad (1.2\text{-}10)$$

$$\Delta \ln x = \ln\left(1 + \frac{h}{x}\right) \quad (1.2\text{-}11)$$

In some respects the role of the number e in the differential calculus is played by the number 2 in the finite difference calculus, since if

$$a^h = 2$$

then

$$\Delta a^x = a^x$$

In particular, if $a = 2$ and $h = 1$, then we get

$$\Delta 2^x = 2^x$$

EXERCISE

1.2-1. Verify Eqs. (1.2-6) through (1.2-11).

1.3 REPEATED DIFFERENCES

Since $\Delta f(x)$ is a function of x, we can apply the Δ operator again to obtain

$$\Delta[\Delta f(x)] = \Delta^2 f(x) \quad (1.3\text{-}1)$$

This notation corresponds to the notation for the second derivative in the ordinary calculus:

$$\frac{d}{dx}\frac{df(x)}{dx} = \frac{d^2f(x)}{(dx)^2} = \frac{d^2f(x)}{dx^2}$$

In general,

$$\Delta^r f(x) = \Delta[\Delta^{r-1} f(x)] \quad (1.3\text{-}2)$$

In the example

$$y(x) = ax^2 + bx + c$$

we had [Eq. (1.2-3)]

$$\Delta y(x) = 2ahx + ah^2 + bh$$

so that

$$\Delta^2 y(x) = 2ah^2$$

and, continuing in this manner, we get

$$\Delta^3 y(x) = 0$$

It is not an accident that $\Delta^3 y(x) = 0$, but rather it is a consequence of the important theorem that follows:

Fundamental Theorem of the Difference Calculus. *The nth difference of a polynomial of degree n*

$$y(x) = a_0 + a_1 x + a_2 x^2 + \cdots + a_n x^n \qquad (a_n \neq 0)$$

is a constant, $a_n n! h^n$, and the $(n+1)$st difference is equal to zero.

The proof is most easily understood if we first examine a lemma.

Lemma. *If $y(x)$ is a polynomial of degree n, then $\Delta y(x)$ is a polynomial of degree $n - 1$.* [That is, the degree of $\Delta y(x)$ is exactly one less than the degree of $y(x)$.]

Proof of the lemma. For the special function $y(x) = x^n$ we find, using the binomial theorem,[1]

$$\begin{aligned}\Delta y(x) = (x + h)^n - x^n &= \sum_{k=0}^{n} C(n,k) h^{n-k} x^k - x^n \\ &= nhx^{n-1} + \frac{n(n-1)h^2}{2} x^{n-2} + \cdots + h^n\end{aligned}$$

Thus, when the operator Δ is applied, the term x^n becomes a polynomial of degree $n - 1$ with the leading coefficient equal to nh. Using the linearity property [Eq. (1.2-2)], we see that the operator Δ decreases the degree of each term of a polynomial by 1. The term $nx^{n-1}ha_n$ cannot cancel out; hence the lemma is proved.

Proof of theorem. We use the lemma n times, which reduces the polynomial to a constant. An examination of the process shows that the coefficient of a_n is $n!h^n$. The next difference vanishes, and the theorem is proved.

This theorem is of great importance in the classical parts of numerical analysis.

EXERCISE

1.3-1. Using $h = 1$, compute the second and fourth differences of

$$y = x^4 - 4x^3 + 6x^2 - 4x + 1$$

Ans. $\Delta^2 y = 12x^2 + 2$; $\Delta^4 y = 24$

[1] We have adopted the older notation $C(n,k)$ for the binomial coefficients in place of the currently popular $\binom{n}{k}$ (or any of the various versions with super and/or subscripts) since the older notation is easier to print, can be produced by a computer, and looks better in the middle of a sentence.

1.4 THE DIFFERENCE TABLE

When using higher differences, it is useful to imagine the work laid out in the form of a difference table (see Table 1.4-1), although the differences are probably not stored in this array in the machine. (We assume $h = 1$.)

TABLE 1.4-1. THE DIFFERENCE TABLE

x	$y(x)$	$\Delta y(x)$	$\Delta^2 y(x)$	$\Delta^3 y(x)$	·
0	$y(0)$				
		$\Delta y(0)$			
1	$y(1)$		$\Delta^2 y(0)$		
		$\Delta y(1)$		$\Delta^3 y(0)$	
2	$y(2)$		$\Delta^2 y(1)$		·
		$\Delta y(2)$		$\Delta^3 y(1)$	
3	$y(3)$		$\Delta^2 y(2)$		·
		$\Delta y(3)$		$\Delta^3 y(2)$	
4	$y(4)$		$\Delta^2 y(3)$		·
		$\Delta y(4)$		·	
5	$y(5)$		·		·
		·		·	
·	·		·		·
		·		·	
·	·		·		
		·			
·	·				

As an example of a difference table, consider the values of the sine-integral function. (We have used a common convention and have written the differences as if the decimal point followed the last digits given for the functional values.) To check the arithmetic, we observe that in Table 1.4-2 the sum of the numbers in any column of differences, when added to the top number in the next column to the left, is equal to the bottom number in the same column on the left.

From the fundamental theorem just proved we know that, if $y(x)$ is a polynomial of degree n in x, then the nth differences will be constant. Reversing this remark, if we wish to tabulate a polynomial of degree n at a set of equally spaced values, then, in principle, we could compute the top number in each column and construct the entire table, using only additions.

As an illustration, let us take the quadratic

$$y(x) = 3x^2 - 6x + 9 \qquad y(0) = 9$$

and evaluate it from $x = 0$ to $x = 10$, using a spacing $h = 1$. We have

$$\Delta y(x) = 6x - 3 \qquad \Delta y(0) = -3$$
$$\Delta^2 y(x) = 6 \qquad \Delta^2 y(0) = 6$$

We now construct the table (Table 1.4-3, where the starting numbers are underlined). Alternatively, we could have computed the first three values and used them to find the differences.

This method requires two additions in this example to compute the *typical* value of the quadratic; in general, n additions are required to compute the typical value of an nth-order polynomial.

TABLE 1.4-2. DIFFERENCE TABLE OF $\text{Si}(x) = \int_0^x \frac{\sin t}{t}\,dt$

x	Si(x)	Δ	Δ^2	Δ^3
0.0	0.0000			
		999		
0.1	0.0999		−2	
		997		−6
0.2	0.1996		−8	
		989		−1
0.3	0.2985		−9	
		980		−5
0.4	0.3965		−14	
		966		−2
0.5	0.4931		−16	
		950		−3
0.6	0.5881		−19	
		931		−3
0.7	0.6812		−22	
		909		−3
0.8	0.7721		−25	
		884		−3
0.9	0.8605		−28	
		856		
1.0	0.9461			

Most scientific computing is now done in "floating-point" notation, but raw data that are automatically recorded are generally in "fixed-point" notation. In order to reduce the raw data, trend lines or even trend polynomials are often removed before the actual processing of the data. Under these circumstances, the above method for evaluating the trend polynomial is especially useful, although some attention must be given to the propagation of errors due to using approximate values of the differences in the computation.

Although Tables 1.4-2 and 1.4-3 started at zero, it is frequently useful to imagine the table extending indefinitely in both directions and, in particular, to imagine the current place of interest as being at zero.

The important Table 1.4-4 shows how a single error of one unit in the function $y(x) \equiv 0$ propagates in the difference table. The error is assumed to be at $x = 0$; thus $y(0) = 1$, and all other $y(x) = 0$.

TABLE 1.4-3. $y(x) = 3x^2 - 6x + 9$

x	y	Δy	$\Delta^2 y$
0	9		
		−3	
1	6		6
		3	
2	9		6
		9	
3	18		6
		15	
4	33		6
		21	
5	54		6
		27	
6	81		6
		33	
7	114		6
		39	
8	153		6
		45	
9	198		6
		51	
10	249		

Check: $y(10) = 300 - 60 + 9 = 249$

It is easy to recognize the numbers in the kth column of differences as $(-1)^r C(k,r)$. To prove this, we introduce the *shift operator* E, which is defined by the rule (we take $h = 1$ in this case)

$$E[f(x)] = f(x + 1) \tag{1.4-1}$$

and is clearly a linear operator since

$$E[af(x) + bg(x)] = aE[f(x)] + bE[g(x)]$$

Repeated use gives

$$E^n[f(x)] = \underbrace{EEE \cdots E}_{n \text{ times}}[f(x)] = f(x + n)$$

In particular, we define for $n = 0$

$$E^0[f(x)] = f(x)$$

and for negative exponents

$$E^{-n}[f(x)] = f(x - n)$$

We have already observed that it is sometimes convenient to imagine the operator as separated from the function upon which it operates, and when we carry this concept so far that we suppress the actual function *in*

TABLE 1.4-4. ERROR PROPAGATION TABLE

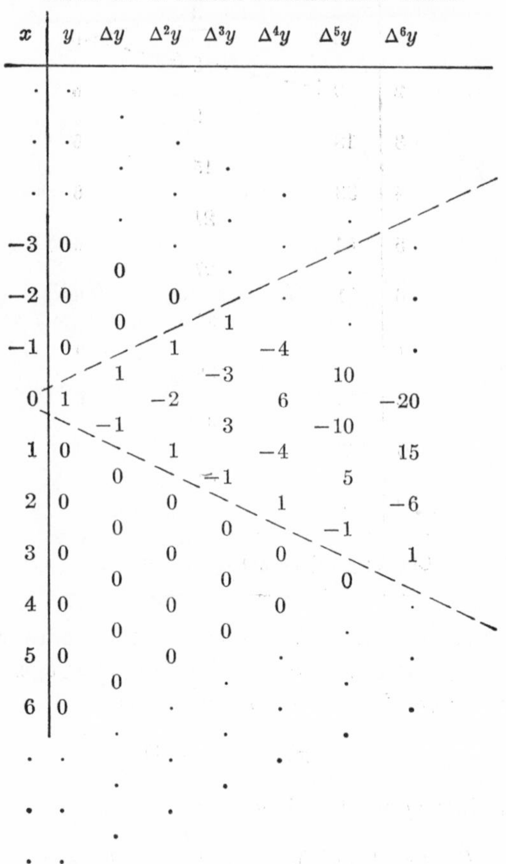

x	y	Δy	$\Delta^2 y$	$\Delta^3 y$	$\Delta^4 y$	$\Delta^5 y$	$\Delta^6 y$
.	.						
		.					
.	.		.				
		.		.			
.	.		.		.		
		.		.		.	
−3	0		.		.		.
		0		.		.	
−2	0		0		.		.
		0		1		.	
−1	0		1		−4		.
		1		−3		10	
0	1		−2		6		−20
		−1		3		−10	
1	0		1		−4		15
		0		−1		5	
2	0		0		1		−6
		0		0		−1	
3	0		0		0		1
		0		0		0	
4	0		0		0		.
		0		0		.	
5	0		0		.		.
		0		.		.	
6	0		.		.		.
		.		.		.	
.	.		.		.		
		.		.			
.	.		.				
		.					
.	.						

an identity we obtain an *operator equation.* An example of such an equation is

$$\Delta = E - 1 \tag{1.4-2}$$

where 1 is the identity operator, $1 \cdot f(x) = f(x)$. The truth of this operator equation follows from the simple observation

$$\Delta y(x) = y(x + 1) - y(x) = E[y(x)] - 1 \cdot y(x) = (E - 1)y(x)$$

In a similar manner, it follows that

$$\Delta^k = (E - 1)^k = \sum_{r=0}^{k} (-1)^{k-r} C(k,r) E^r \tag{1.4-3}$$

We now apply the expansion of Δ^k to Table 1.4-4, where $y(x) = 0$ except for $y(0) = 1$. Only one term remains when Δ^k is applied to $y(x)$, and this gives the corresponding binomial coefficient.

For the general function $y(x)$ the operator equation for Δ^k gives

$$\Delta^k y(n) = \sum_{r=0}^{k} (-1)^{k-r} C(k,r) y(n+r) \tag{1.4-4}$$

which is often called "the Lagrange form of the difference" (the reason for the name will appear later). With Eq. (1.4-4) we can find an isolated difference without computing the entire difference table.

We may reverse formula (1.4-4) when we know the value of the difference and one value of the function is missing. As a specific example, suppose that we know that the sixth differences are either exactly zero or sufficiently close to zero so that we may approximate them by zero, and we are missing one value of the function. Let this value be labeled y_0, and those adjacent be y_{-3}, y_{-2}, y_{-1}, y_1, y_2, y_3. Then we have, using (1.4-4) with $k = 6$ and $n = -3$,

$$y_{-3} - 6y_{-2} + 15y_{-1} - 20y_0 + 15y_1 - 6y_2 + y_3 = 0$$

or

$$y_0 = \tfrac{1}{20}[(y_{-3} + y_3) - 6(y_{-2} + y_2) + 15(y_{-1} + y_1)] \tag{1.4-5}$$

This formula is very useful when one value in a smooth table is missing because of some carelessness. However, a formula like this should be used with caution if the missing value is due to some peculiarity in the function, such as a division by zero.

Note that even-order differences are much easier to use for such purposes than odd-order ones, since then there is a single largest term.

EXERCISES

1.4-1. Compute $f(n) = n^2 - 79n + 1601$ for $n = 0$ to 10. [*Note:* $f(n)$ = prime number for $n = 1, 2, \ldots, 79$.]

1.4-2. Compute a table of cubes from 10 to 20, using a difference table.

1.4-3. Suppose that the value of $x = 0.6$ in Table 1.4-2 were missing; assuming that the fourth differences are zero, compute Si(0.6).

1.4-4. Estimate the value of Si(1.1), using (1.4-4).

1.5 FACTORIALS

In the calculus, x^n plays an important role. In the Taylor series, for example,

$$f(x) = \sum_{n=0}^{\infty} a_n x^n$$

an arbitrary function is expanded in powers of x. The reason that x^n plays this leading role is to a great extent that it satisfies the relation

$$\frac{d}{dx} x^n = nx^{n-1} \qquad (n \geq 1)$$

The natural analogy in the difference calculus requires a set of functions $g_n(x)$ which satisfy the relations

$$\Delta g_n(x) = ng_{n-1}(x) \qquad (n \geq 1) \tag{1.5-1}$$

The *factorial functions*,[1] defined by the equations

$$\begin{aligned} g_n(x) = x^{(n)} = x(x-1)(x-2) \cdots (x-n+1) \qquad (n \geq 1) \\ g_0(x) = x^{(0)} = 1 \end{aligned} \tag{1.5-2}$$

satisfy the relations (1.5-1) since

$$\begin{aligned} \Delta g_n(x) &= g_n(x+1) - g_n(x) \\ &= [(x+1) - (x-n+1)]x(x-1) \cdots (x-n+2) \\ &= nx(x-1) \cdots [x-(n-1)+1] \\ &= nx^{(n-1)} = ng_{n-1}(x) \end{aligned}$$

Note that $x^{(0)} \equiv 1$ leads to $0^{(0)} = 1$ and corresponds to the standard use of $0! = 1$ and that $x^{(n)}$ has exactly n factors.

We now assume that a given function $f(x)$ can be expanded in factorials corresponding to the Taylor expansion in powers of x. Thus

$$f(x) = b_0 + b_1 x + b_2 x^{(2)} + b_3 x^{(3)} + \cdots = \sum_{k=0}^{\infty} b_k x^{(k)}$$

We also assume for the moment that the difference operator is linear for infinite sums and that we may take the difference term by term. In the difference expansion when $x = 0$ we have

$$f(0) = b_0$$

Taking differences on both sides, we get

$$\Delta f(x) = \sum_{k=1}^{\infty} b_k k x^{(k-1)}$$

Again setting $x = 0$, we have

$$\Delta f(0) = b_1$$

Continuing in this way, we have

$$\Delta^n f(x) = \sum_{k=n}^{\infty} b_k k(k-1) \cdots (k-n+1) x^{(k-n)}$$

[1] Not to be confused with $n! = n(n-1)(n-2) \cdots 2 \cdot 1$ which is the value of $x^{(n)}$ when $x = n$.

and, setting $x = 0$,

$$\Delta^n f(0) = n!b_n$$

Hence we get the formal analog of the Taylor series

$$f(x) = \sum_{k=0}^{\infty} \frac{\Delta^k f(0)}{k!} x^{(k)} = \sum_{k=0}^{\infty} C(x,k)\Delta^k f(0) \tag{1.5-3}$$

where $C(x,k)$ is the binomial coefficient which can now be written as $x^{(k)}/k!$.

If we now restrict ourselves to polynomials, then the Taylor series terminates after a finite number of terms, and convergence questions do not arise. Similarly, the expansion (1.5-3) terminates for polynomials of degree n,

$$f(x) = \sum_{k=0}^{n} \frac{\Delta^k f(0)}{k!} x^{(k)} \tag{1.5-4}$$

since by the fundamental theorem (Sec. 1.3) $\Delta^{n+1}f(x) = 0$ and all higher differences also vanish. The expression (1.5-4) is known as *Newton's formula for interpolation.*

The definition of the factorials can be extended in much the same way as is used to extend the laws of exponents in algebra. Using the definition (1.5-2), we clearly have

$$x^{(n)} = x^{(m)}(x - m)^{(n-m)}$$

If we now formally set $n = 0$, we get

$$x^{(0)} = 1 = x^{(m)}(x - m)^{(-m)}$$

or

$$(x - m)^{(-m)} = \frac{1}{x^{(m)}}$$

Replacing $x - m$ by y, we have

$$y^{(-m)} = \frac{1}{(y + m)^{(m)}} = \frac{1}{(y + m)(y + m - 1) \cdots (y + 1)} \tag{1.5-5}$$

which serves to define factorials for negative integer exponents.

Note that, if $m \neq 0$, then

$$x^{(m)}x^{(-m)} \neq x^{(0)}$$

EXERCISES

1.5-1. Using Newton's formula, find a polynomial which takes on the following values [all $f(n)$ are prime numbers $n < 41$]:

n	0	1	2	3	4	5
$f(n)$	41	43	47	53	61	71

Ans. $P(n) = n^2 + n + 41$

1.5-2. Show that

$$3^{(-3)} = 1/120$$

$$1^{(-m)} = \frac{1}{(m+1)!}$$

1.6 SYNTHETIC DIVISION

Synthetic division is sometimes taught in elementary courses in algebra. It will be used frequently, and hence it is worth careful discussion. We shall show how to use synthetic division to go from the usual form of a polynomial in powers of x to the form in which the polynomial is expressed in terms of factorials.

When the polynomial

$$f(x) = x^4 + 6x^2 - 7x + 8$$

is divided by $x - 2$ until only a constant remains, we get the form

$$\begin{array}{rrrrrl}
 & x^3 & 2x^2 & 10x & 13 & = \text{quotient} \\
\text{Divisor} = x - 2 \,\big|\, x^4 & 0x^3 & 6x^2 & -7x & 8 & \\
x^4 & -2x^3 & & & & \\ \hline
 & 2x^3 & +6x^2 & & & \\
 & 2x^3 & -4x^2 & & & \\ \hline
 & & 10x^2 & -7x & & \\
 & & 10x^2 & -20x & & \\ \hline
 & & & 13x & +8 & \\
 & & & 13x & -26 & \\ \hline
 & & & & 34 & = \text{remainder}
\end{array}$$

that is,

$$\frac{f(x)}{x-2} = \frac{x^4 + 6x^2 - 7x + 8}{x - 2} = x^3 + 2x^2 + 10x + 13 + \frac{34}{x-2}$$

Polynomial ↓ Divisor ↓ Quotient ↓ Remainder ↓

$$\text{or} \quad f(x) = x^4 + 6x^2 - 7x + 8 = (x - 2)(x^3 + 2x^2 + 10x + 13) + 34$$

When $x = 2$, we get

$$f(2) = 34$$

which states that the remainder on division by $x - 2$ is exactly $f(2)$. Thus we have the following theorem.

The Remainder Theorem. *If a polynomial $f(x)$ is divided by $x - a$ we get*

$$f(x) = (x - a)Q(x) + R$$

where $Q(x)$ is the quotient and R is the remainder, and we have

$$f(a) = R$$

If $R = 0$, then a is a root of $f(x) = 0$.

This form can be simplified by noting that the powers of x need not be written (provided zero coefficients are supplied for the missing powers). Also, the quotient on the top line is the same as the bottom numbers in each column (except the last) and need not be written. Finally, we need not write the $x - a$ but can simply write a, and add instead of subtract in the body of the form. Thus the form becomes

$$\begin{array}{rl|rrrrl} \text{Divisor} = & 2 & 1 & 0 & 6 & -7 & 8 \\ & & & 2 & 4 & 20 & 26 \\ \hline \text{Quotient} = & & 1 & 2 & 10 & 13 & \lvert 34 = R = \text{remainder} \end{array} \qquad (1.6\text{-}1)$$

To compute

$$y(x) = x^4 + 6x^2 - 7x + 8$$

the synthetic-division algorithm takes the form

$$y(x) = \{[(x + 0)x + 6]x - 7\}x + 8 \qquad (1.6\text{-}2)$$

We note, as an aside, that for $n = 6$ this "factored form" is not the one using the least arithmetic *if* we have repeatedly to evaluate a sixth-degree polynomial[1] (for example, in a library routine). Instead, the pattern

$$\begin{aligned} P_1 &= x(x + a_1) \\ P_2 &= (P_1 + x + a_2)(P_1 + a_3) \\ P &= (P_2 + a_4)(P_1 + a_5) + a_6 \end{aligned}$$

uses one less operation. The coefficients a_i may be found in terms of the A_i in the polynomial

$$P(x) = x^6 + A_1x^5 + A_2x^4 + A_3x^3 + A_4x^2 + A_5x + A_6$$

The price of computing the coefficients a_i in terms of the coefficients A_i is paid only once.

Another aside concerning the use of synthetic division is the observation that if we have a polynomial $P(x)$ with real coefficients to evaluate at a complex number

$$x_1 = a + ib \qquad (i = \sqrt{-1})$$

we can form the quadratic

$$\begin{aligned} (x - x_1)(x - \bar{x}_1) &= (x - a - ib)(x - a + ib) \\ &= x^2 - 2ax + (a^2 + b^2) \end{aligned}$$

and divide the polynomial by this real quadratic factor to get

$$P(x) = [x^2 - 2ax + (a^2 + b^2)]Q(x) + r_1x + r_2$$

[A shortened form similar to (1.6-1) can be devised for quadratic factors.]

[1] An observation due to Motzkin. See J. Todd, Motivation for Working in Numerical Analysis, *Commun. Pure and Appl. Math.*, vol. 8, pp. 97–116, 1955.

Putting $x = x_1$, we get

$$P(x_1) = r_1x_1 + r_2$$

In such situations

$$\begin{aligned} P(x_1) + P(\bar{x}_1) &= r_1(x_1 + \bar{x}_1) + 2r_2 \\ &= 2(r_1a + r_2) \end{aligned}$$

rather than $P(x_1)$ is often the quantity that is to be computed.

We now turn to the problem of writing a given polynomial as a sum of terms in $x^{(n)}$. To do this we need to divide, in turn, the polynomial by x, the quotient by $x - 1$, the next quotient by $x - 2$, etc., for n steps. The division by x being trivial, we get for the particular polynomial

$$P(x) = x^4 + 6x^2 - 7x + 8$$

of the above example (using in turn divisors 0, 1, 2, and 3)

$$\begin{array}{r|rrrrr}
1 & 1 & 0 & 6 & -7 & |+8 \\
 & & 1 & 1 & 7 & \\
\hline
2 & 1 & 1 & 7 & |0 & \\
 & & 2 & 6 & & \\
\hline
3 & 1 & 3 & |13 & & \\
 & & 3 & & & \\
\hline
 & 1 & |6 & & &
\end{array}$$

or, corresponding to (1.6-2),

$$P(x) = (\{[1(x - 3) + 6](x - 2) + 13\}(x - 1) + 0)(x) + 8 \qquad (1.6\text{-}3)$$

$$P(x) = x^{(4)} + 6x^{(3)} + 13x^{(2)} + 0x^{(1)} + 8 \qquad (1.6\text{-}4)$$

Equation (1.6-3) provides a convenient form for numerical work; (1.6-4) exhibits the actual coefficients more clearly.

EXERCISES

1.6-1. Write x^3 as a sum of factorials.

Ans. $x^{(3)} + 3x^{(2)} + x^{(1)}$

1.6-2. If $P(x) = x^5 - 5x^4 + 3x^2 + 1$, compute $P(1 + i) + P(1 - i)$.

Ans. 34

*1.7 STIRLING NUMBERS OF THE FIRST KIND

Since $x^{(n)}$ is important in the difference calculus and x^n is important in the differential calculus, it is clear that relations connecting these two

* Starred sections will not need to be referred to elsewhere in the text, except slightly in Section 1.9.

expressions would be useful. To express $x^{(n)}$ in powers of x, we write

$$x^{(n)} = \sum_{k=0}^{n} S(n,k)x^k \tag{1.7-1}$$

and calculate the coefficients $S(n,k)$. These coefficients are called the *Stirling numbers of the first kind.* For $n = 1$

$$x^{(1)} = x = S(1,0) + S(1,1)x$$

whence $$S(1,0) = 0 \qquad S(1,1) = 1$$

For $n = 2$,

$$x(x-1) = x^2 - x = S(2,0) + S(2,1)x + S(2,2)x^2$$

whence $$S(2,0) = 0 \qquad S(2,1) = -1 \qquad S(2,2) = 1$$

Rather than continuing in this manner, we get the general recurrence relation by writing

$$x^{(n+1)} = (x-n)x^{(n)}$$

Hence, using (1.7-1) twice,

$$\begin{aligned}\sum_{k=1}^{n+1} S(n+1,k)x^k &= (x-n)\sum_{k=1}^{n} S(n,k)x^k \\ &= \sum_{k=1}^{n} [S(n,k-1) - nS(n,k)]x^k \\ &\qquad + S(n,n)x^{n+1} - nS(n,0)\end{aligned} \tag{1.7-2}$$

It is obvious that, for all n, $S(n,n) = 1$, while, for $n > 0$, $S(n,0) = 0$. Equating coefficients in Eq. (1.7-2), we obtain the recurrence relation, which enables us to compute the numbers one at a time,

$$S(n+1,k) = S(n,k-1) - nS(n,k) \qquad (k = 1, 2, \ldots, n) \tag{1.7-3}$$

There is no simple formula for the Stirling numbers, as there is for the binomial coefficients. Table 1.7-1 gives some of the numbers.

TABLE 1.7-1. STIRLING NUMBERS OF THE FIRST KIND $S(n,k)$

n \ k	1	2	3	4	5
1	1				
2	−1	1			
3	2	−3	1		
4	−6	11	−6	1	
5	24	−50	35	−10	1

EXERCISE

1.7-1. Extend Table 1.7-1 one line for $n = 6$.

*1.8 STIRLING NUMBERS OF THE SECOND KIND

Stirling numbers of the second kind express x^n in terms of factorials:

$$x^n = \sum_{k=1}^{n} s(n,k)x^{(k)} \tag{1.8-1}$$

As in the previous section, we first compute a few numbers before we find the general recurrence relation. For $n = 1$,

$$x = s(1,0) + s(1,1)x$$

whence $$s(1,0) = 0 \qquad s(1,1) = 1$$

For $n = 2$,

$$x^2 = s(2,0) + s(2,1)x + s(2,2)x(x-1)$$

whence $$s(2,0) = 0 \qquad s(2,1) = 1 \qquad s(2,2) = 1$$

The recurrence relation follows from

$$x^{n+1} = x \cdot x^n$$

$$\begin{aligned}\sum_{k=1}^{n+1} s(n+1, k)x^{(k)} &= x \sum_{k=1}^{n} s(n,k)x^{(k)} = \sum_{k=1}^{n} s(n,k)(x-k)x^{(k)} \\ &\qquad + \sum_{k=1}^{n} ks(n,k)x^{(k)} \\ &= \sum_{k=1}^{n} s(n,k)x^{(k+1)} + \sum_{k=1}^{n} ks(n,k)x^{(k)} \\ &= \sum_{k=1}^{n} [s(n, k-1) + ks(n,k)]x^{(k)} \\ &\qquad + s(n,n)x^{(n+1)} - s(n,0)\end{aligned}$$

It is clear that, for any n, $s(n,n) = 1$, whereas for $n > 0$, $s(n,0) = 0$. Equating coefficients of similar terms in factorials (since they are linearly independent), we obtain the recurrence relation

$$s(n+1, k) = s(n, k-1) + ks(n,k) \tag{1.8-2}$$

From these observations we can construct a table of Stirling numbers of the second kind (see Table 1.8-1).

TABLE 1.8-1. STIRLING NUMBERS OF THE SECOND KIND $s(n,k)$

$n \backslash k$	1	2	3	4	5
1	1				
2	1	1			
3	1	3	1		
4	1	7	6	1	
5	1	15	25	10	1

EXERCISE

1.8-1. Extend Table 1.8-1 one line for $n = 6$.

1.9 AN EXAMPLE

We now give an example of the use of Newton's formula and the Stirling numbers. The problem is to calculate

$$g(y) = \frac{d}{dy} \int_0^y \frac{f(x)}{\sqrt{y - x}} dx \qquad 0 \le y \le 1$$

when the data for $f(x)$ are given at $x = 0, 0.1, 0.2, \ldots, 1.0$.

We see immediately that at the upper limit, when $x = y$, the integrand becomes infinite. We also note that the result of the integration is to be differentiated with respect to y, but if we differentiate with respect to y under the integral sign, the resulting integral will diverge. Thus we have a problem requiring some care.

We begin with the observation that, if $f(x)$ were x^n, we would have to evaluate the simple expression

$$g(y) = \frac{d}{dy} \int_0^y \frac{x^n\, dx}{\sqrt{y - x}}$$

This form suggests setting

$$x = y \sin^2 \theta$$
$$dx = 2y \sin \theta \cos \theta \, d\theta$$

and we obtain, when $f(x) = x^n$,

$$g(y) = \frac{d}{dy} \frac{2y^{n+1}}{\sqrt{y}} \int_0^{\pi/2} \sin^{2n+1} \theta \, d\theta = \frac{(2n + 1)y^n}{\sqrt{y}} W_{2n+1}$$

where

$$W_{2n+1} = \int_0^{\pi/2} \sin^{2n+1} \theta \, d\theta = \frac{2 \cdot 4 \cdots (2n)}{1 \cdot 3 \cdot 5 \cdots (2n + 1)}$$
$$= \frac{2n}{2n + 1} W_{2n-1}, \qquad W_1 = 1$$

We next observe that if

$$f(x) = \sum_{n=0}^{10} a_n x^n$$

then

$$g(y) = \frac{1}{\sqrt{y}} \sum_{n=0}^{10} (2n + 1) W_{2n+1} a_n y^n$$

If we now assume that the true function $f(x)$ is approximated sufficiently accurately by a tenth-order polynomial through the given samples, then we have merely to find the a_n. One way to determine the a_n is to use Newton's formula (1.5-4) adjusted for spacing $x = 0.1$

$$f(x) = f(0) + 10x\,\Delta f(0) + \frac{10x(10x-1)}{2!}\Delta^2 f(0) + \cdots + (10x)^{(10)}\,\Delta^{10} f(0)$$

(we are using $h = 0.1$ in the differences) and use the Stirling numbers of the first kind [Eq. (1.7-1)] to obtain the coefficients a_n of the powers of x. As will be seen later, it may be dangerous to use tenth differences, but in the specific problem which inspired this example, where the data were crude and the expected answer not much better, it came out satisfactorily as judged by the following facts: (1) The graph of the approximating polynomial $f(x)$ looked reasonable; (2) the results were compatible with other parts of the physical theory being studied; and (3) the results stimulated further work.

1.10 ALTERNATIVE NOTATIONS

The choice [Eq. (1.2-1)]

$$\Delta f(x) = f(x+h) - f(x)$$

was arbitrary. We might just as well have chosen *backward* differences

$$\nabla f(x) = f(x) - f(x-h)$$

What would have happened if we had? In particular, we would have found that the *ascending factorials*

$$^{(n)}x = x(x+h)(x+2h)\cdots[x+(n-1)h]$$

would give, corresponding to (1.5-1),

$$\nabla^{(n)}x = n[^{(n-1)}x]$$

from which we could derive the backward Newton formula corresponding to (1.5-4). Similarly, new Stirling numbers of the first and second kinds, $\Sigma(n,k)$ and $\sigma(n,k)$, would arise. These are related to the old ones by

$$\Sigma(n,k) = (-1)^{n+k}S(n,k) = |S(n,k)|$$
$$\sigma(n,k) = (-1)^{n+k}s(n,k)$$

Thus a theory completely analogous to the one that we have developed can be based on the backward difference. The formulas that result from such a theory are useful to the expert but are likely to be a luxury to the occasional user.

Central differences and *mean differences* are still other notations that are sometimes used, but again they are merely a change in notation and a slight convenience in some situations; hence, they do not seem to be worth the trouble to the occasional user.

1.11 GENERAL REMARKS AND REFERENCES

We have given only a brief introduction to the finite difference calculus —whole books have been written on the subject. One of the best is by Charles Jordan [19].[1]

The finite difference calculus can be used to produce many interesting results such as

$$\sum_{k=0}^{n} S(n,k)s(k,m) = \delta(n,m)$$

where $\delta(n,m)$ is the Kronecker delta function which equals 0 if $m \neq n$ and equals 1 if $m = n$. This comes from the simple observation that, if we expand a factorial in powers of x and then reexpand the powers of x in factorials, then all the coefficients will vanish except that of the original factorial. Such results are seldom useful in practical computing but are interesting mathematically.

Many books make use of elaborate symbolic methods. We have used only the Δ and E operators in a few simple, finite cases and have carefully avoided throughout the book the use of dubious infinite processes with symbolic operators. The symbolic methods are often suggestive, but their use assumes a familiarity which the occasional user does not have. Further, the author believes that the formal manipulation often conceals the meaning of the formulas and hence obscures insight into the problem being investigated.

[1] Numbers in brackets refer to numbered items in the References at the end of the book.

CHAPTER 2

Roundoff Noise

2.1 INTRODUCTION

Even a limited amount of experience with hand calculation raises the problem of roundoff effects due to keeping only a finite number of decimal places in the computation. In hand calculations the problem of roundoff is not serious, for a number of reasons. First, the amount of work that can be done by hand is limited as compared with the amount that a modern high-speed computing machine often produces. Second, during hand calculation a human being can observe many of the effects and take steps where necessary to guard against fatal damage. Third, hand calculation is usually done in a mode which might be called "quasi-fixed, quasi-floating point, variable-length computation," where the length is adjusted to guard against severe roundoff effects, while machine computation is generally in floating point with a fixed length of number. Fourth, in hand calculations it is often practical to use guaranteed maximum bounds on the roundoff, whereas in machine computations such bounds are usually conservative, and to use them may make the computation so expensive that a statistical estimate is necessary. Thus, in large-scale computation it is frequently necessary to adopt the statisticians' attitude that it is better to take action with a known risk than it is to take no action at all.

As a result of these factors, the theories which have been produced in the past for estimating roundoff effects are generally inadequate for modern machine computations. At present there is no really satisfactory theory, and it is necessary to discuss briefly a number of fragments of theories. Actual practice has produced several widely used methods of employing the computer itself to give some indication of the error as well as theoretical estimates of what should happen. The latter are necessary since it is clearly important to have some idea of the expected accuracy *before* starting to write a machine program.

2.2 RANGE ARITHMETIC

Perhaps the simplest and most useful approach to the roundoff problem is *range arithmetic*. In this method each number is, in fact, represented by two numbers, the maximum and the minimum values that it might have. In a sense, each number is replaced by a range in which the correct answer must lie—hence the name. When two numbers are combined, the new range is computed in the appropriate fashion from the given ranges (using proper rounding). Thus at every stage there are safe bounds within which the true answer must lie.

Such a method can be instrumented either with hardware in the machine or by a program which simulates such hardware. In both cases it more than doubles the labor and requires twice as much storage space for the numbers as does the normal computation.

In moderate-sized problems the range-arithmetic approach is very useful and the ranges obtained are practical to use. On the other hand, the bounds produced by the range arithmetic in very large problems are often so far from each other that the answer is almost worthless.

While it is often true, nevertheless it is dangerous to assume that the true answer is near the middle of the range. Suppose that it were generally near the middle and that the last operation to be performed were multiplication. Then, by this assumption, both factors of this final multiplication should also tend to lie in the middle of their ranges. To be specific, suppose that both numbers had a range 0 to 1, with the true value near $\frac{1}{2}$. The final product would then have the same range, but the true value would be near $\frac{1}{4}$—a contradiction. Thus, the feeling that the true value must lie near the middle of the range is not always a reliable one; all that we obtain with certainty is the range, and we have no guarantee as to where in the range the true answer lies.

EXERCISE

2.2-1. Draw a flow diagram for computing the range in the case of a product of two numbers when you know the ranges of the two numbers.

2.3 DOUBLE PRECISION

Another favorite method is to run the problem in both single- and double-precision arithmetic and to draw the conclusion that the number of places which agree in the single- and double-precision answers is the number which are right. Having run several typical computations at single and double precision, one may use single precision for the rest of the work, under the assumption that the same number of places is accurate.

The risks of such a method are evident, but nevertheless it is widely used with fair success. In a sense, it resembles range arithmetic in that it similarly increases the storage requirements and the labor of computation. It also usually produces more accurate answers than does range arithmetic.

2.4 SIGNIFICANT-DIGIT ARITHMETIC

The main idea behind *significant-digit arithmetic* is that great loss of accuracy usually occurs when two numbers close to one another in size are subtracted, thus producing a number of leading zeros which are then removed by the normalization of the floating-point mode of operation. If this final shift is prevented and the leading zeros are kept, then the result tends to indicate the number of significant digits.

In actual practice there are a number of other facts to watch in designing a significant-digit-arithmetic system, but we shall not go into them here.

There are two variants of the method: One[1] keeps the leading zeros and produces numbers that contain only the significant digits, whereas the other[2] actually makes the shift but carries an index of significance much like the exponent in the floating-point form. Both methods are possible to simulate by suitable programs, and both methods are being built into the hardware of machines. Experience with them should throw a good deal of light on their value in scientific computation.

2.5 THE STATISTICAL APPROACH

The statistical approach is based on the tempting thought that roundoff is a random process and consequently we should try to construct a model of roundoff based on probability theory. This approach starts with the following paradox: If the computation is repeated, then exactly the same result will appear (assuming that the machine functions properly), and consequently the result is not a random variable.

The situation is something like that of tossing a penny. In one model we believe that, if we knew enough about the forces and the distribution of weight in the penny, then we could calculate whether it would come up heads or tails. We prefer, however, not to go to so much work and to assume a model in which heads or tails is a random variable. In the same way, while we believe that we can compute the number, we prefer to regard it as a random variable. The number obtained we regard as the true answer plus a random roundoff effect.

In order to apply formal statistics, we need to create, actually or conceptually, a population—technically *an ensemble*—of answers. While there are many ways of creating an ensemble, we shall temporarily adopt the following conceptual model. We imagine that the computation can

[1] R. L. Ashenhurst and N. C. Metropolis, Unnormalized Floating Point Arithmetic, *J. Assoc. Computing Machinery,* vol. 6, no. 3, pp. 415–428, July, 1959.

[2] L. H. Gray and J. C. Harrison, Normalized Floating-point Arithmetic with an Index of Significance, *Proc. Eastern Joint Computer Conf.,* 1959.

be done with infinite precision, and we choose to add *roundoff noise* after each arithmetic step—*noise* analogous to noise in information theory, where it is usually considered as being added to the desired signal.

Our problem is now to find various statistics of the ensemble or at least estimates of them. If our particular answer is typical of the ensemble, then we shall be able to draw valid conclusions; if not, then we cannot. However, with fair regularity our particular case will have properties differing from the majority of the ensemble. Indeed, because of our sample having the peculiar roundoff that exactly produces all zeros past some point, it is rather likely also to have some other peculiar properties.

The several theories of roundoff differ mainly in how they attempt to find the statistics of the ensemble that we have imagined. It is clear that we shall not be able actually to form the ensemble and shall have to estimate it from what little we can produce.

2.6 RANDOM ROUNDOFF

One obvious method of producing some samples from the ensemble is to construct a random roundoff, either by a simulation program or by actual hardware, and to repeat the computation several times. From the distribution of the answers obtained we hope to make valid estimates of the true distribution of the ensemble.

The main objection to this approach is that it costs a great deal of time and money to produce enough answers to form a reasonably reliable estimate of the sampling fluctuations and thus to be able to judge what confidence we may have in the answer. Even to estimate the average at all accurately requires quite a few cases.

2.7 VARIABLE PRECISION

A recent proposal has been to run the problem a number of times, each run with one less place in the accuracy of the computation than the last. From this sample of solutions the estimate of the accuracy of the most accurate solution is formed.

It would seem to be intuitively evident that the same number of runs using random roundoff as in Sec. 2.6 would give a more reliable estimate (since the numbers being computed are more closely related to the desired result) than the variable-precision method, but the question has not been adequately investigated either experimentally or theoretically.

2.8 ESTIMATING THE NOISE IN A TABLE

The method of this section is applicable in situations where a table of numbers has been produced *and* each number has a roundoff that is independent of the other roundoffs. This is often the situation in practical work, where many cases of a problem are run for an equal spacing in the independent variable. Thus we try to answer the following question: Given a table of answers to a computation, x versus $f(x)$, for a set of equally spaced x, what is the expected roundoff noise in the answers? Note that we in no way try to find any systematic errors, but rather from the results obtained, plus a little more calculation, we try to estimate the noise in the answers.

Table 1.4-4 shows how a single error propagates in a difference table. Using the linearity property of the Δ operator, we can imagine the difference table of the tabulated function as being the sum of the difference table of the true function and the difference table of the noise. The noise difference table can, in turn, be imagined as being a sum of elementary difference tables, the rth table being due to the noise ϵ_r in the rth place in the noise table. The total noise in the kth difference, using Eq. (1.4-4), is

$$\Delta^k \epsilon_n = \sum_{r=0}^{k} (-1)^{k-r} C(k,r) \epsilon_{n+r} \tag{2.8-1}$$

An extreme bound can easily be found from a bound on the individual noises in the table. If the ϵ_r are bounded by ϵ, that is,

$$|\epsilon_r| \leq \epsilon$$

then

$$|\Delta^k \epsilon_n| \leq \sum_{r=0}^{k} C(k,r)\epsilon = 2^k \epsilon \tag{2.8-2}$$

For $k = 10$, for example, this represents a thousandfold amplification of the noise.

Since the extreme bound is so pessimistic, we turn to the statistical approach. It is reasonable to suppose that negative and positive errors ϵ_n are equally likely, and therefore the average over the ensemble is zero. We wish to measure the variance of the distribution. Let us assume that the variance of the roundoff noise is σ^2, that is,

$$\text{Variance } (\epsilon_n) = \text{average } (\epsilon_n{}^2) - [\text{average } (\epsilon_n)]^2 = \text{average } (\epsilon_n{}^2) = \sigma^2$$

where the average is, of course, over the ensemble. The variance of the kth differences is, using (2.8-1) and the fact that the average is zero,

$$\text{Variance } (\Delta^k \epsilon_n) = \text{average} \left[\sum_{r=0}^{k} (-1)^{k-r} C(k,r) \epsilon_{n+r} \right]^2$$

When we multiply this out, we get

$$\text{Average} \left[\sum_{r,s} (-1)^r (-1)^s C(k,r) C(k,s) \epsilon_{n+r} \epsilon_{n+s} \right]$$
$$= \text{average} \left[\sum_{r} C^2(k,r) \epsilon_{n+r}^2 + 2 \sum_{r<s} (-1)^{r+s} C(k,r) C(k,s) \epsilon_{n+r} \epsilon_{n+s} \right]$$

Under our assumptions (1) that the roundoff noises ϵ_{n+r} and ϵ_{n+s} (for $r \neq s$) are independent (over the ensemble) and (2) that the average, or mean, is zero and using the known result (Exercise 2.8-2)

$$\sum_{r=0}^{k} C^2(k,r) = C(2k,k)$$

and

$$\text{Average } (\epsilon_{n+r}\epsilon_{n+s}) = 0$$

we have the following result:

$$\text{Average}\left[\sum_r C^2(k,r)^2\epsilon_r{}^2\right] = \sum_r C^2(k,r) \text{ average } (\epsilon_r{}^2)$$

$$= C(2k,k)\sigma^2 = \frac{(2k)!}{(k!)^2}\sigma^2 \qquad (2.8\text{-}3)$$

Table 2.8-1 gives an idea of the size of the *noise amplification factor* which is simply $\sqrt{C(2k,k)}$.

TABLE 2.8-1. ROUNDOFF AMPLIFICATION FACTORS

Order of difference	Variance $C(2k,k)$	Root-mean-square noise $\sqrt{C(2k,k)}$	Maximum noise 2^k
1	2	1.414	2
2	6	2.449	4
3	20	4.472	8
4	70	8.367	16
5	252	15.875	32
6	924	30.397	64

The theoretical result that we have just obtained requires that we average the quantities over the ensemble. In practice we cannot, of course, do this, and we resort to the usual trick in such situations: We replace the average over the ensemble by the average over the function (the standard trick used in ergodic theory).

It might be thought that the average over the kth differences in Table 1.4-1 would give a moderately good estimate of the average over the ensemble (assuming that the ergodic principle applies) but it is unfortunately a fact that the kth differences are highly correlated. Indeed, the coefficient of correlation for adjacent values is

$$\sum_r \frac{(-1)^r C(k,r)}{\sqrt{C(2k,k)}} \frac{(-1)^{r-1}C(k, r-1)}{\sqrt{C(2k,k)}} = -\sum_r \frac{C(k,r)C(k, r-1)}{C(2k,k)}$$

$$= -\frac{C(2k, k+1)}{C(2k,k)} = -\frac{k}{k+1} \qquad (2.8\text{-}4)$$

(see Exercise 2.8-3).

As $k \to \infty$ this correlation coefficient approaches -1 and shows that the entries in the kth column are random but not independent, as is so often stated; they have a strong tendency to alternate in sign, $+, -, +, -, \ldots$.

This strong negative correlation between successive kth differences enables us to locate isolated errors in a table. The method is best understood by an example. Consider Table 2.8-2 where, as usual, we carry only the last digits in the differences. The behavior of the fourth differences centered about 35° makes us suspect an error in the entry at 35°. We therefore assume that the fourth differences are constant and that there is an error at 35°. Since we know the pattern of the fourth difference [Eq. (1.4-3)]

$$1, \quad -4, \quad 6, \quad -4, \quad 1$$

we try to fit the following equations approximately:

$$\begin{aligned} 41 &= -4\epsilon + C \\ -59 &= 6\epsilon + C \\ 37 &= -4\epsilon + C \end{aligned}$$

Solving these, we get, approximately, $\epsilon = -10$, so that the correct entry at 35° is 6,421. Using 6,421 in place of 6,411, we have the difference table

$$\begin{array}{ccccc} & & & & -4 \\ & & & 8 & \\ & & -50 & & 1 \\ & -468 & & 9 & \\ 6{,}421 & & -41 & & 1 \\ & -509 & & 10 & \\ & & -31 & & -3 \\ & & & 7 & \\ & & & & 0 \end{array}$$

Having shown the strong correlation that exists between the successive entries in the kth difference column and given an illustration of how to take advantage of it, we return to our main problem of estimating the noise in a table. We assumed previously that we knew the variance of the noise in the table values was σ^2; now we wish to estimate the variance from the noise in the kth column. If the true table values were from a polynomial, then their differences would vanish at some order, and what would be left in the difference table would be due to the noise. Our troubles are that the function is not a polynomial and that we would not know what order of differences to take even if it were, unless we knew the degree of the polynomial. In practice, in favorable situations, the differences first tend to decrease in size and then to increase while at the same time showing the strong oscillation due to the negative correlation.

TABLE 2.8-2. TABLE OF $C_n(z)$

Deg.	$C_n(z)$	Δ	Δ^2	Δ^3	Δ^4	Theoretical fourth difference due to error ϵ
0	8,346					
		−44				
5	8,302		−87			
		−131		4		
10	8,171		−83		3	
		−214		7		
15	7,957		−76		−1	
		−290		6		
20	7,667		−70		6	
		−360		12		
25	7,307		−58		−14	$+\epsilon$
		−418		−2		
30	6,889		−60		41	-4ϵ
		−478		39		
35	6,411		−21		−59	$+6\epsilon$
		−499		−20		
40	5,912		−41		37	-4ϵ
		−540		17		
45	5,372		−24		−10	$+\epsilon$
		−564		7		
50	4,808		−17		−2	
		−581		5		
55	4,227		−12		0	
		−593		5		
60	3,634		−7		−2	
		−600		3		
65	3,034		−4		−2	
		−604		1		
70	2,430		−3		2	
		−607		3		
75	1,823		−0		−4	
		−607		−1		
80	1,216		−1		2	
		−608		1		
85	0,608		0			
		−608				
90	0					

In this situation it is customary to take the first column of differences just past the minimum as being due to noise, although, so far as the author knows, adequate research on this matter has not been done. We thus take the average of the squares of the differences in this column as an estimate of the squares of the differences averaged over the ensemble.

Dividing this average by $C(2k,k)$, we have an estimate of the square of the noise level in the original table.

As an example, consider a table of rounded values of $\Gamma(x)$, as in Table 2.8-3. The average $[\Delta^4\Gamma(x)]^2 = (9 + 1 + 4 + 4 + 4 + 4)/6 = 13/3$. We divide this by $C(8,4) = 70$:

$$\frac{13}{3 \cdot 70} = 0.062 = \sigma_1{}^2 = \text{estimate of noise in table}$$

TABLE 2.8-3. TABLE OF THE GAMMA FUNCTION $\Gamma(x)$

x	$\Gamma(x)$	Δ	Δ^2	Δ^3	Δ^4
1.0	1.000				
		−49			
1.1	0.951		+16		
		−33		−4	
1.2	0.918		+12		3
		−21		−1	
1.3	0.897		+11		−1
		−10		−2	
1.4	0.887		+9		+2
		−1		0	
1.5	0.886		+9		−2
		+8		−2	
1.6	0.894		+7		+2
		+15		0	
1.7	0.909		+7		+2
		+22		+2	
1.8	0.931		+9		
		+31			
1.9	0.962				

A theoretical estimate based on a flat distribution of random roundoff is

$$\sigma_2{}^2 = \int_{-\frac{1}{2}}^{\frac{1}{2}} x^2\,dx = \tfrac{1}{12} = 0.0833 \tag{2.8-5}$$

In view of the smallness of the sample, the estimated noise levels

$$\sigma_1 = 0.25$$
$$\sigma_2 = 0.29$$

provide a fair check on the theory. Exact calculation[1] of the roundoff error gives

$$\sigma_3 = 0.32$$

[1] My thanks to H. C. Thacher, Jr., for pointing this out to me.

EXERCISES

2.8-1. Expand both sides of $(1 + t)^{a+b} = (1 + t)^a(1 + t)^b$ and equate like powers of t to get

$$C(a + b, r) = \sum_{s=0}^{a} C(a,s)C(b, r - s)$$

2.8-2. In Exercise 2.8-1 set $a = b = r$ to get

$$C(2r,r) = \sum_{s=0}^{r} C^2(r,s)$$

2.8-3. In Exercise 2.8-1 let $a = b = n$, $r = n + 1$, to get

$$C(2n, n + 1) = \sum_{s=0}^{n} C(n,s)C(n, s - 1)$$

2.9 "LEAST-SIGNIFICANT-DIGIT" THEORY

We now examine the first of two theories whose aims are to show how the roundoff effects develop. They may be called "roundoff theories in the small." We are going to drop the statistical model of Sec. 2.5 and produce our ensemble by letting the numbers that enter into the arithmetic operations vary. In this way we can study how roundoff propagates in a typical small piece of computation.

When a computed number is rounded off, it seems reasonable to suppose that the error committed (actually the change due to the rounding) is uniformly distributed in the range $-\frac{1}{2}$ to $\frac{1}{2}$ times the least significant digit kept.[1] In other words, any one size of error is as likely as any other size in the range. In probability language, this becomes

$$\text{Probability } (x_1 \leq \epsilon \leq x_2) = x_2 - x_1 \qquad (x_1 \leq x_2 \text{ in } -\tfrac{1}{2}, \tfrac{1}{2})$$

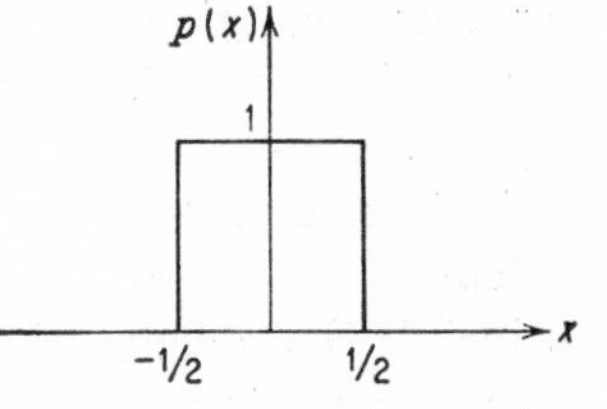

FIG. 2.9-1. Roundoff distribution for one number.

(see Fig. 2.9-1). In this model we are assuming a smooth distribution and ignoring the obvious fact that the actual distribution of machine roundoff must be granular since only certain roundoffs can occur (because only certain numbers can occur in the input).

We now examine what happens when several such numbers are added together, under the twin hypotheses that all the numbers have the same

[1] Some machines merely drop the extra digits with no attempt to round, and the distribution for such machines is taken as uniform from 0 to 1, but beware!

exponent in their floating-point form and that at no time does the sum (or difference) cause a change in the exponent. As a first step, consider the sum of two such numbers. The range of the sum of the roundoff errors is clearly -1 to 1 times the least significant digit kept. Let ϵ_1 be the error in the first number, ϵ_2 be the error in the second, and ϵ the error in the sum. Consider, now, the probability of finding $\epsilon = -a$ $(a \geq 0)$. The choice of ϵ_1 is restricted to the range $-\frac{1}{2}$ to $\frac{1}{2} - a$, and ϵ_2 is uniquely determined by the choice of ϵ_1. Thus the probability density is

$$P'(a) = p(a) = \int_{-\frac{1}{2}}^{\frac{1}{2}-a} d\epsilon_1 = 1 - a$$

A similar argument applies to errors greater than zero. Thus we get the triangular distribution shown in Fig. 2.9-2.

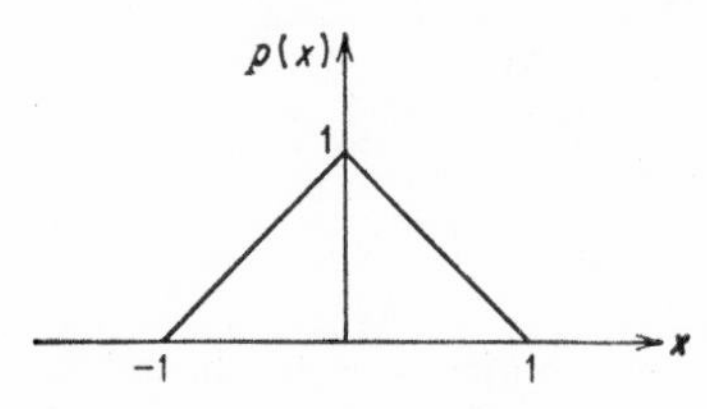

FIG. 2.9-2. Roundoff distribution for the sum of two numbers.

Rather than continue by considering the sum of three, and then four, and then five, etc., we turn to the *central-limit theorem* which is fairly widely known, at least in an intuitive way. This theorem states, for our case when the average value of the error is zero, that, as we increase the number of rounded numbers being added, the distribution function for the error of the sum approaches the normal distribution

$$\text{Probability } (x \leq \epsilon \leq x + dx) = \frac{1}{\sigma\sqrt{2\pi}} e^{-\frac{1}{2}(x/\sigma)^2}\, dx$$

for some suitable σ (which we shall later determine). The actual approach is so rapid, in fact, that frequently the sum of 10 or 12 random numbers chosen from a flat distribution is used as a good approximation to a single random number chosen from a normal distribution (see Sec. 32.4).

The range of the sum of n roundoff errors is $-n/2$ to $n/2$ while in principle the range of the normal distribution is $-\infty$ to $+\infty$, and therefore we cannot use the range to determine the value of σ. The average of both distributions is zero, and we are forced to consider choosing σ so that the variances of the two distributions agree. We have already computed [Eq. (2.8-5)]

$$\text{Variance } (\epsilon) = \int_{-\frac{1}{2}}^{\frac{1}{2}} \epsilon^2\, d\epsilon = \tfrac{1}{12}$$

For the sum of n independent roundoff errors $\epsilon = \epsilon_1 + \epsilon_2 + \cdots + \epsilon_n$ we have

$$\text{Variance } (\epsilon) = \text{variance}\left(\sum_i \epsilon_i\right) = \iint \cdots \int \left(\sum_i \epsilon_i\right)^2 d\epsilon_1\, d\epsilon_2 \cdots d\epsilon_n$$

Using both

$$\int_{-1/2}^{1/2} d\epsilon_i = 1 \qquad \text{and} \qquad \int_{-1/2}^{1/2} \epsilon_i \, d\epsilon_i = 0$$

we get

$$\text{Variance } (\epsilon) = \sum_{i,j} \iint \epsilon_i \epsilon_j \, d\epsilon_i \, d\epsilon_j$$

$$= \sum_i \int \epsilon_i^2 \, d\epsilon_i = \frac{n}{12} = \sigma^2$$

Thus, matching the variances of the two distributions, we have $\sigma^2 = n/12$. If n is moderately large, the normal distribution is a good approximation, and we have the symmetric interval in which there is a 50 per cent chance of the error lying as

$$-0.6745\sigma \text{ to } +0.6745\sigma$$

Thus we see that a 50 per cent range varies as $n^{1/2}$ while the full range varies as n. This is one of the reasons that the range-arithmetic approach, which naturally gives the full range, is so pessimistic as compared with statistical estimates.

This model should not be taken too seriously, in spite of the somewhat elaborate mathematical treatment, since the hypothesis that no shifts occur during the additions is highly unrealistic. Each time a shift does occur, there is a tendency to go back toward the uniformly flat distribution. Thus for addition and subtraction there are two opposing forces acting on the least significant digits; one pushes in the direction of the normal distribution while the other pushes toward the flat distribution.

EXERCISE

2.9-1. Combine two numbers having random roundoff errors from the triangular distribution of Fig. 2.9-2 to get the distribution of the sum of four random roundoffs.

2.10 "MOST-SIGNIFICANT-DIGIT" THEORY

When we consider the operations of multiplication and division, it is necessary to examine the distribution of the most significant digits in order to understand the propagation of roundoff errors. Let x_1 and x_2 be the two numbers with roundoff errors ϵ_1 and ϵ_2. When these numbers are multiplied, we get

$$(x_1 + \epsilon_1)(x_1 + \epsilon_2) = x_1x_2 + x_1\epsilon_2 + x_2\epsilon_1 + \epsilon_1\epsilon_2$$

If the noise due to roundoff is fairly high, the additional roundoff errors that occur are small in comparison with the roundoff noise in the products

$x_1\epsilon_2$ and $x_2\epsilon_1$. Thus the leading digits in x_1 and x_2 tend to control the roundoff propagation through multiplication and, by a similar argument, through division.

We are going to show that the leading digits are not uniformly distributed in frequency of occurrence. Since this is usually considered surprising, we first appeal to some experimental evidence. One hundred physical constants having a physical dimension were selected in a random fashion (no dimensionless ratios or pure numbers were accepted) from the "Handbook of Physics and Chemistry" [15]. The observed distribution of the leading digit was as shown in Table 2.10-1.

TABLE 2.10-1. DISTRIBUTION OF LEADING DIGITS OF RANDOMLY SELECTED PHYSICAL CONSTANTS

Leading digit	Recorded number	Theoretical number	Difference
1	34	30	+4
2	12	18	−6
3	13	12	+1
4	15	10	+5
5	7	8	−1
6	3	7	−4
7	4	6	−2
8	4	5	−1
9	8	4	+4
Total	100	100	0

The theoretical values in column 3 are from a log distribution, that is, in row n we have $100[\log (n + 1) - \log n]$. The small differences show that there is something in the theory which states that the leading digits are distributed logarithmically. A number of different arguments may be given on physical grounds to justify this result. First, consider the distribution of the leading digits of all physical constants. This distribution covers many decades in the size of the numbers. Now consider a change in the unit of length. This change will produce changes in the individual physical constants, but it is rather difficult to believe that the shape of the distribution of the leading digits of all the physical constants will change very much. If there is no change at all, then it can be shown that it must have been a log distribution.

Another intuitive argument for the distribution not being a flat one goes as follows: Consider a one-digit floating-point computing machine, and assume that the frequency is uniform. If we now examine all 81 sums or products that could be formed, we find a strong bias toward the numbers with the smaller leading digits.

This argument can be made more rigorous. Suppose that the initial density distribution of z is continuous, $1 \leq z \leq 10$, and of height $1/9$. The following results can be obtained.[1]

1. The product, in floating-point form, of two numbers selected independently from the above distribution has the density distribution

$$\frac{10 \ln 10 - 9 \ln z}{81} \tag{2.10-1}$$

2. The quotient similarly has the density distribution

$$\frac{1}{18}\left(1 + \frac{10}{z^2}\right) \tag{2.10-2}$$

3. A long sequence of independent multiplications and/or divisions of numbers from any (reasonable) initial distribution always produces a distribution of numbers whose density distribution function rapidly approaches the limiting form

$$\frac{1}{z \ln 10} \tag{2.10-3}$$

The curves for these three distributions are shown in Fig. 2.10-1 along with the original flat distribution from which the numbers are first selected.

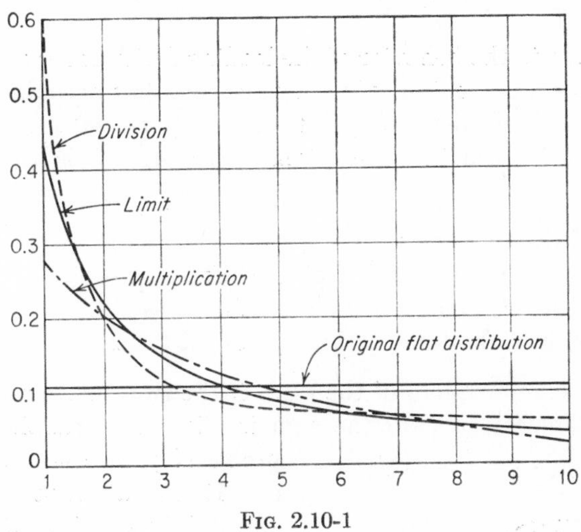

FIG. 2.10-1

Let us consider how we might prove the first statement. How can the number z occur in the product, neglecting the exponent? Let us choose

[1] Private communication from R. C. Prim, III.

the first factor in any way we please. The probability density is

$$\tfrac{1}{9}dx$$

Now if $x_1 \leq z$, then the second factor is uniquely determined as

$$\frac{z}{x_1}$$

while if $x_1 \geq z$ the second factor is

$$\frac{10z}{x_1}$$

and the probability requires another factor of $\tfrac{1}{9}$. Thus we have the total probability density

$$\frac{1}{9}\left(\int_1^z \frac{dx_1}{9x_1} + \int_z^{10} \frac{10dx_1}{9x_1}\right) = \frac{1}{81}(\ln z - \ln 1 + 10 \ln 10 - 10 \ln z)$$
$$= \frac{10 \ln 10 - 9 \ln z}{81}$$

As a check we integrate over all z:

$$\tfrac{1}{81}[(10 \ln 10)z - 9(z \ln z - z)]_1^{10}$$
$$= \tfrac{1}{81}(100 \ln 10 - 10 \ln 10 - 90 \ln 10 + 90 - 9)$$
$$= 1$$

as it should.

We now turn to the behavior of the leading digits under addition. We need to assume a model indicating how large a shift occurs when the two numbers have their decimal points lined up *before* the addition. A reasonable model that seems to match experience is to assume the following:

One-half of the time, no shift
One-quarter of the time, a one-place shift
One-eighth of the time, a two-place shift
One sixteenth of the time, a three-place shift etc.

In the first case, one-half of the time, if the two numbers come from a flat distribution $1 \leq x < 10$ the sum is triangular, running from zero probability at $z = 2$ to a peak at $z = 11$ and down to zero again at $z = 20$ (Fig. 2.10-2). Since the area must be 1, we have the peak of $\tfrac{1}{9}$. All the numbers from 10 to 20 have leading digit 1, and this amounts to an area of $\tfrac{1}{2} + \tfrac{19}{81} = \tfrac{119}{162}$.

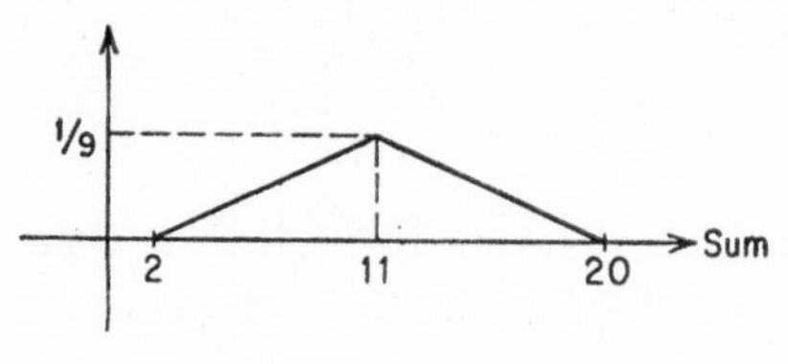

FIG. 2.10-2. Distribution for a sum.

Evidently, from the assumed flat distribution, we are rapidly building up an excess of 1 as the leading digit. Without wasting more time and space on a detailed analysis of the other cases, we can still see that the greater the shift the less the original distribution will be disturbed, but in all cases some excess of 1's will occur. Thus even for the process of addition we see that the flat distribution of leading digits is an unrealistic assumption.

Thus in floating-point arithmetic the propagation of errors through long chains of operations is not as severe as the flat-distribution model predicts, since the favorable distribution of the leading digits influences the propagation of error through the operations of multiplication and division.

EXERCISES

2.10-1. Prove statement 2 above.

2.10-2. Discuss the theory for the process of subtraction.

2.11 ON ANALYZING ERROR PROPAGATION IN A SHORT CALCULATION

The analysis of error propagation in a calculation is quite simple in elementary cases, but in the more elaborate developments it is quite complex. The simple theory is based on the use of the obvious approximation

$$\delta f = \frac{df}{dx}\,\delta x$$

and we shall make use of this occasionally.

The more elaborate developments will be treated in various places in this book, but a general treatment lies outside the scope of a first course. This should not be interpreted to mean that the topic is not important.

2.12 GENERAL REMARKS AND REFERENCES

The last three theories, if they may be dignified by the term, try to examine how roundoff grows in individual operations. Such theories, developed further than we have undertaken, are necessary if we are ever to be in a position to make reasonable estimates of the accuracy of a computation *before* the computation is done, using only the general magnitude and form of the problem as guides for making the estimate. In a sense, therefore, the various fragments of theories that we have given do not compete but rather complement each other.

It should be evident that any theory which purports to study roundoff in a statistical sense and which ignores the distribution of the leading digits is not likely to be very accurate. The nonuniform distribution of leading digits is not usually discussed and hence, though long known,[1] has not influenced numerical error analysis. Thus there are no good references to standard texts.

The beginner in the field of computing always has the feeling that if he is lucky he will be able to ignore roundoff effects, and thus he tends to ignore roundoff theories. And if he is lucky, he can; but as the problems become larger and more involved and more dependence is placed on the results obtained, he is gradually forced to think about the topic which, at present, is messy and poorly developed. Hopefully, better and more complete theories will be found in the near future. Certainly when one faces the problem of estimating the accuracy of a projected computation *before* money is spent on coding, debugging, etc., he is forced to think about the last two models of roundoff, and he will quite likely wish that they were more highly developed. Experience has shown that all too frequently the roundoff errors in a computation are correlated and that the assumption of independence is not correct; hence we have not pursued the above models to the limit.

[1] Frank Benford, The Law of Anomalous Numbers, *Proc. Am. Phil. Soc.*, vol. 78, no. 4, pp. 551–572, March, 1938; S. A. Goudsmit and W. H. Furry, Significant Figures of Numbers in Statistical Tables, *Nature*, vol. 154, pp. 800–801, 1944; W. H. Furry and H. Hurwitz, Distribution of Numbers and Distribution of Significant Figures, *Nature*, vol. 155, pp. 52–53, 1945; Roger S. Pinkham, On the Distribution of First Significant Digits, *Ann. Math. Statist.*, vol. 34, no. 4, pp. 1223–1230, December, 1961.

CHAPTER 3

The Summation Calculus

3.1 INTRODUCTION AND NOTATION

The difference calculus was introduced in Chap. 1 and was applied in Chap. 2 to the important problem of roundoff. In this chapter we return to the main theme of Part I and take up the summation calculus. The summation calculus is related to the difference calculus as the integral calculus is to the differential. As in the integral calculus, finding the inverse operation is essentially a guessing process.

The most convenient notation for the summation calculus is to define $\sum_{x=a}^{b} f(x)$ as the sum

$$f(a) + f(a + 1) + \cdots + f(b - 1)$$

which is the notation used by Boole [3], Jordan [19], and many others.[1] (Note that we are using $h = 1$ throughout this chapter.) This is not the usual notation in other parts of mathematics, and its adoption would lead to confusion. It seems best to face the troubles that arise from using the standard, inconvenient notation

$$\sum_{x=a}^{b} f(x) = f(a) + f(a + 1) + \cdots + f(b)$$

The summation methods will be based entirely on the forward difference operator

$$\Delta f(x) = f(x + 1) - f(x)$$

rather than on the backward difference operator ∇ or on the central difference operator. If we add the above equation from $x = a$ to $x = b - 1$, we get

$$\sum_{x=a}^{b-1} \Delta f(x) = f(b) - f(a) \tag{3.1-1}$$

[1] J. W. Tukey suggests $\sum_{x=a}^{<b}$ as a more suitable notation.

This corresponds to

$$\int_a^b \frac{df(x)}{dx}\, dx = f(b) - f(a)$$

in the integral calculus.

The basic result of the summation calculus is that, if two functions have the same first differences, then they differ at most by a constant (at the discrete set of points). This leads to the concept of an indefinite sum corresponding to the indefinite integral and to the concept of the additive constant in the indefinite summation table.

In the infinitesimal calculus the indefinite integration table is based on the corresponding table of derivatives; likewise the indefinite summation table is based on the corresponding difference table. From

$$\Delta x^{(n)} = nx^{(n-1)}$$

we get, using (3.1-1),

$$\sum_{x=a}^{b-1} x^{(n)} = \frac{b^{(n+1)} - a^{(n+1)}}{n+1} \qquad (n \neq -1) \tag{3.1-2}$$

As a simple example, we set $n = 0$:

$$\sum_{a}^{b-1} x^{(0)} = \sum_{a}^{b-1} 1 = \frac{b^{(1)} - a^{(1)}}{1} = b - a$$

Using the general formula (3.1-2) and the obvious linearity of the operator Σ, we can find sums of polynomials by simply converting the powers of x into factorials, using either the Stirling numbers of the second kind (Sec. 1.8) or repeated synthetic division (Sec. 1.6). By this method we can show that

$$\sum_{x=1}^{n} x = \sum_{1}^{n} x^{(1)} = \frac{(n+1)n}{2}$$

$$\sum_{x=1}^{n} x^2 = \sum_{1}^{n} [x^{(2)} + x^{(1)}] = \frac{(n+1)(n)(2n+1)}{6}$$

$$\sum_{x=1}^{n} x^3 = \sum_{1}^{n} [x^{(3)} + 3x^{(2)} + x^{(1)}] = \left[\frac{(n+1)n}{2}\right]^2$$

$$\sum_{x=1}^{n} x^4 = \frac{(n+1)n(2n+1)}{6}\,\frac{3n^2+3n-1}{5} \tag{3.1-3}$$

$$\sum_{x=1}^{n} x^5 = \left[\frac{(n+1)n}{2}\right]^2 \frac{2n^2+2n-1}{3}$$

$$\sum_{x=1}^{n} x^6 = \frac{(n+1)n(2n+1)}{6} \frac{3n^4+6n^3-3n+1}{7}$$

$$\sum_{x=1}^{n} x^7 = \left[\frac{(n+1)n}{2}\right]^2 \frac{3n^4+6n^3-n^2-4n+2}{6} \qquad (3.1\text{-}3\ \textit{cont.})$$

EXERCISES

3.1-1. Verify the formulas for x^3 and x^5, using the synthetic-division method.

3.1-2. Using $\Delta C(n,k) = C(n+1, k) - C(n,k) = C(n, k-1)$, show that

$$\sum_{x=k}^{m} C(x,k) = C(m+1, k+1)$$

3.1-3. Show that

$$\sum_{x=1}^{n} (2x-1) = n^2$$

$$\sum_{x=1}^{n} (2x-1)^2 = \frac{n(4n^2-1)}{3}$$

$$\sum_{1}^{n} (2x-1)^3 = n^2(2n^2-1)$$

3.2 SUMMATION FORMULAS

The summation formula for $x^{(n)}$ is also valid for negative exponents ($n \neq -1$); for example,

$$\sum_{x=1}^{m} \frac{1}{x(x+1)} = \sum_{x=1}^{m} (x-1)^{(-2)} = \frac{m^{(-1)} - 0^{(-1)}}{-1}$$

$$= -\frac{1}{m+1} + \frac{1}{1} = \frac{m}{m+1} \qquad (3.2\text{-}1)$$

Similarly,

$$\sum_{1}^{m} \frac{1}{x(x+1)(x+2)} = \sum_{1}^{m} (x-1)^{(-3)} = \frac{m^{(-2)} - 0^{(-2)}}{-2}$$

$$= \frac{1}{2}\left[\frac{1}{1 \cdot 2} - \frac{1}{(m+1)(m+2)}\right]$$

The difference formula

$$\Delta a^x = (a - 1)a^x$$

leads to the summation of a geometric progression

$$\sum_{0}^{m} a^x = \frac{a^{m+1} - 1}{a - 1}$$

The formulas for the differences of sin and cos

$$\Delta \sin (ax + b) = 2 \sin \frac{a}{2} \cos \left[a\left(x + \frac{1}{2}\right) + b \right]$$

$$\Delta \cos (ax + b) = -2 \sin \frac{a}{2} \sin \left[a\left(x + \frac{1}{2}\right) + b \right]$$

lead to the useful formulas

$$\sum_{x=0}^{m} \sin (ax + b) = -\frac{\cos [a(m + \frac{1}{2}) + b] - \cos [-a/2 + b]}{2 \sin (a/2)} \tag{3.2-2}$$

$$= \frac{\sin [a(m + 1)/2] \sin [(am/2) + b]}{\sin (a/2)}$$

$$\sum_{x=0}^{m} \cos (ax + b) = \frac{\sin [a(m + \frac{1}{2}) + b] - \sin [-(a/2) + b]}{2 \sin (a/2)} \tag{3.2-3}$$

$$= \frac{\sin [a(m + 1)/2] \cos [(am/2) + b]}{\sin (a/2)}$$

Many other formulas can be generated, but we shall not burden the reader with them.

EXERCISES

3.2-1. Compute $1 + 2 + 4 + 8 + 16 + \cdots + 2^{64}$.

3.2-2. Show that

$$\sum_{x=0}^{2N-1} \cos \frac{\pi x}{N} = 0$$

$$\sum_{x=0}^{2N-1} \sin \frac{\pi x}{N} = 0$$

3.2-3. Show that

$$\sum_{x=0}^{2N-1} \cos \frac{\pi k x}{N} \cos \frac{\pi m x}{N} = N\,\delta(m,k) \qquad (0 < m + k < 2N)$$

$$\sum_{x=0}^{2N-1} \cos \frac{\pi k x}{N} \sin \frac{\pi m x}{N} = 0$$

$$\sum_{x=0}^{2N-1} \sin \frac{\pi k x}{N} \sin \frac{\pi m x}{N} = N\,\delta(m,k) \qquad (0 < m + k < 2N)$$

3.2-4. Compute

$$\sum_{x=1}^{N} \frac{1}{x(x+1)(x+2)(x+3)}$$

3.2-5. Compute

$$\sum_{m=0}^{N-1} \sin^m \theta \qquad (\theta \neq 0, \pm\pi, \pm 2\pi, \ldots)$$

Ans. $(1 - \sin^N \theta)/(1 - \sin \theta)$

3.3 SUMMATION BY PARTS

A bit of reflection reveals that, besides a table of integrals, two methods are used to perform integrations:

1. Change of variable
2. Integration by parts

Of these two methods the first is not available in the summation calculus, which depends on equally spaced arguments, usually $h = 1$. This makes computing a sum in an analytic, closed form more difficult, in general, than computing an integral.

On the other hand, there is an analog to integration by parts in the summation calculus, and it is a very powerful tool. In the calculus, integration by parts is based on the formula for the derivative of a product

$$d(uv) = u\,dv + v\,du$$

From this equation the formula for integration by parts

$$\int u\,dv = uv - \int v\,du$$

is found by integration. Similarly, from the formula for the difference of a product,

$$\Delta(uv) = u\,\Delta v + v(x+1)\,\Delta u$$

we obtain by summation

$$\sum_{x=0}^{m-1} u\,\Delta v = u(m)v(m) - u(0)v(0) - \sum_{x=0}^{m-1} v(x+1)\,\Delta u(x)$$

We can choose v so that $v(0) = 0$ (or any other prescribed value).

As an example of using summation by parts, consider

$$\sum_{x=0}^{m-1} xa^x = \frac{xa^x}{a-1}\Bigg|_0^m - \sum_0^{m-1} \frac{a^{x+1}}{a-1}\cdot 1$$

$$= \frac{xa^x}{a-1} - \frac{a^{x+1}}{(a-1)^2}\Bigg|_0^m = \frac{a}{(1-a)^2}[(m-1)a^m - ma^{m-1} + 1]$$

In general, the use of summation by parts closely resembles the use of integration by parts. For example, summations such as

$$\sum \frac{1}{2^x} \sin \theta x$$

are accomplished by using summation by parts twice and then combining like terms. There are no new ideas in the process, only tedious algebra.

EXERCISES

3.3-1. Show that

$$\sum_{x=1}^{n} \frac{1}{2^x} \sin \theta x = \frac{2 \sin \theta - 2^{-n}[2 \sin \theta(n+1) - \sin n\theta]}{1 + 8 \sin^2 (\theta/2)}$$

3.3-2. Compute

$$\sum_{x=1}^{n} x^2 \cos x$$

3.4 GENERAL REMARKS

Comparatively few finite series can be summed in a nice, closed form. On the other hand, special series, such as those which involve binomial coefficients, can be summed surprisingly often. The reader is advised to try his hand at summing a finite series before resorting to machine computation; success often illuminates the original problem. Special methods for summing series are too numerous to be discussed here.

Perhaps the best reference for the summation of series is L. B. W. Jolley [18]. See also Ref. 1.

CHAPTER 4

Evaluation of Infinite Series

4.1 INTRODUCTION

Most books on infinite series discuss convergence, divergence, and "summability" of series at great length but almost completely neglect the actual computation (summation) of series. One reason for this is that there are very few methods for summing a series in closed form. Of course, if the indefinite summation can be done in closed form and if the series converges, then the infinite series can also be summed by letting the upper limit go to infinity. As an example, we had [Eq. (3.2-1)]

$$\sum_{x=1}^{n} \frac{1}{x(x+1)} = 1 - \frac{1}{n+1}$$

Hence

$$\sum_{x=1}^{\infty} \frac{1}{x(x+1)} = 1$$

In general,

$$\sum_{x=1}^{\infty} \frac{1}{x(x+1)\cdots(x+k-1)} = \frac{1}{k-1}\frac{1}{(k-1)!} \qquad (k \geq 2) \qquad (4.1\text{-}1)$$

Frequently a problem in analysis can be reduced to that of evaluating an infinite series. Many times it is more work to evaluate the series than to do the original problem, but sometimes the series representation is an advantage. One of the reasons that the series is preferable is that the error committed, when only a finite number of terms are used, is more easily controlled than the errors made in the direct approach. Thus in the problem of evaluating

$$\int_0^x e^{-t^2}\,dt$$

for values of x, say less than 1, we can expand the exponential in an infinite series and integrate term by term to obtain

$$\int_0^x e^{-t^2}\,dt = \int_0^x \sum_{k=0}^{\infty} \frac{(-1)^k t^{2k}}{k!}\,dt = \sum_{k=0}^{\infty} \frac{(-1)^k x^{2k+1}}{(2k+1)k!}$$

If we are interested in values of x less than 1 and want eight figures correct, then taking 11 terms will suffice. This follows from the facts that the series is alternating in sign, the terms are monotone, the series is convergent, and the first neglected term has a denominator about 9.2×10^8. Were we to try to estimate the value of the integral by some approximate method of integration taken up later in the book, then the problem of estimating the error would be more difficult.

EXERCISES

4.1-1. Prove (4.1-1).
4.1-2. Write the series for

$$Si(x) = \int_0^x \frac{\sin t}{t}\,dt$$

How many terms do you need in order to compute $Si(2\pi)$ to eight decimal places?

4.2 KUMMER'S COMPARISON METHOD

If a given series converges rapidly, there is no serious problem in planning the computation of its sum. If the series converges slowly, then we look for a series with a known sum which converges at about the same rate as the given series. The words "about the same rate" really mean that the difference approaches zero rather rapidly. When we find a suitable series, we are reduced to computing the sum of the difference of the two series, which by definition converges more rapidly. This is the idea behind Kummer's method.

To fix notation, suppose that the given series is

$$S = \sum_{x=r}^{\infty} a_x$$

and suppose that we know the sum

$$S' = \sum_{x=r}^{\infty} C_x a_x$$

where $C_x \to C$ as $x \to \infty$. Then

$$S = \frac{S'}{C} + \sum_{x=r}^{\infty} \left(1 - \frac{C_x}{C}\right) a_x$$

As an example, consider the series

$$S = \sum_{x=1}^{\infty} \frac{x}{(x^2 + 1)^2}$$

which converges as $1/x^3$ and suggests taking

$$S' = \sum_{x=2}^{\infty} \frac{1}{(x-1)(x)(x+1)} \qquad (C = 1)$$

as a comparison series, provided that we take care of the term $x = 1$ separately.

$$\begin{aligned} S &= S' + (S - S') \\ &= S' + \frac{1}{4} + \sum_{2}^{\infty} \left[\frac{x}{(x^2+1)^2} - \frac{1}{(x-1)(x)(x+1)} \right] \\ &= \frac{1}{4} + \frac{1}{4} + \sum_{2}^{\infty} \frac{x^4 - x^2 - (x^4 + 2x^2 + 1)}{x(x^2-1)(x^2+1)^2} \\ &= \frac{1}{2} - \sum_{2}^{\infty} \frac{3x^2 + 1}{x(x^2-1)(x^2+1)^2} = 0.39711677 \cdots \end{aligned}$$

The new series converges as $1/x^5$ [and could be treated in a similar fashion, if desired, by using (4.1-1) with $k = 5$, $C = 3$ for the $3x^2$ part of the numerator].

EXERCISES

4.2-1. Use $\sum_{1}^{\infty} [1/x(x+1)]$ to approximate $\sum_{1}^{\infty} (1/x^2)$.

4.2-2. Approximate the result of Exercise 4.2-1 by

$$\sum_{1}^{\infty} \frac{1}{x(x+1)(x+2)}$$

Continue for k steps.

4.3 SOME SPECIAL SUMS

Kummer's method requires a supply of comparison series and their sums. One of the most useful sequences of comparison series, besides those of (4.1-1), is the sequence of sums

$$S_k = \sum_{x=1}^{\infty} \frac{1}{x^k} \qquad (k = 2, 3, \ldots) \tag{4.3-1}$$

which are really the values of the Riemann zeta function

$$\zeta(z) = \sum_{n=1}^{\infty} \frac{1}{n^z}$$

(see Table 8.7-1 for $z = 2, 3, \ldots, 17$). The values for the even integers are known in closed form,

$$S_2 = \zeta(2) = \frac{\pi^2}{6} \qquad S_4 = \frac{\pi^4}{90}$$

$$S_6 = \frac{\pi^6}{945} \qquad S_8 = \frac{\pi^8}{9{,}450}$$

but those for the odd integers are not.

Many other series having known sums are available in numerous books and tables,[1] and the reader undoubtedly is familiar with those which arise by expanding familiar functions in Maclaurin series.

4.4 EULER'S METHOD

Another method of summing series numerically is Euler's method. It can be derived as follows:

Consider the finite series

$$\sum_{k=0}^{n-1} a_k t^k \tag{4.4-1}$$

We apply summation by parts and recall that

$$\sum_{a}^{b-1} u(k)\,\Delta v(k) = [u(k)v(k)]_a^b - \sum_{a}^{b-1} v(k+1)\,\Delta u(k)$$

[1] See Refs. 18 and 1.

We pick $u(k) = a_k$, $\Delta v(k) = t^k$. Since we may use any additive constant, we choose

$$v(k) = \sum_{x=0}^{k-1} t^x = \frac{1-t^k}{1-t}$$

Then

$$\sum_{k=0}^{n-1} a_k t^k = \left[a_k \frac{1-t^k}{1-t} \right]_0^n - \sum_{k=0}^{n-1} \frac{1-t^{k+1}}{1-t} \Delta a_k$$

$$= a_n \left(\frac{1-t^n}{1-t} \right) - \frac{1}{1-t} \sum_0^{n-1} \Delta a_k + \frac{t}{1-t} \sum_0^{n-1} t^k \Delta a_k \qquad (4.4\text{-}2)$$

but $\displaystyle\sum_{k=0}^{n-1} \Delta a_k = (a_1 - a_0) + (a_2 - a_1) + \cdots + (a_n - a_{n-1})$

$$= a_n - a_0$$

and so (4.4-2) becomes

$$\sum_{k=0}^{n-1} a_k t^k = a_n \left(\frac{1-t^n}{1-t} \right) - \frac{a_n - a_0}{1-t} + \frac{t}{1-t} \sum_0^{n-1} t^k \Delta a_k$$

$$= \frac{a_0}{1-t} - \frac{a_n t^n}{1-t} + \frac{t}{1-t} \sum_{k=0}^{n-1} t^k \Delta a_k \qquad (4.4\text{-}3)$$

We apply summation by parts to the third term, noting that it is of the same form as the original series, with a_k replaced by Δa_k in (4.4-3),

$$\frac{t}{1-t} \sum_{k=0}^{n-1} t^k \Delta a_k = \frac{t}{1-t} \left(\frac{\Delta a_0}{1-t} - \frac{\Delta a_n t^n}{1-t} + \frac{t}{1-t} \sum_{k=0}^{n-1} t^k \Delta^2 a_k \right)$$

Thus, as a result of the application of summation by parts twice, we have obtained

$$\sum_{k=0}^{n-1} a_k t^k = \frac{a_0}{1-t} + \frac{t\,\Delta a_0}{(1-t)^2} + \frac{t^2}{(1-t)^2} \sum_{k=0}^{n-1} t^k \Delta^2 a_k$$

$$- \frac{a_n t^n}{1-t} - \frac{1}{1-t} \frac{\Delta a_n t^{n+1}}{1-t}$$

After r applications, the expression (4.4-1) has the form

$$\sum_{k=0}^{n-1} a_k t^k = \frac{1}{1-t} \sum_{i=0}^{r-1} \left(\frac{t}{1-t}\right)^i \Delta^i a_0 + \left(\frac{t}{1-t}\right)^r \sum_{k=0}^{n-1} t^k \Delta^r a_k - \frac{t^n}{1-t} \sum_{i=0}^{r-1} \left(\frac{t}{1-t}\right)^i \Delta^i a_n$$

Since the original series converges, given an $\epsilon > 0$, there exists an n_0 such that, for $n > n_0$, $|a_n| < \epsilon/2^r$. Hence, by (1.4-3), $|\Delta^i a_n| \leq \epsilon$. The last term in the above expression therefore goes to zero as $n \to \infty$, and we have

$$\sum_{k=0}^{\infty} a_k t^k = \frac{1}{1-t} \sum_{i=0}^{r-1} \left(\frac{t}{1-t}\right)^i \Delta^i a_0 + \left(\frac{t}{1-t}\right)^r \sum_{k=0}^{\infty} t^k \Delta^r a_k \qquad (4.4\text{-}4)$$

We now suppose that the given series (4.4-1) is smooth enough so that the second term on the right approaches zero as $r \to \infty$. We are then left with the series

$$\sum_{k=0}^{\infty} a_k t^k = \frac{1}{1-t} \sum_{i=0}^{\infty} \left(\frac{t}{1-t}\right)^i \Delta^i a_0 \qquad (4.4\text{-}5)$$

As an example of using the above process, given a series $\sum_{k=0}^{\infty} u_k$, if u_{k+1}/u_k approaches t, we can write

$$\sum_0^{\infty} u_k = \sum_0^{\infty} a_k t^k$$

where a_{k+1}/a_k approaches 1, and apply Euler's method.

The most frequent case of application is when $t = -1$. Euler's method gives, from (4.4-5),

$$\sum_{k=0}^{\infty} (-1)^k a_k = \frac{1}{2} \sum_{i=0}^{\infty} \frac{(-1)^i}{2^i} \Delta^i a_0 \qquad (4.4\text{-}6)$$

Sometimes Euler's transformation makes the series converge faster and sometimes it does not. Consider the following examples:

Example 1

$$\sum_{k=0}^{\infty} \frac{(-1)^k}{2^k}$$

We have

$$a_n = \frac{1}{2^n}$$

$$\Delta a_n = \frac{1}{2^{n+1}} - \frac{1}{2^n} = \frac{-1}{2^{n+1}}$$

$$\Delta^i a_0 = \frac{(-1)^i}{2^i}$$

and so, by (4.4-6),

$$\sum_{k=0}^{\infty} \frac{(-1)^k}{2^k} = \frac{1}{2} \sum_{i=0}^{\infty} \frac{(-1)^i}{2^i} \frac{(-1)^i}{2^i} = \frac{1}{2} \sum_{i=0}^{\infty} \frac{1}{4^i}$$

which converges more quickly.

Example 2

$$\sum_{k=0}^{\infty} \frac{(-1)^k}{3^k}$$

It is easy to see that

$$\Delta^i a_0 = \frac{(-2)^i}{3^i}$$

and so

$$\sum_{k=0}^{\infty} \frac{(-1)^k}{3^k} = \frac{1}{2} \sum_{i=0}^{\infty} \frac{(-1)^i}{2^i} \frac{(-2)^i}{3^i} = \frac{1}{2} \sum_{i=0}^{\infty} \frac{1}{3^i}$$

which converges somewhat more slowly.

Example 3. Similarly,

$$\sum_{k=0}^{\infty} \frac{(-1)^k}{4^k} = \frac{1}{2} \sum_{i=0}^{\infty} \left(\frac{3}{8}\right)^i$$

which converges more slowly.

The "break point" for the application of Euler's method to such alternating series is seen to be at ratios of $|a_i|$ between $\frac{1}{2}$ and $\frac{1}{3}$.

In practice, what is usually done is to sum the first few (say 10) terms directly and then apply Euler's transformation to the rest.[1]

EXERCISES

4.4-1. Apply Euler's method to

$$\log 2 = \sum_{k=1}^{\infty} \frac{(-1)^{k-1}}{k} = \sum_{i=1}^{\infty} \frac{1}{k2^k}$$

[1] Reference 4 and J. B. Rosser, Transformations to Speed the Convergence of Series, *J. Research Natl. Bur. Standards*, vol. 46, 1951.

4.4-2. Show that

$$\arctan 1 = \sum_{k=0}^{\infty} \frac{(-1)^k}{2k+1} = \frac{1}{2} \sum_{k=0}^{\infty} \frac{k!}{1 \cdot 3 \cdot 5 \cdots (2k+1)}$$

4.4-3. Add the first eight terms of log 2 and apply Euler's method to the rest.
Ans. $1 - \frac{1}{2} + \frac{1}{3} - \cdots - \frac{1}{8} = 0.63452381$; Euler part gives $= \frac{1}{2}(0.11724674)$

4.5 A NONLINEAR TRANSFORMATION

Most transformations used in mathematics are linear, but the nonlinear transformation on the partial sums S_n of a series

$$T(S_n) = \frac{S_{n+1}S_{n-1} - S_n^2}{S_{n+1} - 2S_n + S_{n-1}} \tag{4.5-1}$$

is often very useful. If T were linear, we would have

$$T(CS_n) = CT(S_n)$$

and

$$T(S_n + U_n) = T(S_n) + T(U_n)$$

The second is not true in general. However, we do have the weaker rule

$$T(S_n + C) = T(S_n) + C$$

This transformation is useful in situations where, loosely speaking, the behavior is like that of a geometric progression.

As an example of the power of a single application, consider the Leibnitz series

$$\pi = 4 - \frac{4}{3} + \frac{4}{5} - \frac{4}{7} + \cdots$$

The convergence is so slow that it is practically valueless, but a single application gives the following:

n	S_n	$T(S_n)$
0	4.00000	
1	2.66667	3.16667
2	3.46667	3.13333
3	2.89524	3.14524
4	3.33968	3.13968
5	2.97605	3.14271
6	3.28374	3.14088
7	3.01707	3.14207
8	3.25237	3.14125
9	3.04184	

The theory of this transformation and of many related ones which are often even more powerful has been studied extensively[1] and will not be discussed here.

This transformation is useful not only for handling slowly convergent and divergent series but also for many iterative sequences which converge slowly. It is also easy to express (4.5-1) in terms of the individual terms of a series, and this provides an alternate approach.

EXERCISE

4.5-1. Using (4.5-1), compute $\sum_{k=1}^{\infty} [(-1)^{k-1}/k]$ to four decimal places.

4.6 POWER SERIES

Power series are widely used in mathematics, especially in linear problems. Even in nonlinear problems the power series is a useful tool. Since in nonlinear problems the computation of the successive coefficients is frequently quite laborious, the computer itself should be used to find them. Routines to add, subtract, multiply, divide, substitute one power series in another, and to invert a series, all to a given number of coefficients, are easy to write.

The advantage of the power-series approach, or the polynomial approach as in the problem of Sec. 1.9, is that many of the analytic operations of the calculus can be performed on the coefficients individually one at a time. An abstract mathematical description of what was done in the problem of Sec. 1.9 goes as follows: The 11 data points can be viewed as a single point in an 11-dimensional data space. The first operation of fitting a Newton factorial polynomial transformed the point in the data space to a corresponding point in an 11-dimensional coefficient space. The operation of using the Stirling numbers to get to the direct polynomial representation amounts to a change in the basis of representing the coefficient space. The subsequent operations were transformations in the coefficient space, while the final evaluation was a transformation back to the data space. Thus most of the operations were in a transform space rather than in the actual space of the data and the answer. This is characteristic of the power-series approach; we operate on the coefficients of a representation, or transformation, of the data, and only at the end do we return to the data space.

[1] D. Shanks, Non-linear Transformations of Divergent and Slowly Convergent Sequences, *J. Math. and Phys.*, vol. 34, pp. 1–42, 1955.

Similar remarks apply to asymptotic series as well as to many of the methods discussed in this chapter.

EXERCISES

4.6-1. Make a flow diagram for a routine to multiply two power series of k terms to get a k-term product.

4.6-2. The same as Exercise 4.6-1, but for division.

4.6-3. The same as Exercise 4.6-1, but substitute one series in another.

4.6-4. Given the integral

$$g(T) = \int_0^\infty \frac{f(x)(x/T)^2 e^{x/T}}{(e^{x/T} - 1)^2}\, dx$$

use the method of Sec. 1.9 and put $f(x) = x^k$. Show $g(T) = (n + 2)!T^{n+1}\zeta(n + 2)$. Hence show that, if $g(T)$ is given and can be approximated by

$$g(T) = \sum_{n=0}^{N} a_n T^{n+1}$$

then

$$f(x) = \sum_{n=0}^{N} \frac{a_n x^n}{(n + 2)!\zeta(n + 2)}$$

4.7 EXPANSIONS IN SPECIAL FUNCTIONS

In addition to expansions in power series, analysis often makes use of expansions in special functions such as Legendre polynomials $P_n(x)$, Laguerre polynomials $L_n(x)$, Hermite polynomials $H_n(x)$, Bessel functions $J_n(x)$, etc. A superficial approach may suggest that these expansions are not useful in numerical computation because of the labor of computing the values of the special functions. It is a known fact that most of the families of special functions satisfy a three-term recurrence relation of the form

$$f_{n+1}(x) = A(n,x)f_n(x) + B(n)f_{n-1}(x)$$

In the case of the polynomials, the zero and first-order polynomials are particularly easy to compute for a specific value of x. Thus the labor of computing expansions in families of special functions is not much greater than the labor of computing power series expansions.[1]

Some consideration must be given to the propagation of errors when using a recurrence relation to compute successive functions, but usually

[1] See also I. Stegun and M. Abramowitz, Generation of Bessel Functions on a High Speed Computer, *Math. Tables and Aids to Comp.*, vol. 11, pp. 255–257, 1957, and M. C. Gray, Bessel Functions of Integral Order and Complex Argument, *Commun. Assoc. Computing Machinery*, vol. 4, no. 4, p. 169, April, 1961.

the growth of the error is not severe for moderate (say 15) values of the index n.

4.8 INTEGRALS AS APPROXIMATIONS TO SUMS

The definite integral is defined as the limit of a sum. It is evident, therefore, that integrals can be used to approximate sums. One such actual formula for doing this appears in Sec. 12.3 and may be written as

$$\begin{aligned} f_0 + f_1 + \cdots + f_n = \int_0^n f(x)\,dx + \tfrac{1}{2}f_0 + \tfrac{1}{2}f_n \\ - \tfrac{1}{12}(\Delta f_0 - \Delta f_{n-1}) + \tfrac{1}{24}(\Delta^2 f_0 + \Delta^2 f_{n-2}) \\ - \tfrac{19}{720}(\Delta^3 f_0 - \Delta^3 f_{n-3}) + \tfrac{3}{160}(\Delta^4 f_0 + \Delta^4 f_{n-4}) + \cdots \qquad (4.8\text{-}1) \end{aligned}$$

A closely related formula is the Euler-Maclaurin formula [43, pp. 127–128]

$$\begin{aligned} f_0 + f_1 + \cdots + f_n = \int_0^n f(x)\,dx + \frac{1}{2}f_0 + \frac{1}{2}f_n - \frac{1}{12}[f_0' - f_n'] \\ + \frac{1}{720}[f_0''' - f_n'''] - \frac{1}{30{,}240}[f_0^{(5)} - f_n^{(5)}] + \cdots \qquad (4.8\text{-}2) \end{aligned}$$

*4.9 THE DIGAMMA FUNCTION

For the final topic of this chapter, recall the formula

$$\int x^m\,dx = \frac{x^{m+1}}{m+1} + C$$

which does not apply when $m = -1$. Indeed, when $m = -1$ the integral can be said to define a new function, $\log x$.

Similarly, the summation formula

$$\sum_{t=1}^{x-1} t^{(m)} = \frac{x^{(m+1)}}{m+1} + C$$

does not apply when $m = -1$ and correspondingly can be said to define a new function, called the digamma function and denoted by $F(x)$:

$$F(x) = \sum_{r=1}^{x} \frac{1}{r} - \gamma \qquad (4.9\text{-}1)$$

where γ = Euler's constant $= 0.5772156649 \cdots$ and hence $F(0) = -\gamma$.

This formula applies when x is an integer. Another form,

$$\digamma(x) = \sum_{r=1}^{\infty} \frac{x}{r(r+x)} - \gamma \qquad \digamma(0) = -\gamma \tag{4.9-2}$$

provides a means of extending the definition of $\digamma(x)$ to noninteger values. We need to show, however, that this new definition is consistent with the old one. This amounts to showing that $\digamma(x)$ satisfies the equation

$$\Delta \digamma(x) = \frac{1}{x+1}$$

(This is analogous to the process of showing that a function satisfies an integration formula by showing that the derivative of the function is the integrand.) We have, using (4.9-2),

$$\begin{aligned} \Delta \digamma(x) &= \sum_{r=1}^{\infty} \left[\frac{x+1}{r(r+x+1)} - \frac{x}{r(r+x)} \right] \\ &= \sum_{r=1}^{\infty} \left[\frac{1}{(r+x)(r+x+1)} \right] \\ &= \sum_{r=1}^{\infty} \left[\frac{1}{r+x} - \frac{1}{r+x+1} \right] \\ &= \frac{1}{x+1} \end{aligned}$$

The following relation exists between the digamma function and the natural logarithm:

$$\digamma(x) = \lim_{n \to \infty} \left(\log(x+n+1) - \frac{1}{x+1} - \frac{1}{x+2} - \cdots - \frac{1}{x+n} \right)$$

Another relationship between known functions is

$$\frac{d}{dx} \log [\Gamma(1+x)] = \digamma(x)$$

If we differentiate again, we get the so-called trigamma function

$$\frac{d^2}{dx^2} \log [\Gamma(1+x)] = \sum_{1}^{\infty} \frac{1}{(r+x)^2} = \digamma'(x)$$

the tetragamma function

$$\frac{d^3}{dx^3}\log\left[\Gamma(1+x)\right] = -2\sum_{1}^{\infty}\frac{1}{(r+x)^3} = F''(x)$$

the pentagamma function

$$\frac{d^4}{dx^4}\log\left[\Gamma(1+x)\right] = 6\sum_{1}^{\infty}\frac{1}{(r+x)^4} = F'''(x)$$

etc.

These functions have been tabulated [35] and can be used to sum series whose general coefficient is a rational function. For example, suppose that we have

$$S = \sum_{x=1}^{\infty}\frac{P_{n-2}(x)}{P_n(x)}$$

where $P_n(x)$ is a polynomial of degree n and $P_{n-2}(x)$ is at least two degrees less than $P_n(x)$. Using partial fractions, we can write [where the a_i are the zeros of $P_n(x)$]

$$S = \sum_{x=1}^{\infty}\left[\frac{A_1}{x-a_1} + \frac{A_2}{x-a_2} + \cdots + \frac{A_k}{x-a_k} + \frac{B_1}{(x-a_1)^2} + \frac{B_1}{(x-a_2)^2} + \cdots + \frac{B_k}{(x-a_k)^2} + \cdots + \frac{M_1}{(x-a_1)^m} + \frac{M_2}{(x-a_2)^m} + \cdots + \frac{M_k}{(x-a_k)^m}\right]$$

We cannot rearrange the first group of terms since individually the series diverge. However, it is easy to see that $\sum_{i=1}^{k} A_i = 0$; hence we can write

$$\left[\left(\frac{A_1}{x-a_1} - \frac{A_1}{x}\right) + \left(\frac{A_2}{x-a_2} - \frac{A_2}{x}\right) + \cdots + \left(\frac{A_k}{x-a_k} - \frac{A_k}{x}\right)\right]$$

without changing the value of the term. We now have the series in a form that we can rearrange easily. As an example, consider (see Ref. 35)

$$\begin{aligned} S &= \sum_{1}^{\infty}\frac{1}{(4x+2)(4x+1)(4x+3)^2} \\ &= \sum_{1}^{\infty}\left[-\frac{1}{4x+2} + \frac{1}{4}\cdot\frac{1}{4x+1} + \frac{3}{4}\cdot\frac{1}{4x+3} + \frac{1}{2}\cdot\frac{1}{(4x+3)^2}\right] \\ &= \tfrac{1}{4}F(\tfrac{1}{2}) - \tfrac{1}{16}F(\tfrac{1}{4}) - \tfrac{3}{16}F(\tfrac{3}{4}) + \tfrac{1}{32}F'(\tfrac{3}{4}) \end{aligned}$$

CHAPTER 5

Finite Difference Equations

5.1 NOTATION

Corresponding to differential equations in the differential calculus, there are difference equations in the difference calculus. Thus, for example, one might have a differential equation for the unknown function y,

$$y' + 2y = x$$

or a difference equation

$$\Delta y + 2y = x \tag{5.1-1}$$

It is a curious fact that in this form the analogy between differential and difference equations is not close. But if we use

$$\Delta y = y(x + 1) - y(x)$$

in (5.1-1), we obtain

$$y(x + 1) + y(x) = x \tag{5.1-2}$$

which is the more useful form and which may be treated in much the same way as we treat

$$y' + y = x$$

It is evident that we can pass from one form of the difference equation to the other so that we can always use the second form in solving difference equations.

One reason for using the form (5.1-2) is that it more clearly reveals the nature of the problem. Thus if we have

$$\Delta^2 y + 2\Delta y + y = f(x)$$

we might suppose that we had a second-order difference equation. But this reduces to

$$\begin{aligned} f(x) &= y(x + 2) - 2y(x + 1) + y(x) + 2[y(x + 1) - y(x)] + y(x) \\ &= y(x + 2) \end{aligned}$$

which is trivial. It is the maximum difference of the arguments of the unknown function that determines the order of the difference equation.

Since there is a close analogy between difference and differential equations, it is worth going over some of the details of how differential equations are solved. Again, all tricks except change of the independent variable generally apply in the two cases.

5.2 AN EXAMPLE OF A FIRST-ORDER DIFFERENCE EQUATION

To solve the differential equation

$$y' + y = x \tag{5.2-1}$$

we first consider the homogeneous equation

$$y' + y = 0 \tag{5.2-2}$$

and look for the "complementary function." In this equation we *guess* that $y = e^{mx}$ will be a solution, provided that we choose m properly. Substituting in the equation, we get

$$\begin{aligned} me^{mx} + e^{mx} &= 0 \\ e^{mx}(m + 1) &= 0 \\ m + 1 &= 0 \qquad (e^{mx} \neq 0) \\ m &= -1 \end{aligned}$$

whence

$$y = Ce^{-x} \tag{5.2-3}$$

where C is an arbitrary constant. Next we seek a particular solution of the complete equation. In view of the right-hand side, we *guess* at

$$y = ax + b$$

The differential equation (5.2-1) gives

$$a + ax + b = x$$

so that

$$a = 1$$
$$b = -1$$

and

$$y = x - 1 \tag{5.2-4}$$

is a particular solution of (5.2-1).

The general solution is now (5.2-3) plus (5.2-4):

$$y = Ce^{-x} + (x - 1)$$

which is the sum of the complementary function and the particular solution. Of course this method applies only to linear equations.

For the difference equation

$$y(x + 1) + y(x) = x$$

we do the same. In the homogeneous equation

$$y(x + 1) + y(x) = 0$$

we substitute, instead of e^{mx},

$$y = \rho^x$$

which is merely a notational change, since we can set $e^m = \rho$. We obtain

$$\rho^x(\rho + 1) = 0$$

so that

$$\rho = -1$$

and

$$y = C(-1)^x$$

For the complete equation we again *guess* that

$$y = ax + b$$

The difference equation gives

$$a(x + 1) + b + ax + b = x$$

$$2a = 1 \qquad a = \tfrac{1}{2}$$

so that

$$a + 2b = 0 \qquad b = -\tfrac{1}{4}$$

and

$$y = \frac{x}{2} - \frac{1}{4}$$

Then the general solution is

$$y = C(-1)^x + \frac{x}{2} - \frac{1}{4}$$

Thus we see the close analogy for first-order linear equations.

5.3 AN EXAMPLE OF A SECOND-ORDER EQUATION

For higher-order linear equations we use the same technique. Consider, for example, the Fibonacci numbers which are defined by

$$y_{n+1} = y_n + y_{n-1}$$

with

$$y_0 = 0 \qquad y_1 = 1$$

We first find the general solution. Trying

$$y = \rho^n$$

we obtain

$$\rho^2 - \rho - 1 = 0$$

or

$$\rho = \frac{1 \pm \sqrt{5}}{2}$$

Thus

$$y_n = C_1\left(\frac{1 + \sqrt{5}}{2}\right)^n + C_2\left(\frac{1 - \sqrt{5}}{2}\right)^n$$

Using the condition $y_0 = 0$ gives

$$0 = C_1 + C_2$$

and using the condition $y_1 = 1$,

$$y_1 = 1 = C_1\left(\frac{1+\sqrt{5}}{2}\right) + C_2\left(\frac{1-\sqrt{5}}{2}\right)$$

These equations lead to

$$C_1 = \frac{1}{\sqrt{5}} = -C_2$$

and

$$y_n = \frac{1}{\sqrt{5}}\left[\left(\frac{1+\sqrt{5}}{2}\right)^n - \left(\frac{1-\sqrt{5}}{2}\right)^n\right]$$

One reason for choosing this example is the fact that the difference equation itself enables one to calculate quite easily the successive numbers in turn. Thus

$$\begin{array}{lll} y_0 = 0 & y_4 = 3 & y_8 = 21 \\ y_1 = 1 & y_5 = 5 & y_9 = 34 \\ y_2 = 1 & y_6 = 8 & y_{10} = 55 \\ y_3 = 2 & y_7 = 13 & y_{11} = 89 \\ & & \cdots\cdots \end{array}$$

No question of existence can arise, and the general solution, while perhaps useful to compute some isolated number, is obviously too difficult to use in practice.

5.4 LINEAR DIFFERENCE EQUATIONS WITH CONSTANT COEFFICIENTS

Let us continue the analogy to linear differential equations. In the case of double roots of the characteristic equation, we try both e^{mx} and xe^{mx}. For example, consider

$$y'' - 2y' + y = 0$$

Letting $y = e^{mx}$, we obtain

$$m^2 - 2m + 1 = 0$$

or

$$m = 1, 1$$

Then e^x and xe^x each satisfy the equation, and the general solution is

$$y = C_1e^x + C_2xe^x$$

Consider now the difference equation

$$y(x+2) - 2y(x+1) + y(x) = 0$$

Substituting $y = \rho^x$, we obtain

$$\rho^2 - 2\rho + 1 = 0$$
$$\rho = 1, 1$$

Then $(1)^x = 1$, $x(1)^x = x$ each satisfy the equation, and the general solution is

$$y = C_1 + C_2 x$$

It will be recalled that, if the right-hand side of a nonhomogeneous equation has a term that appeared in the solution of the homogeneous equation, then there is a similar effect. Thus in the case

$$y' + y = e^{-x}$$

the homogeneous equation

$$y' + y = 0$$

has the solution $y = Ce^{-x}$. Therefore, in guessing at a particular solution of the whole equation, we try

$$y = kxe^{-x}$$
$$ke^{-x} - kxe^{-x} + kxe^{-x} = e^{-x} \qquad k = 1$$

We obtain, finally,

$$y = Ce^{-x} + xe^{-x} = (C + x)e^{-x}$$

Again, in the difference equation

$$y(x + 1) - y(x) = 1$$

we form the homogeneous equation

$$y(x + 1) - y(x) = 0$$

This has the solution $y = C$, where C is an arbitrary constant. Hence, when we guess the particular solution, we try

$$y = kx$$

Substituting, we see that $k = 1$, so that the general solution of the original equation is

$$y = C + x$$

Milne-Thomson [27] discusses the solution of difference equations in more detail. He shows, in particular, how the knowledge of one solution of a homogeneous linear difference equation enables one to reduce the order of the equation by 1.

5.5 AN EXAMPLE

The purpose of this section is to give an example of the use of difference equations and to show how some ideas and methods in the field of ordinary differential equations can be applied in an obvious way to difference equations.

The example is to find the integrals

$$J(n) = \int_0^\infty x^n e^{-x} \sin x \, dx \qquad (n \geq 0)$$
$$K(n) = \int_0^\infty x^n e^{-x} \cos x \, dx$$

and in particular to show that $J(4k + 3) = 0$.

Integration by parts gives the simultaneous difference equations

$$J(n) = \frac{n}{2} [J(n-1) + K(n-1)]$$
$$K(n) = \frac{n}{2} [-J(n-1) + K(n-1)] \qquad (5.5\text{-}1)$$

Standard integration tables give $J(0) = K(0) = \frac{1}{2}$.

These difference equations have variable coefficients $(n/2)$; that is, $J(n)$ has a factor n, $J(n + 1)$ has a factor $n + 1$, etc., suggesting a behavior like $n!$. Thus we are led to consider the transformation

$$J(n) = n!j(n)$$
$$K(n) = n!k(n)$$

This leaves the factor $\frac{1}{2}$ in the coefficients, and a little more thought suggests the transformation

$$J(n) = \frac{n!}{2^n} j(n)$$
$$K(n) = \frac{n!}{2^n} k(n) \qquad (5.5\text{-}2)$$

which leads to the system

$$j(n) = j(n-1) + k(n-1) \qquad j(0) = \tfrac{1}{2}$$
$$k(n) = -j(n-1) + k(n-1) \qquad k(0) = \tfrac{1}{2} \qquad (5.5\text{-}3)$$

If this were a system of ordinary differential equations, we would either try to recall how to solve systems or reduce the problem to a single second-order equation by differentiating one equation and eliminating one variable. The latter seems to be easier to recall and adapt to difference equations. Thus we write the top equation (5.5-3) as

$$j(n+1) = j(n) + k(n) \qquad (5.5\text{-}4)$$

and, using this with the second of (5.5-3), we eliminate $k(n)$. We then use the first of (5.5-3) to eliminate $k(n-1)$ to get

$$j(n+1) - 2j(n) + 2j(n-1) = 0 \tag{5.5-5}$$

This is a second-order linear difference equation with constant coefficients and initial conditions, and, using (5.5-4),

$$\begin{aligned} j(0) &= \tfrac{1}{2} \\ j(1) &= 1 \end{aligned} \tag{5.5-6}$$

The general solution of (5.5-5) is (where $i = \sqrt{-1}$)

$$j(n) = C_1(1+i)^n + C_2(1-i)^n$$

and the initial conditions (5.5-6) give

$$\begin{aligned} C_1 + C_2 &= \tfrac{1}{2} \\ C_1(1+i) + C_2(1-i) &= 1 \end{aligned}$$

which leads to

$$C_1 = \frac{1-i}{4} \qquad C_2 = \frac{1+i}{4}$$

Thus,

$$j(n) = \tfrac{1}{2}[(1+i)^{n-1} + (1-i)^{n-1}]$$

and

$$J(n) = \frac{n!}{2^{n+1}}[(1+i)^{n-1} + (1-i)^{n-1}]$$

(which is actually a real number).

To show the second half of the problem, we observe that

$$(1+i)^4 = 1 + 4i + 6i^2 + 4i^3 + i^4 = -4 = (1-i)^4$$

Hence

$$j(4k+3) = (-4)^k j(3) = \frac{(-4)^k}{2}[(1 + 2i + i^2) + 1 - 2i + i^2] = 0$$

Thus

$$J(4k+3) = 0$$

EXERCISE

5.5-1. Let

$$I_k(\Phi) = \int_0^\pi \frac{\cos k\theta - \cos k\Phi}{\cos\theta - \cos\Phi}\, d\theta$$

Show that for $k =$ integer

$$\begin{aligned} I_{k+2}(\Phi) - (2\cos\Phi)I_{k+1}(\Phi) + I_k(\Phi) &= 0 \\ I_0 &= 0 \\ I_1 &= \pi \end{aligned}$$

and hence

$$I_k(\Phi) = \frac{\pi \sin k\Phi}{\sin\Phi}$$

CHAPTER 6

The Finite Fourier Series

6.1 INTRODUCTION

The Fourier series

$$f(x) = \frac{a_0}{2} + \sum_{k=1}^{\infty} (a_k \cos kx + b_k \sin kx) \tag{6.1-1}$$

where

$$a_k = \frac{1}{\pi} \int_0^{2\pi} f(x) \cos kx \, dx$$

$$b_k = \frac{1}{\pi} \int_0^{2\pi} f(x) \sin kx \, dx \tag{6.1-2}$$

has played an important role in the development of mathematics.[1] Corresponding to the usual theory which is based on knowing the function in an entire interval of length 2π, there is a theory based on knowing the function at a discrete set of equally spaced points. We shall adopt as a basis of discussion $2N$ equally spaced points and leave the case of an odd number of points, $2N + 1$, as an exercise for the reader.

The basis for the expansion as a Fourier series is the *orthogonality* of the functions, 1, $\cos kx$, $\sin kx$, with respect to integration over the interval $[0,2\pi]$

$$\int_0^{2\pi} \cos kx \cos mx \, dx = \begin{cases} 0 & k \neq m \\ \pi & k = m \neq 0 \\ 2\pi & k = m = 0 \end{cases}$$

$$\int_0^{2\pi} \sin kx \sin mx \, dx = \begin{cases} 0 & k \neq m \\ \pi & k = m \neq 0 \end{cases} \tag{6.1-3}$$

$$\int_0^{2\pi} \sin kx \cos mx \, dx = 0$$

6.2 ORTHOGONALITY OVER A DISCRETE SET OF POINTS

It is a remarkable fact that if we use summation in place of integration the set of functions

$$\begin{matrix} 1, & \cos x, & \cos 2x, & \ldots, & \cos (N-1)x, & \cos Nx \\ & \sin x, & \sin 2x, & \ldots, & \sin (N-1)x & \end{matrix}$$

[1] For an elementary treatment of the Fourier series, see Ref. 17.

is orthogonal over the discrete set of points

$$0, \frac{\pi}{N}, \frac{2\pi}{N}, \ldots, \frac{(2N-1)\pi}{N} \tag{6.2-1}$$

The transformation

$$x \to \frac{\pi}{N} x$$

reduces us to unit spacing with the functions

$$\begin{aligned} 1, \quad \cos\frac{\pi}{N}x, \quad \cos\frac{2\pi}{N}x, \quad \ldots, \quad \cos\frac{(N-1)\pi}{N}x, \quad \cos\frac{\pi}{N}Nx \\ \sin\frac{\pi}{N}x, \quad \sin\frac{2\pi}{N}x, \quad \ldots, \quad \sin\frac{(N-1)\pi}{N}x \end{aligned} \tag{6.2-2}$$

The orthogonality relations are ($k \leq N$, $m \leq N$)

$$\begin{aligned} &\sum_{x=0}^{2N-1} \sin\frac{\pi}{N}kx \sin\frac{\pi}{N}mx = \begin{cases} 0 & \text{if } k \neq m \\ N & \text{if } k = m \neq 0 \end{cases} \\ &\sum_{x=0}^{2N-1} \sin\frac{\pi}{N}kx \cos\frac{\pi}{N}mx = 0 \\ &\sum_{x=0}^{2N-1} \cos\frac{\pi}{N}kx \cos\frac{\pi}{N}mx = \begin{cases} 0 & \text{if } k \neq m \\ N & \text{if } k = m \neq 0, N, \ldots \\ 2N & \text{if } k = m = 0, N, \ldots \end{cases} \end{aligned} \tag{6.2-3}$$

which were proved in Exercise 3.2-3. These may be viewed as analogs of the condition for perpendicularity (orthogonality) in the usual two- and three-dimensional spaces. (See Sec. 17.6.)

If we now assume that an arbitrary function $f(x)$ can be written in the form

$$f(x) = \frac{a_0}{2} + \sum_{k=1}^{N-1}\left(a_k \cos\frac{\pi}{N}kx + b_k \sin\frac{\pi}{N}kx\right) + \frac{a_N}{2}\cos\pi x \tag{6.2-4}$$

we can use the orthogonality relations in the usual way to determine the coefficients of the expansion. To obtain a_k ($k = 1, \ldots, N-1$), multiply both sides by $\cos(\pi/N)mx$ and sum over all x. We get

$$\sum_{x=0}^{2N-1} f(x)\cos\frac{\pi}{N}mx = Na_m \qquad (1 \leq m \leq N-1) \tag{6.2-5}$$

since by (6.2-3) all the other terms are zero. Using $\sin (\pi/N)mx$ in place of $\cos (\pi/N)mx$, we obtain

$$\sum_{x=0}^{2N-1} f(x) \sin \frac{\pi}{N} mx = Nb_m \qquad (1 \leq m \leq N-1) \tag{6.2-6}$$

Finally we have

$$\begin{aligned} \sum_{x=0}^{2N-1} f(x) &= 2N\left(\frac{a_0}{2}\right) = Na_0 \\ \sum_{x=0}^{2N-1} f(x) \cos \pi x &= 2N\left(\frac{a_N}{2}\right) = Na_N \end{aligned} \tag{6.2-7}$$

These last two formulas are in the same forms as (6.2-5) and show why it is advantageous to assume the expansion with $a_0/2$ and $a_N/2$ rather than a_0 and a_N.

EXERCISE

6.2-1. Discuss the expansion for an odd number, $2N + 1$, of sample points.

$$\textit{Ans. } f(x) = \frac{a_0}{2} + \sum_{k=1}^{N} \left(a_k \cos \frac{2\pi k}{2N+1} x + b_k \sin \frac{2\pi k}{2N+1} x\right)$$

6.3 ACCURACY OF THE EXPANSION

In this manner the $2N$ values of $f(x)$ $(x = 0, 1, \ldots, 2N-1)$ determine $2N$ coefficients, a_k, b_k, and it is reasonable to expect that the series will exactly reproduce the original function at the given points (technically speaking, that the set of functions is complete with respect to the set of points used). To prove this reasonable assumption, we fix $x = \bar{x}$ in our mind and rearrange the sum, using Eqs. (6.2-5) to (6.2-7) to eliminate the a_k and b_k:

$$\begin{aligned} f(\bar{x}) &= \frac{1}{N} \sum_x f(x) \left[\frac{1}{2} + \sum_{k=1}^{N-1} \left(\cos \frac{\pi}{N} kx \cos \frac{\pi}{N} k\bar{x} \right.\right. \\ &\qquad \left.\left. + \sin \frac{\pi}{N} kx \sin \frac{\pi}{N} k\bar{x}\right) + \frac{1}{2} \cos \pi x \cos \pi\bar{x}\right] \\ &= \frac{1}{N} \sum_x f(x) \left[\frac{1}{2} + \sum_{k=1}^{N-1} \cos \frac{\pi}{N} k(x - \bar{x}) + \frac{1}{2} \cos \pi x \cos \pi\bar{x}\right] \end{aligned}$$

We now write

$$\sum_{k=1}^{N-1} \cos\frac{\pi}{N} k(x - \bar{x}) = \frac{1}{2}\sum_{k=1}^{N-1} \cos\frac{\pi}{N} k(x - \bar{x}) + \frac{1}{2}\sum_{k=1}^{N-1} \cos\frac{\pi}{N} k(x - \bar{x})$$

and in the second sum use the fact that

$$\cos\frac{\pi}{N} k(x - \bar{x}) = \cos\frac{\pi}{N}(2N - k)(x - \bar{x})$$

to get

$$\sum_{k=1}^{N-1} \cos\frac{\pi}{N} k(x - \bar{x}) = \frac{1}{2}\sum_{k=1}^{N-1} \cos\frac{\pi}{N} k(x - \bar{x}) + \frac{1}{2}\sum_{k=N+1}^{2N-1} \cos\frac{\pi}{N} k(x - \bar{x})$$

Using this, we now have

$$f(\bar{x}) = \frac{1}{2N}\sum_{x} f(x)\left[\sum_{k=0}^{2N-1} \cos\frac{\pi}{N} k(x - \bar{x})\right]$$

But the term in the brackets

$$\sum_{k=0}^{2N-1} \cos\frac{\pi}{N} k(x - \bar{x}) = \begin{cases} 2N & \text{if } x = \bar{x} \\ 0 & \text{if } x \neq \bar{x} \end{cases}$$

so that we have only the term $x = \bar{x}$ from the sum, and

$$f(\bar{x}) = \frac{1}{2N}\cdot f(\bar{x})\cdot 2N = f(\bar{x})$$

Thus both sides of the equation give the same number.

There is a well-known relationship between the sum of the squares of the coefficients and the sum of the squares of the function. This relation is

$$\sum_{x}[f(x)]^2 = N\left[\frac{a_0{}^2}{2} + \sum_{k=1}^{N-1}(a_k{}^2 + b_k{}^2) + \frac{a_N{}^2}{2}\right] \tag{6.3-1}$$

To prove this, we merely form the sum of the squares of the function

$$\sum_{x}[f(x)]^2 = \sum_{x}\left[\frac{a_0}{2} + \sum_{k=1}^{N-1}\left(a_k\cos\frac{\pi}{N}kx + b_k\sin\frac{\pi}{N}kx\right) + \frac{a_N}{2}\cos\pi x\right]^2$$

and multiply it out. Doing the sum over x, the orthogonality relations (6.2-5) to (6.2-7) give

$$\sum_x [f(x)]^2 = \left(\frac{a_0}{2}\right)^2 \cdot 2N + \sum_k (a_k^2 \cdot N + b_k^2 \cdot N) + \left(\frac{a_N}{2}\right)^2 \cdot 2N$$

which is (6.3-1).

This result may be used to compute the sum of the squares of the errors in the fit when terms only up to $k = M < N - 1$ are used. Let $f_M(x)$ be the sum of the first M harmonics, that is,

$$f_M(x) = \frac{a_0}{2} + \sum_{k=1}^{M} \left(a_k \cos\frac{\pi}{N} kx + b_k \sin\frac{\pi}{N} kx\right)$$

Then

$$\sum_x (f - f_M)^2 = \sum_x \left[\sum_{k=M+1}^{N-1} \left(a_k \cos\frac{\pi}{N} kx + b_k \sin\frac{\pi}{N} kx\right) + \frac{a_N}{2}\cos \pi x\right]^2$$

$$= N\left[\sum_{k=M+1}^{N-1} (a_k^2 + b_k^2) + \frac{a_N{}^2}{2}\right]$$

$$= \sum_x f^2 - N\left[\frac{a_0{}^2}{2} + \sum_{k=1}^{M} (a_k^2 + b_k^2)\right] \tag{6.3-2}$$

In other words, using only terms up to $k = M$ in the expansion (6.2-4) of the function gives us an approximation to $f(x)$ whose least-square error is given by (6.3-2). By examining how (6.3-2) decreases as M increases, we can estimate the value of taking additional terms in the Fourier expansion.

6.4 NUMERICAL COMPUTATION OF THE COEFFICIENTS

The coefficients of the expansion (6.2-4) are given by

$$a_k = \frac{1}{N}\sum_{x=0}^{2N-1} f(x)\cos\frac{\pi}{N} kx \qquad (k = 0, 1, 2, \ldots, N)$$

$$b_k = \frac{1}{N}\sum_{x=0}^{2N-1} f(x)\sin\frac{\pi}{N} kx \qquad (k = 1, 2, \ldots, N - 1) \tag{6.4-1}$$

It has been shown[1] that these coefficients may be computed readily from $\cos (\pi/N)k$ and $\sin (\pi/N)k$ without directly computing the other values of the sine and cosine. The method is as follows: Let

$$\begin{aligned} U_0 &= 0 \\ U_1 &= f(2N - 1) \\ U_m &= \left(2 \cos \frac{\pi}{N} k\right) U_{m-1} - U_{m-2} + f(2N - m) \\ & \qquad (m = 2, 3, \ldots, 2N - 1) \end{aligned} \tag{6.4-2}$$

$$\begin{aligned} \text{Then} \quad Na_k &= \sum_{x=0}^{2N-1} f(x) \cos \frac{\pi}{N} kx = \left(\cos \frac{\pi}{N} k\right) U_{2N-1} - U_{2N-2} + f(0) \\ Nb_k &= \sum_{x=0}^{2N-1} f(x) \sin \frac{\pi}{N} kx = \left(\sin \frac{\pi}{N} k\right) U_{2N-1} \end{aligned} \tag{6.4-3}$$

A direct proof of this can be given, but it obscures the reason why the method works, so that we shall adopt a slightly more indirect proof that yields an additional useful result.

We begin by defining

$$\begin{aligned} V_0 &= 0 \\ V_1 &= 1 \\ V_m &= (2 \cos t) V_{m-1} - V_{m-2} \qquad (m = 2, 3, \ldots) \end{aligned} \tag{6.4-4}$$

and first show that

$$V_m = \frac{\sin mt}{\sin t} \tag{6.4-5}$$

Since (6.4-5) is true for $m = 0, 1$, it suffices to check that

$$\begin{aligned} (2 \cos t) V_{m-1} - V_{m-2} &= \frac{2 \cos t \sin (m - 1)t}{\sin t} - \frac{\sin (m - 2)t}{\sin t} \\ &= \frac{\sin mt + \sin (m - 2)t - \sin (m - 2)t}{\sin t} \\ &= \frac{\sin mt}{\sin t} = V_m \end{aligned}$$

[1] G. Goertzel, *Am. Math. Monthly*, January, 1958, p. 34. We have used a notation different from his.

as required. We also note that

$$\begin{aligned}(\cos t)V_m - V_{m-1} &= \frac{\cos t \sin mt - \sin (m-1)t}{\sin t}\\ &= \frac{\tfrac{1}{2}[\sin (m+1)t + \sin (m-1)t] - \sin (m-1)t}{\sin t}\\ &= \frac{\sin (m+1)t - \sin (m-1)t}{2 \sin t}\\ &= \cos mt \qquad (6.4\text{-}6)\end{aligned}$$

Thus, from only the values of $\sin t$ and $\cos t$ we can compute, using (6.4-4) to (6.4-6), $\sin 2t$, $\cos 2t$, $\sin 3t$, $\cos 3t$, . . . , $\sin kt$, $\cos kt$. In the process we do less than

$$\begin{Bmatrix}3k \text{ multiplications}\\ 2k \text{ additions}\end{Bmatrix} \quad \text{or} \quad \begin{Bmatrix}2k \text{ multiplications}\\ 3k \text{ additions}\end{Bmatrix}$$

There is, of course, a gradual loss in accuracy in computing the values in this manner, but usually this is not significant.

We now show how Eqs. (6.4-4) to (6.4-6) apply to the proof of (6.4-3), using (6.4-2). Consider first the case where all the $f(x) = 0$ except $f(j)$ and set $t = (\pi/N)k$. We see at once, from (6.4-2), that

$$U_0 = U_1 = \cdots = U_{2N-j-1} = 0 \qquad U_{2N-j} = f(j)$$

Thus U_{2N-j} corresponds to V_1, except that all the U numbers that we compute will be $f(j)$ times as large as their equivalent V numbers. By the time we reach U_{2N-1} we shall have gone exactly j steps and be at V_j. By (6.4-5) and (6.4-6), the results that we find in (6.4-3) are

$$f(j) \sin j \frac{\pi}{N} k \qquad f(j) \cos j \frac{\pi}{N} k$$

We next note that in the general case, where the $f(x)$ may be any values, the linearity of (6.4-2) causes each term to be added at the proper time so that it has exactly the correct number of multiplications to produce the proper harmonic

$$f(x) \sin x \frac{\pi}{N} k \qquad f(x) \cos x \frac{\pi}{N} k$$

and that, owing to the addition in (6.4-2), we get the required sums (6.4-3).

In this process, for each pair a_k, b_k we do about $6N$ arithmetic operations, so that for all $2N$ coefficients we do about $6N^2$ arithmetic operations. This may be contrasted to the case where we have the matrix of sines and cosines in storage and merely perform the matrix times vector product to get the coefficients. In the latter case we do about $8N^2$ arithmetic operations. Thus we may say that the proposed method saves machine time over the very favorable situation in which the matrix of sines and cosines is given.

From the two values $\sin \pi/N$, $\cos \pi/N$ we may use the V notation method to generate

$$\sin \frac{\pi}{N} k \qquad \cos \frac{\pi}{N} k \qquad (k = 2, 3, \ldots, N - 1)$$

The main process in the U notation now produces the required coefficients. By suitably interlacing the V and U processes, only the $2N$ data points plus a few registers are needed at any one time for numbers. The programs for computing the numbers are also fairly direct and consequently use a small amount of storage.

6.5 FOURIER ANALYSIS USING 12 POINTS

The special case of 12 points ($N = 6$) is of interest because it occurs frequently and is easy to do, even by hand! We have

$$6a_k = \sum_{x=0}^{11} f(x) \cos \frac{\pi}{6} kx$$

$$6b_k = \sum_{x=0}^{11} f(x) \sin \frac{\pi}{6} kx$$

If we break up each sum into the ranges 0 to 6 and 7 to 11 in x, and in the second range put $x = 12 - x'$, we get

$$6a_k = \sum_{x=0}^{6} f(x) \cos \frac{\pi}{6} kx + \sum_{x'=1}^{5} f(12 - x') \cos \frac{\pi}{6} kx'$$

$$6b_k = \sum_{x=0}^{6} f(x) \sin \frac{\pi}{6} kx - \sum_{x'=1}^{5} f(12 - x') \sin \frac{\pi}{6} kx'$$

This suggests writing

	$f(0)$	$f(1)$	$f(2)$	$f(3)$	$f(4)$	$f(5)$	$f(6)$
		$f(11)$	$f(10)$	$f(9)$	$f(8)$	$f(7)$	
Add	$s(0)$	$s(1)$	$s(2)$	$s(3)$	$s(4)$	$s(5)$	$s(6)$
Subtract		$t(1)$	$t(2)$	$t(3)$	$t(4)$	$t(5)$	

We then have

$$6a_k = \sum_{x=0}^{6} s(x) \cos \frac{\pi}{6} kx$$

$$6b_k = \sum_{x=1}^{5} t(x) \sin \frac{\pi}{6} kx$$

We again break up the range into two parts, this time 0 to 3 and 4 to 6, setting $x = 6 - x'$ in the second range.

$$6a_k = \sum_{x=0}^{3} s(x) \cos \frac{\pi}{6} kx + (-1)^k \sum_{x'=0}^{2} s(6 - x') \cos \frac{\pi}{6} kx'$$

$$6b_k = \sum_{x=1}^{3} t(x) \sin \frac{\pi}{6} kx - (-1)^k \sum_{x'=1}^{2} t(6 - x') \sin \frac{\pi}{6} kx'$$

This suggests writing

	$s(0)$	$s(1)$	$s(2)$	$s(3)$	$t(1)$	$t(2)$	$t(3)$
	$s(6)$	$s(5)$	$s(4)$		$t(5)$	$t(4)$	
Add	$u(0)$	$u(1)$	$u(2)$	$u(3)$	$p(1)$	$p(2)$	$p(3)$
Subtract	$v(0)$	$v(1)$	$v(2)$		$q(1)$	$q(2)$	

The result, when written out, is

$$6a_0 = u(0) + u(1) + u(2) + u(3) \qquad = [u(0) + u(3)] + [u(1) + u(2)]$$

$$6a_1 = v(0) + \frac{\sqrt{3}}{2} v(1) + \frac{1}{2} v(2) \qquad = \left[v(0) + \frac{1}{2} v(2)\right] + \frac{\sqrt{3}}{2} v(1)$$

$$6a_2 = u(0) + \tfrac{1}{2}[u(1) - u(2)] - u(3) = [u(0) - u(3)] + \tfrac{1}{2}[u(1) - u(2)]$$

$$6a_3 = v(0) - v(2)$$

$$6a_4 = u(0) - \tfrac{1}{2}u(1) - \tfrac{1}{2}u(2) + u(3) = [u(0) + u(3)] - \tfrac{1}{2}[u(1) + u(2)]$$

$$6a_5 = v(0) - \frac{\sqrt{3}}{2} v(1) + \frac{1}{2} v(2) \qquad = \left[v(0) + \frac{1}{2} v(2)\right] - \frac{\sqrt{3}}{2} v(1)$$

$$6a_6 = u(0) - [u(1) - u(2)] - u(3) \qquad = [u(0) - u(3)] - [u(1) - u(2)]$$

$$6b_1 = \frac{1}{2} p(1) + \frac{\sqrt{3}}{2} p(2) + p(3) \qquad = \left[\frac{1}{2} p(1) + p(3)\right] + \frac{\sqrt{3}}{2} p(2)$$

$$6b_2 = \frac{\sqrt{3}}{2} [q(1) + q(2)]$$

$$6b_3 = p(1) - p(3)$$

$$6b_4 = \frac{\sqrt{3}}{2} [q(1) - q(2)]$$

$$6b_5 = \frac{1}{2} p(1) - \frac{\sqrt{3}}{2} p(2) + p(3) \qquad = \left[\frac{1}{2} p(1) + p(3)\right] - \frac{\sqrt{3}}{2} p(2)$$

In the process we do less than 60 arithmetic operations, and most of them are simple additions. This compares favorably with the $6N^2 = 6^3 = 216$ needed by the method of Sec. 6.4. The programming, of course, is a bit longer to write but is very easy as there is no complex logic and only one nonsimple number $\sqrt{3}/2 = 0.8660254075$ is required. It is remarkably easy to do by hand.

These general methods can be, and have been, applied in cases other than $N = 6$.

EXERCISE

6.5-1. If $f(x) = x(12 - x)$ for $x = 0, 1, \ldots, 11$, find, by hand, the Fourier expansion of $f(x)$.

6.6 MINIMUM MULTIPLICATION METHODS

In general, it is not the purpose of this book to go into the details of how to organize a computation to save machine time. Since there is normally such a strong preference for polynomial representation at the expense of other forms, and because the finite Fourier summations sometimes consume so much actual machine time, it has been necessary to show how Fourier series can be computed reasonably efficiently. The methods given should not be taken as necessarily the best.

The number of multiplications can be kept to a minimum by arranging to perform all the additions first, making full use of the reduction of the trigonometric functions to values in the first quadrant. This is why the method of Sec. 6.5 works. Whenever the order of the harmonic being found divides the number $2N$, then some of the possible set of values of the trigonometric function do not occur. It is for this reason that values $2N = 12$, 24, and 60 are so often used in practice.

6.7 COSINE EXPANSIONS

If we assume that the function $f(x)$ is periodic of period $2N$, then whether we use $x = 0, 1, \ldots, 2N - 1$ or $-(N - 1), -(N - 2), \ldots, 0, 1, \ldots, N$, the result is the same. Now suppose that we have a function defined for $x = 0, 1, \ldots, N$ and we wish to expand it in a series of cosine terms alone. Let us *define* $f(-x) = f(x)$ for $x = 1, 2, \ldots, N - 1$. The function $f(x)$ is now an *even function*. When we make the Fourier expansion, we find that all the $b_k = 0$, since

$$b_k = \frac{1}{N} \sum_{x=-N+1}^{N} f(x) \sin \frac{\pi k}{N} x$$

and the sine is an odd function. Thus we have a cosine expansion. The formulas for the a_k can be simplified:

$$a_k = \frac{1}{N} \sum_{x=-N+1}^{N-1} f(x) \cos \frac{\pi k}{N} x$$

$$= \frac{2}{N} \left[\frac{1}{2} f_0 + \sum_{x=1}^{N-1} f(x) \cos \frac{\pi k}{N} x + \frac{1}{2} f_N (-1)^k \right]$$

By defining $f(x)$ to be an odd function, we can get sine expansion in a similar fashion.

We also note that if we have a range $0 \leq x \leq R$ the substitution

$$x = \frac{R}{2} (1 + \cos t)$$

makes the function periodic in t as well as an even function of t. This substitution has a number of advantages in matters of speed of convergence which will be discussed in Chap. 22.

EXERCISES

6.7-1. If $f(x)$ is defined for the interval $0 \leq x \leq \pi/2$, show how the function may be extended to $-\pi < x \leq \pi$ so that only odd cosine harmonics occur. Only even cosine harmonics.

6.7-2. Discuss in detail the problem of finding a Fourier series in sines only. In odd (or even) sine harmonics.

6.8 LOCAL FOURIER SERIES

In many problems the phenomena change slowly, and the idea of a "local Fourier series" whose coefficients change slowly with time is quite natural. It is also natural to center the range about the origin and use the values from $-N + 1$ to N rather than from 0 to $2N - 1$. Thus we have for Eqs. (6.2-5) to (6.2-7)

$$\begin{aligned} a_m(t) &= \frac{1}{N} \sum_{x=-N+1}^{N} f(x+t) \cos \frac{\pi}{N} mx \qquad m = 0, 1, \ldots, N \\ b_m(t) &= \frac{1}{N} \sum_{x=-N+1}^{N} f(x+t) \sin \frac{\pi}{N} mx \qquad m = 1, 2, \ldots, N-1 \end{aligned} \tag{6.8-1}$$

It is easy to obtain $a(t+1)$ and $b(t+1)$ from $a(t)$ and $b(t)$. For example,

$$a_m(t+1) = \frac{1}{N} \sum_{x=-N+1}^{N} f(x+t+1) \cos \frac{\pi}{N} mx$$

Set $x + 1 = x'$:

$$\begin{aligned} a_m(t+1) &= \frac{1}{N} \sum_{x'=-N}^{N+1} f(x'+t) \cos \frac{\pi}{N} m(x'-1) \\ &= \frac{1}{N} \left\{ \sum_{x'=-N+1}^{N} f(x'+t) \cos \frac{\pi}{N} m(x'-1) \right. \\ &\qquad \left. + (-1)^m [f(N+1+t) - f(-N+t)] \right\} \end{aligned}$$

Expanding the cosine term,

$$\begin{aligned} a_m(t+1) = &\left[a_m(t) \cos \frac{\pi}{N} m + b_m(t) \sin \frac{\pi}{N} m \right] \\ &+ \frac{(-1)^m}{N} [f(N+1+t) - f(-N+t)] \end{aligned}$$

Note that we use fixed (for a given frequency) multipliers $\cos \pi m/N$ and $\sin \pi m/N$ and add or subtract the two new values in the range.

Similarly for the sine term,

$$b_m(t+1) = \left[-a_m(t) \sin \frac{\pi}{N} m + b_m(t) \cos \frac{\pi}{N} m \right]$$

PART II

Polynomial Approximation—Classical Numerical Analysis

CHAPTER 7

Introduction to Polynomial Approximations

7.1 ORIENTATION

The first experience in numerical computation is usually in the area of function evaluation. The main techniques used in this field are "common sense" and "low cunning." The purpose of this book is to present the main ideas in the field of computing, and we shall only occasionally illustrate the "low cunning" aspect, although this is still an important part of the art of computing.

The second type of experience is usually interpolation for missing values in a table, such as a logarithmic or trigonometric table; "reading between the lines" it has been called. Abstractly speaking, in interpolation we are given some samples of the function and are asked to compute an estimate of some value that is not in the table. In most tables the assumption is made that the function behaves as a straight line between successive sample points, although occasionally quadratic polynomials and even higher powers are assumed.

The main problems of numerical analysis are often thought to be centered around interpolation, but this is not so. Rather they involve analytic operations such as integration, differentiation, finding zeros, maximization, etc., of a function, when all that we have, or can compute, are some samples of the function, and these are usually not known exactly but merely approximately, being contaminated by roundoff noise.

The classical numerical-analysis approach is to use some samples of the function to produce an approximating polynomial and then perform the analytic operation on this approximating polynomial. This process may be called [38, p. 51] "analytic substitution," that is, for a function we could not handle we substitute another function on which we can perform the analytic operation. For example, in Newton's iterative method for finding a zero of a function

$$y = f(x)$$

we are given an approximate value x_1, and in place of the curve we use the line

$$y - y_1 = y_1'(x - x_1)$$

which is tangent to the curve at the point (x_1, y_1). The zero of the line is given by $y = 0$, or

$$x = x_1 - \frac{y_1}{y_1'}$$

This new value of x is used as the next approximate value of the zero.

Because polynomials are easy to handle, much of classical numerical analysis is based on polynomial approximation. However, for many purposes other classes of functions are preferable, and we shall study them after we examine polynomial approximation.

Having selected the samples and the class of approximating functions, we still must choose one particular member of the class by means of some criterion—some measure of approximation or "goodness of fit." The most widely used criterion is that of requiring the approximating function to match exactly the samples of the function at the sample points. Another rather common criterion is "least squares," meaning that "the sum of the squares of the deviations between the given samples and the approximating function at the sample points is to be made a minimum." Still other criteria are sometimes used.

Before we begin a computation we must also decide what accuracy we want in the answer and what criterion we shall adopt for measuring this accuracy. This approach may be summarized in the form of four questions:

1. What samples shall we use?
2. What class of approximating functions shall we use?
3. What criterion of goodness of fit shall we use?
4. What accuracy do we want?

EXERCISE

7.1-1. Using the three-term Taylor series

$$y = y(x_1) + (x - x_1)y'(x_1) + \frac{(x - x_1)^2 y''(x_1)}{2}$$

to approximate a function at the point x_1, develop an extension of Newton's formula for approximating zeros of a function.

Ans. $x = x_1 - \dfrac{y(x_1)}{y'(x_1)}\left[1 + \dfrac{y(x_1)y''(x_1)}{(2y'(x_1))^2}\right]$

7.2 ALTERNATIVE FORMULATIONS

In the process just described, having found the approximating function, we then apply the analytic operation and evaluate the result at some point or points. Usually the final result is found to be a linear combination of the values of the original samples. This suggests that we can go directly from the samples to the answer without producing the approximating function along the way.

In order to make these remarks more precise, suppose that we have some linear operator L operating on some class of functions, that is,

$$L[af(x) + bg(x)] = aL[f(x)] + bL[g(x)] \tag{7.2-1}$$

Integration, differentiation, and interpolation are typical linear operations. Finding a zero is not a linear operation.

The numerical estimate of the analytic operation $L[f(x)]$ will be estimated from samples of the function and its various derivatives (occasionally we use means, moments, etc., but they are not used in this formula), since they can usually be computed easily when the function is given. Thus we seek a formula of the class

$$\begin{aligned} L[f(x)] = {} & a_1 f(x_1) + a_2 f(x_2) + \cdots + a_n f(x_n) \\ & + b_1 f'(x_1) + b_2 f'(x_2) + \cdots + b_n f'(x_n) + \cdots \\ & + k_1 f^{(k-1)}(x_1) + k_2 f^{(k-1)}(x_2) + \cdots + k_n f^{(k-1)}(x_n) \end{aligned} \tag{7.2-2}$$

where many of the coefficients may be zero. To simplify matters of presentation, we shall assume for the moment that we have only values of the function (b's, c's, . . . , k's $= 0$).

Under the assumptions that we are using polynomial approximation and exact matching at the given sample points, there are three equivalent ways to determine the coefficients a_i. The first method, which we have already discussed, is to find the approximating polynomial and operate on the polynomial with the operator L; it is called *the analytic substitution method.*

The second method is to expand every term in a suitable *Taylor series* and equate coefficients of like derivatives. As an example of this method, consider finding a formula such as Simpson's

$$\int_{-1}^{1} f(x)\,dx = a_{-1}f(-1) + a_0 f(0) + a_1 f(1) \tag{7.2-3}$$

Since

$$f(x) = f(0) + xf'(0) + \frac{x^2}{2}f''(0) + \frac{x^3}{3!}f'''(0) + \frac{x^4}{4!}f^{\text{IV}}(0) + \cdots$$

we get for the left-hand side of the equation

$$\int_{-1}^{1} f(x)\,dx = 2f(0) + \frac{2}{3!}f''(0) + \frac{2}{5!}f^{\text{IV}}(0) + \cdots$$

while on the right-hand side we get

$$a_{-1}f(-1) = a_{-1}f(0) - a_{-1}f'(0) + \frac{a_{-1}}{2!}f''(0) - \frac{a_{-1}}{3!}f'''(0) + \frac{a_{-1}}{4!}f^{\text{IV}}(0) + \cdots$$

$$a_0 f(0) = a_0 f(0)$$

$$a_1 f(1) = a_1 f(0) + a_1 f'(0) + \frac{a_1}{2!}f''(0) + \frac{a_1}{3!}f'''(0) + \frac{a_1}{4!}f^{\text{IV}}(0) + \cdots$$

Equating the coefficients of $f(0)$, $f'(0)$, . . . as far as we can, we get

$$\begin{aligned}
2 &= a_{-1} + a_0 + a_1 \\
0 &= -a_{-1} \qquad + a_1 \\
2/3 &= a_{-1} \qquad + a_1 \\
0 &= -a_{-1} \qquad + a_1 \\
2/5 &= a_{-1} \qquad + a_1 + E
\end{aligned}$$

where E is the error. The solution of these equations is $a_{-1} = a_1 = 1/3$, $a_0 = 4/3$ which gives the usual formula,

$$\int_{-1}^{1} f(x)\,dx = 1/3 f_{-1} + 4/3 f_0 + 1/3 f_1 \tag{7.2-4}$$

The third method requires that the formula be *exact for a sequence of functions* $f(x) = 1, x, x^2, \ldots$ as far as we can go. This method is most useful for deriving specific formulas, and we shall use it for this purpose most of the time. However, for purposes of computing a specific problem, the method of analytic substitution, which first produces the approximating function to the original data, is often preferable because it enables the user of the results to see with his eye what kind of an approximation is being made. Thus in the example of Sec. 1.9, when we found the approximating polynomial, we evaluated this polynomial at the points midway between the sample points and plotted the results to see what kind of a pattern they made.

We need to show that these three methods are equivalent and give the same answer. The third method, which makes the formula exact for the various powers of x up to some order, say n, also makes the formula exact for any polynomial of degree less than or equal to n. Thus if the function is a polynomial of degree n (or less), the formula is exact. But if the function is a polynomial, then the Taylor series also ends at the nth-order term and fits the result of the second method. The converse is also true, and the second method is equivalent to the third method.

In order to show that the first method, that of finding the interpolating polynomial and then operating on it, is equivalent to the other two methods, it is necessary to have available some of the theory of interpolating polynomials, and this we shall develop in the next chapter. Meanwhile, suppose that given the samples $f_i(i = 1, 2, \ldots, n)$ of the function, we can find a polynomial[1]

$$P(x) = L_1(x)f_1 + L_2(x)f_2 + \cdots + L_n(x)f_n \qquad (7.2\text{-}5)$$

By setting $f(x) = 1, x, x^2, \ldots, x^{n-1}$, we can find corresponding polynomials $P(x)$. If we make the analytic substitution of these functions, one at a time, into the formula and carry out the operations involved in $L[f(x)]$, we obtain a formula which is exact for the functions $1, x, x^2, \ldots, x^{n-1}$.

Conversely, since $L[f(x)]$ is linear, if it is exact for $1, x, x^2, \ldots, x^{n-1}$, then it is exact for any linear combination; hence it is exact for the linear combination (7.2-5). Thus the third method is equivalent to the first method.

The third method which we shall use for finding formulas is one of broad, general application. However, it is often more work to find a formula by this general method than it is to use various tricks usually found in textbooks. In defense of our approach, we say, "In view of the vast amount of knowledge now available, it is better to use one general, somewhat inefficient method than to learn many special tricks." The resulting formulas are the same regardless of the approach. The method, being quite general, easily produces new formulas and provides a useful tool for further research. Lastly, the adoption of such a uniform method is in the direction of gradually putting the labor of derivation on the machine where, in the long run, it belongs.

Another feature of this book is the use of the ideas of information theory. It may be merely the author's prejudice, but it seems reasonable to regard a large-scale computer as processing information. Many of the ideas of information theory are closely related to band-limited functions which are treated in Part III of this book. Although, as we have seen in Chap. 2, a theory that considers noise as a fundamental fact has not yet been developed, there is a significant trend in this direction, both in recent books and published papers. The best that we can do at present is to remember that almost every number that we get will have some noise contaminating it and to examine the formulas that we produce in the light of how they will propagate the noise.

EXERCISE

7.2-1. Derive the trapezoid rule by the first, second, and third methods.

[1] The $L_i(x)$ are defined in Sec. 8.3 and are not to be confused with the linear operator L in Eq. (7.2-1).

7.3 SAMPLES; INFORMATION

The first of the four questions listed at the end of Sec. 7.1 was "What samples shall we use?" In principle, this is a question in statistics, especially in the field known as "design of experiments." In practice, we often have the samples given to us by outside circumstances, or we use a set of equally spaced points. If we use the equally spaced points, then we come face to face with the sampling theorem of information theory, which falls in Part III but lies behind, and really "explains," much of the material in Part II. Thus in some respects much of Part II is somewhat superficial.

To see that there is a real problem in choosing which samples to use, imagine that you are asked to compute

$$\int_0^1 f(x)\, dx$$

for some difficult function which you cannot integrate analytically. You are being asked, really, to estimate the area under the curve based on a number of samples (say n of them) or, what is the same thing, to estimate the average value of $f(x)$. Ideally, we would need to know a great deal about the function to place our samples to best advantage. As we assume various properties of the function, we get various sampling plans. Because of the author's prejudices, the words "samples" and "information" will often be used interchangeably.

As another example of asking "Where does the information lie?" suppose that we are asked to estimate a derivative at a point (which is a local property, as contrasted with integration). Common sense dictates the use of information close to the point where the derivative is to be estimated. On the other hand, if the samples are taken too near, then the noise in the computing, or in the measuring, of the samples will prevent an accurate estimate from being made. Thus an adequate answer to the question of where to take our samples requires an assessment of the noise that contaminates all our information. As remarked before, an adequate theory of noise is not available, but progress is slowly being made in that direction.

7.4 THE CLASS OF FUNCTIONS

The second of the four questions was "What class of approximating functions shall we use?"

There are three classes, or sets, of functions widely used in numerical analysis. The first set is generated by linear combinations of

$$1, \quad x, \quad x^2, \quad \ldots, \quad x^n$$

and is the same as the class of all polynomials of degree n (or less). The second set is generated by

$$\cos a_i x \qquad \sin a_i x$$

which is related to the Fourier series and Fourier integral. These are treated in Part III. The third set is generated by

$$e^{-a_i x}$$

which are the growth and decay functions, so common in actual situations, and is treated in Part III, Chap. 26.

Each of these three sets has this important property: The finite set, as a set, goes into itself when x is replaced by $x + k$. For example, if $P(x)$ is a polynomial of degree n, then $P(x + k)$ is also a polynomial of degree n in x, although, of course, the coefficients are different. The same applies to the other two sets.

This property is important since it implies that when we choose the set of approximating functions we do not need to have any idea of where the origin is. Any other finite set of approximating functions (not a combination of these three) implies that we have a natural origin, since the location of the origin will affect the answer.[1] If there is a singularity in the problem, this automatically defines a natural origin, and usually the nature of the singularity suggests the proper set of approximating functions to use (see Part III, Chap. 27). But in most problems there is no natural origin, and we are forced to select one of these three sets, or combinations of them, if the choice of the origin is not to influence the answer being computed; it is unfortunate enough that the set itself influences the answer.

The set of linear combinations of $1, x, x^2, \ldots, x^n$ has a second important property, namely, that as a set it is also unchanged by replacing x by kx. The other two sets do not have this property. Thus if there is no natural scale in the problem, we are forced to use polynomials or to influence the answer by the very choice of the scale. Although this is a convenient property to have, many problems have a natural scale, especially those depending on time, and the dominance of polynomials over the other two has been unfortunate for the computing field. The comparative neglect of the other two sets means that their theories have not been developed as far as that of polynomials, and as a result we tend to regard their theories as being more difficult than they really are.

One combination of these functions which has this scale-free property

[1] We do not prove this statement here but merely use it to explain why these three sets are mainly discussed.

is the ratio of two polynomials or, more simply, rational functions, and there is gradually developing an adequate theory for them (see Chap. 20). We do not normally come across constants raised to a polynomial power or polynomials raised to a polynomial power, each of which would also be scale-free.

7.5 GOODNESS OF FIT

The third of the four questions was "What criterion of goodness of fit shall we use?" The classical answer has been "exact matching at the sample points," which has the advantage of a simple theory and ease of computing and also the disadvantage of ignoring the noise. Much of Part II is devoted to exact matching, but it should be remembered that this criterion neglects fundamentally important aspects of the typical computing situation. We have already observed that the criterion of exact matching at the sample points is the same as making the formula exact for $1, x, \ldots, x^n$ in the polynomial case.

Another well-regarded criterion is "least squares," meaning that the sum of the squares of the errors of the fit at the sample points is to be made the least possible, or, more simply, minimized. This method uses redundant information to obtain some smoothing of the noise. The criterion is popular with mathematicians because of the beauty of the mathematical theory which goes with it and with physicists because they believe that least squares is a principle of nature, but a person in the computing field is entitled to view the criterion with suspicion in any particular application (see Chaps. 17 and 18).

A third criterion is associated with the name Chebyshev.[1] Basically the idea is to reduce the maximum error to a minimum, a criterion that is becoming quite popular these days (see Chap. 19).

Obviously other criteria can be used, and the proper one to select normally comes from the physical problem from which the computation develops. We shall try to give a little feeling for the various criteria as we introduce them, but since this is a text on numerical methods we cannot spend the time to develop a real feeling for the art of selecting the proper criterion. The choice of the criterion can frequently affect the computation, both in the amount to be done and in the results obtained, and therefore it deserves careful consideration before starting a computation.

[1] This has many Americanized spellings, among which are

$$\begin{Bmatrix}\text{Tsch}\\ \text{Tch}\\ \text{Ch}\\ \text{\v{C}}\end{Bmatrix} \text{eb} \begin{Bmatrix}\text{y}\\ \text{i}\end{Bmatrix} \begin{Bmatrix}\text{sch}\\ \text{ch}\\ \text{sh}\\ \text{\v{s}}\end{Bmatrix} \begin{Bmatrix}\text{\"{e}}\\ \text{e}\\ \text{o}\end{Bmatrix} \begin{Bmatrix}\text{f}\\ \text{ff}\\ \text{v}\end{Bmatrix}$$

7.6 ACCURACY

The last of the four questions was "What accuracy do we want?" The superficial type of answer usually given is "Three or four decimal places in the final answer."

In many problems this is not adequate; it ignores fundamental aspects of the question. Consider, for example, solving a system of linear algebraic equations for the unknowns x_i. We must answer the following questions:

1. Is it the x_i that are to be accurate?
2. Is it the residuals of the equations when the computed x_i are substituted in them which are to be small?
3. Or is it the given set of equations which are to be close to the set of equations for which the x_i are the exact answers?

Evidently still other measures of accuracy can be devised to suit particular situations.

The first one is the usual answer given, and only occasionally is it the most suitable one.

The second criterion is often the proper one, because the vanishing of the equations is being sought, and what is being computed are the x_i that will produce this vanishing.

The third criterion is also very useful and is gradually being used more and more in applications as well as in formal error analysis. In a sense it is saying, "Since neither your measurements nor your physical theory are believed to be exact, how close to them should I come when I am computing?" In many respects this is the most basic criterion, the one from which guesses are made for the other criteria. Errors in the known weight of a missile or of the planets or even in the amount of the light pressure can be converted into equivalent errors allowed at each step of the computation of a trajectory, and this often makes more sense than giving an arbitrary "allowable error" to be achieved at the end of the trajectory. How was the "allowable error" determined?

Thus we see that the way in which we answer the four questions

1. What samples shall we use?
2. What class of approximating functions shall we use?
3. What criterion of goodness of fit shall we use?
4. What accuracy do we want?

can greatly influence the answers that we obtain from the computation. It should also be clear that the answers to these questions must be found in the original problem and not in mathematical treatises or even in books on numerical analysis. Sound computing practice requires constant examination of the problem being studied, not only before the computing is organized but also as it progresses and especially during the stage when the numbers obtained are being translated back to, and interpreted in terms of, the original problem.

CHAPTER 8

Polynomial Interpolation—Arbitrarily Spaced Data

8.1 PHILOSOPHY

There are two main uses for interpolation formulas. In the first place, they are used in the process of analytic substitution of a tractable function for an intractable one. They are also used occasionally to separate out the dependence of an answer on a parameter, as in the example of Sec. 1.9.

The second main use is for interpolating in tables, something that in these days, with large-scale digital computers, is not done very much. Computers usually use a subroutine to compute the functional values rather than to try to find them from a table stored in the machine. The formulas for computing the values are often found semiempirically and do not depend on the technique of polynomial interpolation. On the other hand, the highly developed art of table making[1] does make use of elaborate interpolating methods. The prospective maker of a table is strongly advised not to enter into such a project without first making a careful study of the known techniques and pitfalls.

8.2 INTERPOLATION POLYNOMIALS

A polynomial of degree n

$$y(x) = a_0 + a_1x + a_2x^2 + \cdots + a_nx^n = \sum_{k=0}^{n} a_kx^k \qquad (8.2\text{-}1)$$

has $n + 1$ coefficients, and it is reasonable to expect that $n + 1$ conditions imposed on this general form will determine the coefficients uniquely. In particular, we can require the polynomial to pass through the $n + 1$ points (x_i,y_i) $(i = 1, 2, \ldots, n + 1)$ with $x_i \neq x_j$. To say that the

[1] See, for example, Ref. 10.

polynomial passes through the points (x_i,y_i) means that

$$y_i = \sum_{k=0}^{n} a_k x_i^k \qquad (i = 1, 2, \ldots, n+1)$$

The determinant of the unknowns a_k of these $n + 1$ linear equations is the Vandermonde determinant

$$\Delta = \begin{vmatrix} 1 & x_1 & x_1^2 & \cdots & x_1^n \\ 1 & x_2 & x_2^2 & \cdots & x_2^n \\ \cdots & \cdots & \cdots & \cdots & \cdots \\ 1 & x_{n+1} & x_{n+1}^2 & \cdots & x_{n+1}^n \end{vmatrix} = f(x_1,x_2, \ldots, x_{n+1})$$

which we now show is not zero if $x_i \neq x_j$ for $i \neq j$. The determinant is clearly a function of $x_1, x_2, \ldots, x_{n+1}$. If we regard it first as a function of x_{n+1}, then it is a polynomial of degree n and vanishes whenever $x_{n+1} = x_j$ (for $j = 1, 2, \ldots, n$). Thus $f(x_1,x_2, \ldots, x_{n+1})$ has the factors

$$\prod_{i=1}^{n} (x_{n+1} - x_i) = (x_{n+1} - x_1)(x_{n+1} - x_2) \cdots (x_{n+1} - x_n)$$

Regarding the determinant now as a function of x_n, we see by the same argument that there are factors

$$\prod_{i=1}^{n-1} (x_n - x_i) = (x_n - x_1)(x_n - x_2) \cdots (x_n - x_{n-1})$$

and, in general, all factors

$$\prod_{j>i=1}^{n+1} (x_j - x_i)$$

occur in the determinant. The product of all these factors is a polynomial of degree

$$n + (n-1) + (n-2) + \cdots + 1 = \frac{n(n+1)}{2}$$

But the determinant is also a polynomial of the same degree; hence

$$|x_i^k| = C \prod_{j>i=1}^{n} (x_j - x_i) \tag{8.2-2}$$

where C is some constant still to be found. To determine C, consider the term that comes from multiplying the main diagonal terms

$$1 \cdot x_2 \cdot x_3^2 \cdots x_{n+1}^n$$

Expanding the product, we find exactly the same term, and hence $C = 1$. Thus the Vandermonde determinant is not zero if $x_i \neq x_j$ for $i \neq j$.

Returning to our main problem of finding the polynomial through the $n + 1$ points (x_i, y_i), we see that we can always solve for the coefficients a_k by Cramer's rule or by any other method. Putting these values into the general form of the polynomial (8.2-1), we can rearrange the result in the form

$$\begin{vmatrix} y & 1 & x & x^2 & \cdots & x^n \\ y_1 & 1 & x_1 & x_1^2 & \cdots & x_1^n \\ y_2 & 1 & x_2 & x_2^2 & \cdots & x_2^n \\ \cdots & \cdots & \cdots & \cdots & \cdots & \cdots \\ y_{n+1} & 1 & x_{n+1} & x_{n+1}^2 & \cdots & x_{n+1}^n \end{vmatrix} = 0 \qquad (8.2\text{-}3)$$

Alternatively, we could have written this last expression by inspection, reasoning as follows: First, (8.2-3) must be a polynomial of degree n in x since we can expand the determinant by the elements of the top row, and, second, it passes through the $n + 1$ points (x_i, y_i), since putting these values in the top row makes two rows of the determinant the same.

It is possible that the coefficient of x^n is zero (the condition for this to happen is obvious) and that the polynomial is of degree less than n. To cover this detail, the statement that a polynomial is of degree n often means of degree n or less. In the trivial case when all the y_i are equal the polynomial is of degree zero, and $y(x) \equiv C$.

This may be summarized by saying that, if we have $n + 1$ sample points of a function, we can find a polynomial of degree n which agrees (neglecting roundoff errors) with the function at the sample points. How closely the two functions agree between the sample points will be investigated in Sec. 8.6, but, assuming that they are close, we are in a position to use the polynomial in place of the function in future analytical processes such as integration, differentiation, finding zeros, etc.

Not only may we require a polynomial to pass through certain given points; we may also require it to have given slopes at some points. Thus the determinant

$$\begin{vmatrix} y & 1 & x & x^2 & x^3 \\ y_1 & 1 & x_1 & x_1^2 & x_1^3 \\ y_2 & 1 & x_2 & x_2^2 & x_2^3 \\ y_1' & 0 & 1 & 2x_1 & 3x_1^2 \\ y_2' & 0 & 1 & 2x_2 & 3x_2^2 \end{vmatrix} = 0$$

is a cubic in x and passes through (x_1, y_1) with slope y_1' and through (x_2, y_2) with slope y_2' (since differentiating the determinant with respect to x is the same as differentiating the top line term by term).

It is not always necessary to specify the y value when giving a y' value at a point, and the same applies to higher derivatives. The restraint on the choice of conditions requires that the minor of the y term must not be zero; otherwise there would be no polynomial. This means that at least one function value y_i must be given, at least two conditions on the y_i and y_i', three on the y_i, y_i', y_i'', etc., up to the highest-order derivative given.

The condition just stated is not sufficient. Consider, for example, three equally spaced points, which we label -1, 0, 1. At each point we are given the function y_i and the second derivative, the position and acceleration in physical terms. The minor of the y term is

$$\begin{vmatrix} 1 & -1 & 1 & -1 & 1 & -1 \\ 1 & 0 & 0 & 0 & 0 & 0 \\ 1 & 1 & 1 & 1 & 1 & 1 \\ 0 & 0 & 2 & -6 & 12 & -20 \\ 0 & 0 & 2 & 0 & 0 & 0 \\ 0 & 0 & 2 & 6 & 12 & 20 \end{vmatrix} = 0$$

Thus these six conditions do not suffice to determine a quintic polynomial.

EXERCISES

8.2-1. Write the polynomial passing through (x_1,y_1) with slope y_1' and second derivative y_1'' and having slope y_2' at x_2.

$$Ans. \quad \begin{vmatrix} y & 1 & x & x^2 & x^3 \\ y_1 & 1 & x_1 & x_1^2 & x_1^3 \\ y_1' & 0 & 1 & 2x_1 & 3x_1^2 \\ y_1'' & 0 & 0 & 2 & 6x_1 \\ y_2' & 0 & 1 & 2x_2 & 3x_2^2 \end{vmatrix} = 0$$

8.2-2. Show that, if we use the conditions on y, y', y'', . . . , $y^{(n-1)}$ at one point x_1, then the determinant becomes the truncated Taylor series.

8.2-3. Find a nontrivial polynomial $y(x)$ such that $y(-1) = y(0) = y(1) = y''(-1) = y''(0) = y''(1) = 0$. Discuss its significance. One possible answer: $y(x) = 3x^5 - 10x^3 + 7x$. (*Note:* $\sin \pi x$ also has this property.)

8.3 THE LAGRANGE METHOD OF INTERPOLATION

Another approach to the interpolation problem is the *Lagrange method.* The basic idea behind the method is first to find a polynomial which takes on the value 1 at a particular sample point and the value 0 at all the other sample points. It is easy to see that the function

$$L_j(x) = \frac{(x - x_1)(x - x_2) \cdots (x - x_{j-1})(x - x_{j+1}) \cdots (x - x_{n+1})}{(x_j - x_1)(x_j - x_2) \cdots (x_j - x_{j-1})(x_j - x_{j+1}) \cdots (x_j - x_{n+1})}$$

$$= \frac{\prod_{i=1}^{n+1}{}' (x - x_i)}{\prod_{i=1}^{n+1}{}' (x_j - x_i)}$$

(where the prime on the product means "excluding the jth value") is such a polynomial of degree n; it is 1 when $x = x_j$ and 0 when $x = x_i$, $i \neq j$.

The polynomial $L_j(x)y_j$ takes on the value y_j at the sample point x_j and is zero at all other sample points. It then follows that

$$y(x) = \sum_{j=1}^{n+1} L_j(x)y_j$$

is a polynomial of degree n passing through the $n + 1$ points (x_i,y_i).

We may now ask if the two polynomials, the Lagrange and the determinant, are the same. Their difference is at most a polynomial of degree n, and it vanishes at the $n + 1$ sample points $x = x_i$; hence it is identically zero. This is an important observation, namely, *given the* $n + 1$ *sample points, the corresponding* nth *degree polynomial passing through these points is uniquely* (within roundoff errors) *determined, regardless of how it is constructed or the particular notation used.* It is necessary to emphasize this, because some books on numerical analysis would lead the reader to think that the various formulas represent different polynomials, either because of how they were found or because of notation. If different sample points are used, then, of course, the polynomials may be different, but *the same sample points must lead to the same polynomial* (within roundoff errors).

The Lagrange interpolation formula is not very practical for computations. Its form can be modified somewhat to make it more tractable. Consider the special case when all the $y_i = 1$. Then $y(x) = 1$ for all x, that is,

$$1 = \sum_{j=1}^{n+1} L_j(x)$$

is an identity. We may now divide the right-hand side of the Lagrange formula by this expression, and defining A_j as

$$A_j = \frac{1}{\prod_i{}' (x_j - x_i)}$$

we get, when we divide numerator and denominator by $\prod_j (x - x_j)$,

$$y(x) = \frac{\sum_{j=1}^{n+1} [A_j y_j/(x - x_j)]}{\sum_{j=1}^{n+1} [A_j/(x - x_j)]}$$

This is sometimes called the "barycentric formula"[1] and is easier to use than the Lagrange formula.

[1] W. J. Taylor, *J. Research Natl. Bur. Standards*, vol. 35, pp. 151–159, 1945.

We now consider the problem of finding an interpolation polynomial which at each point x_i takes the value y_i and has the derivative y'_i ($i = 1, \ldots, n+1$). This is called the *Hermite polynomial* and clearly is of degree $2n + 1$ (since there are $2n + 2$ conditions to be satisfied).

We are more concerned with the method used to construct the polynomial than we are with the result. The approach is a natural extension of the Lagrange approach. We seek polynomials

$$H_i(x) \qquad \text{and} \qquad h_i(x)$$

of degree (at most) $2n + 1$ such that

$$\begin{aligned} H_j(x_i) &= 0 & h_j(x_i) &= 0 \\ H'_j(x_i) &= 0 & h'_j(x_i) &= 0 & \quad (i \neq j) \\ H_j(x_j) &= 1 & h_j(x_j) &= 0 \\ H'_j(x_j) &= 0 & h'_j(x_j) &= 1 \end{aligned}$$

Once we find these polynomials, it is then clear that the solution is

$$y(x) = \sum_{j=1}^{n+1} [H_j(x)y_j + h_j(x)y'_j]$$

We first construct $h_j(x)$. If $h_j(x_i) = 0$ and $h'_j(x_i) = 0$, then clearly $h_j(x)$ has a multiple factor

$$(x - x_i)^2 \qquad (\text{for } i \neq j)$$

If $h_j(x_j) = 0$, but $h'_j(x_j) \neq 0$, then there is a simple factor

$$(x - x_j)$$

In order to have $h'_j(x_j) = 1$ we must have

$$h_j(x) = \frac{(x - x_1)^2(x - x_2)^2 \cdots (x - x_{j-1})^2(x - x_j)(x - x_{j+1})^2 \cdots (x - x_{n+1})^2}{(x_j - x_1)^2(x_j - x_2)^2 \cdots (x_j - x_{j-1})^2(x_j - x_{j+1})^2 \cdots (x_j - x_{n+1})^2}$$

The reader should convince himself that $h_j(x)$ has all the required properties.

The construction of $H_j(x)$ is much the same. Again we have the factors

$$(x - x_i)^2 \qquad (\text{for } i \neq j)$$

But this time we require

$$H_j(x_j) = 1 \qquad H'_j(x_j) = 0$$

We try an arbitrary linear factor

$$p(x) = ax + b$$

$$H_j(x) = \frac{(x - x_1)^2(x - x_2)^2 \cdots (x - x_{j-1})^2(ax + b)(x - x_{j+1})^2 \cdots (x - x_{n+1})^2}{(x_j - x_1)^2(x_j - x_2)^2 \cdots (x_j - x_{j-1})^2(x_j - x_{j+1})^2 \cdots (x_j - x_{n+1})^2}$$

But $H_j(x_j) = 1$ requires that

$$ax_j + b = 1$$

Next, taking the derivatives one term at a time and putting $x = x_j$ in each, we get

$$H'_j(x_j) = \frac{2}{x_j - x_1} + \frac{2}{x_j - x_2} + \cdots + \frac{2}{x_j - x_{j-1}} + a + \frac{2}{x_j - x_{j+1}} + \cdots + \frac{2}{x_j - x_{n+1}} = 0$$

Thus

$$a = -2 \sum_{i=1}^{n+1}{}' \frac{1}{x_j - x_i}$$

$$b = 1 - ax_j$$

and we have constructed the required formulas.

This method of constructing polynomials having specific properties occurs in many places in the theory of numerical computation and should be mastered. We have used it in two specific situations, Lagrange and Hermite interpolation, but the general method of matching $y_i, y'_i, \ldots, y^{(n_i)}$ at x_i should be clear. As we shall see when we examine the error in the interpolation process, the use of derivatives at nearby points greatly increases the accuracy of a formula.

EXERCISE

8.3-1. Show how to obtain an interpolation formula which has the values y_i, y'_i, and y''_i at each x_i.

8.4 NEWTON'S INTERPOLATION FORMULA

The first two methods for finding an interpolating polynomial through $n + 1$ points tacitly assumed that the number of points to be used was known. Often what is known is the accuracy desired, and the number of points to be used is determined as information about the function is computed. Newton's interpolation formula, which we now develop, is simply another way of writing the interpolating polynomial. It is useful because the number of points being used can easily be increased or decreased without repeating all the computation. The Newton's formula of Sec. 1.5 is really the special case when the sample points are equally spaced.

As before, let the polynomial passing through the $n + 1$ points $(x_i,y_i)(i = 1, 2, \ldots, n + 1)$ be labeled $P_n = P_n(x)$. We can write

$$P_n(x) = y_1 + (x - x_1)P_{n-1}(x)$$

where $P_{n-1}(x)$ is some polynomial of degree $n - 1$. It is clear that $P_n(x_1) = y_1$, and we are reduced to taking care of only n points $(i = 2, 3, \ldots, n + 1)$. The above equation can be written

$$P_{n-1}(x) = \frac{P_n(x) - y_1}{x - x_1}$$

and hence we want

$$P_{n-1}(x_i) = \frac{P_n(x_i) - P_n(x_1)}{x_i - x_1} = \frac{y_i - y_1}{x_i - x_1} \qquad (i = 2, 3, \ldots, n + 1)$$

That is to say, we want $P_{n-1}(x)$ to pass through the points

$$\left(x_i, \frac{y_i - y_1}{x_i - x_1}\right) \qquad (i = 2, 3, \ldots, n + 1)$$

The quantities

$$\frac{y_i - y_1}{x_i - x_1} = [x_i,x_1] = [x_1,x_i]$$

are called "divided differences" and are customarily written with brackets.[1]

The next step is, of course, to write

$$P_{n-1}(x) = [x_1,x_2] + (x - x_2)P_{n-2}(x)$$

and require $P_{n-2}(x)$ to take on the values of the divided differences of the divided differences,

$$[[x_i,x_1],[x_2,x_1]] = \frac{[x_i,x_1] - [x_2,x_1]}{x_i - x_2} = [x_i,x_2,x_1]$$

It was easy to see that the first-order divided differences were independent of the order of the arguments in the brackets. We now show that the same is true of the second-order divided differences. If we start with three points

$$(x_1,y_1) \quad (x_2,y_2) \quad (x_3,y_3)$$

[1] Many other notations are used, for example, $f[x_i,x_1]$ and $\rho[x_i,x_1]$.

we get a unique polynomial (quadratic) through the three points. This may be written

$$y = y_1 + (x - x_1)\{[x_2,x_1] + (x - x_2)[x_3,x_2,x_1]\}$$

If we now choose the points in the order x_a, x_b, x_c we have

$$y = y_a + (x - x_a)\{[x_b,x_a] + (x - x_b)[x_c,x_b,x_a]\}$$

Since these are both the same quadratic, the coefficients of the x^2 term must be the same, that is, the two symbols

$$[x_3,x_2,x_1] = [x_c,x_b,x_a]$$

are in fact the same thing in alternative forms.

In general, we define

$$[x_1,x_2,x_3, \ . \ . \ . \ ,x_n] = \frac{[x_1,x_2, \ . \ . \ . \ ,x_{n-2},x_{n-1}] - [x_1,x_2, \ . \ . \ . \ ,x_{n-2},x_n]}{x_{n-1} - x_n}$$

and in exactly the same manner show that it is independent of the order of the x_i. Note that the denominator is the difference of the non-repeated x's in each $[\cdot \ \cdot \ \cdot]$ taken in the same order.

One way to make a table of the values needed is as follows:

$$\begin{array}{c|ccccc}
x_1 & y_1^* & & & \\
 & & [x_2,x_1]^* & & \\
x_2 & y_2 & & [x_3,x_2,x_1]^* & \\
 & & [x_3,x_1] & & [x_4,x_3,x_2,x_1]^* \\
x_3 & y_3 & & [x_4,x_2,x_1] & \cdot \\
 & & [x_4,x_1] & \cdot & \\
x_4 & y_4 & \cdot & & \cdot \\
\cdot & \cdot & & \cdot & \\
 & & \cdot & & \cdot \\
\cdot & \cdot & & \cdot & \\
 & & \cdot & & \\
\cdot & \cdot & & &
\end{array}$$

The asterisks indicate entries used as *pivots* in calculating succeeding values.

From this table we can write

$$y(x) = y_1 + (x - x_1)([x_2,x_1] + (x - x_2)\{[x_3,x_2,x_1] + (x - x_3)\{\cdot \ \cdot \ \cdot\}\})$$

As a specific example, consider the log table

x	$\log x$	[,]	[,,]	[,,,]
1	0.0000*			
		0.30100*		
2	0.3010		−0.06245*	
		0.23855		+0.01230
3	0.4771		−0.05015	
		0.20070		
4	0.6021			

The asterisks indicate the pivotal values.

Hence we get

$$y(x) = 0 + (x - 1)\{0.3010 + (x - 2)[(-0.06245) + (x - 3)(0.01230)]\}$$

In particular,

$$\begin{aligned} y(2.5) &= \tfrac{3}{2}\{0.3010 + \tfrac{1}{2}[(-0.06245) + (-\tfrac{1}{2})(0.01230)]\} \\ &= 0.40001 \end{aligned}$$

The correct value for log 2.5 is 0.3979.

EXERCISE

8.4-1. Given a divided difference table, show how to add one more data point at the bottom of the table. Also show how to add one at the top. In the first case, how does one change the Newton polynomial?

8.5 AN ALTERNATIVE FORM FOR THE DIVIDED DIFFERENCE TABLE

The divided difference table which lies at the heart of Newton's interpolation formula may be written in an alternative and sometimes more useful form. This is based on the observation that

$$\begin{aligned} [x_1,x_2, \ldots ,x_n] &= \frac{[x_1,x_2, \ldots ,x_{n-1}] - [x_1,x_2, \ldots ,x_{n-2},x_n]}{x_{n-1} - x_n} \\ &= \frac{[x_1,x_2, \ldots ,x_{n-1}] - [x_2,x_3, \ldots ,x_n]}{x_1 - x_n} \end{aligned}$$

In particular,

$$\begin{aligned} [x_1,x_2,x_3] &= \frac{[x_1,x_2] - [x_1,x_3]}{x_2 - x_3} \\ &= \frac{[x_1,x_2] - [x_2,x_3]}{x_1 - x_3} \end{aligned}$$

Thus we can write

		[,]	[,,]	[,,,]
x_1	y_1			
		$[x_2,x_1]$		
x_2	y_2		$[x_3,x_2,x_1]$	
		$[x_3,x_2]$		$[x_4,x_3,x_2,x_1]$
x_3	y_3		$[x_4,x_3,x_2]$	
		$[x_4,x_3]$		
x_4	y_4			

While the entries in this table differ from those in the previous section, the entries in the top row, which are the ones used in Newton's formula, are the same.

Using the same example of log x, we get

x	log x	[,]	[,,]	[,,,]
1	0.0000			
		0.3010		
2	0.3010		−0.06245	
		0.1761		+0.01230
3	0.4771		−0.02555	
		0.1250		
4	0.6021			

In this form, we may easily add a line at either end of the table and still expect the entries in the body of the table to vary smoothly. While there is no theoretical need to choose the points in numerical order, the smoothness in the table is lost if they are not. The first divided differences are secant lines and hence approximations to the first derivative in the interval (x_n,x_{n+1}). Similarly, the second divided differences are local approximations to the second derivatives, etc.

EXERCISES

8.5-1. Using the table

x	2	0	3	1
y	8	0	27	1

compute the Newton interpolating polynomial. Show that $y = x^3$.

8.5-2. Compute the difference table both ways (Secs. 8.4 and 8.5) for $y = \sin x$ at a spacing of 30°:

x	$\sin x$
0	0.0000
30°	0.5000
60°	0.8660
90°	1.0000

Find the approximating polynomial.

8.5-3. If you have only the top row of the divided difference table in storage in a computer, how can you add one point at the bottom of the table and compute the next divided difference? How do you add one point at the top and find the new line?

8.6 ERROR OF POLYNOMIAL APPROXIMATIONS

Given a function $f(x)$, we have been taking $n + 1$ sample points (x_i,y_i) $(i = 1, 2, \ldots, n + 1)$ and finding a polynomial $P_n(x)$ through these points. We then intend to use this polynomial in place of the original function, and it is therefore important to examine the question of how much the function and the polynomial can differ at points other than the sample points (where they agree within roundoff error).

As an example, consider the function $y(x) = \log x$. In Secs. 8.4 and 8.5 we found the polynomial approximating the function. As a measure of the difference $\log x - P(x)$, let us examine the values at the sample points and the mid-points between them where we would expect the error to be fairly large.

x	$\log x$	$P(x)$	$\log x - P(x)$
1.0	0.0000	0.0000	0.0000
1.5	0.1761	0.1707	+0.0054
2.0	0.3010	0.3010	0.0000
2.5	0.3979	0.4000	−0.0021
3.0	0.4771	0.4771	0.0000
3.5	0.5441	0.5414	+0.0027
4.0	0.6021	0.6021	0.0000

A theoretical expression for the difference between the original function $f(x)$ and the approximating polynomial $P(x)$ can be found if we observe that the difference is zero at all the sample points and write

$$y(x) - P(x) = (x - x_1)(x - x_2) \cdots (x - x_{n+1})K(x)$$

where $K(x)$ is suitably chosen. We now choose an arbitrary x^*. We have

$$y(x^*) - P(x^*) - (x^* - x_1)(x^* - x_2) \cdots (x^* - x_{n+1})K(x^*) = 0$$

Now consider the function

$$\Phi(x) = y(x) - P(x) - (x - x_1)(x - x_2) \cdots (x - x_{n+1})K(x^*)$$

If $y(x)$ has an $(n + 1)$st derivative, we can differentiate $n + 1$ times, and, since $P(x)$ is a polynomial of degree n and $K(x^*)$ is a constant, we get

$$\Phi^{(n+1)}(x) = y^{(n+1)}(x) - (n + 1)!K(x^*)$$

But $\Phi(x)$ vanishes $n + 2$ times (at x^* and $x_1, x_2, \ldots, x_{n+1}$). Hence, by the mean-value theorem, $\Phi'(x)$ vanishes at least $n + 1$ times in the interval containing all the x values (including x^*). Continuing to apply this theorem, we find that $\Phi^{(k)}(x)$ vanishes at least $n + 2 - k$ times and that $\Phi^{(n+1)}(x)$ vanishes at least once. Thus there is an $\bar{x}$ in the interval of the x values such that

$$y^{(n+1)}(\bar{x}) = (n + 1)!K(x^*)$$

We now have a value for the constant $K(x^*)$ and can put this back in the original expression to get

$$y(x^*) = P(x^*) + \frac{(x^* - x_1)(x^* - x_2) \cdots (x^* - x_{n+1})y^{(n+1)}(\bar{x})}{(n + 1)!}$$

But, since x^* was arbitrary, we may write x^* as x.

$$y(x) = P(x) + \frac{(x - x_1)(x - x_2) \cdots (x - x_{n+1})y^{(n+1)}(\bar{x})}{(n + 1)!} \tag{8.6-1}$$

If we use this error term to estimate the error made in interpolating in the log table, we get

$$\frac{(x - 1)(x - 2)(x - 3)(x - 4)}{4!} y^{(4)}(\bar{x})$$

To estimate the error at $x = 3/2$, we obtain

$$\frac{(1/2)(-1/2)(-3/2)(-5/2)}{4!}\left(-\frac{6}{\bar{x}^4}\right)\log e = \frac{0.102}{\bar{x}^4}$$

All that we know is that $1 \leq \bar{x} \leq 4$. At worst the error is about 0.102, while at best it is 0.0004—in fact, it is about 0.0054, which shows how little accuracy such an error term actually gives in this case.

It should be noted that the mean value $\bar{x} = \bar{x}(x)$ depends on x, and in fact there may be several $\bar{x}$'s for some values of x. Also, $\bar{x}$ is not necessarily a continuous function of x. This latter effect is illustrated by applying the mean-value theorem to the function $y(x) = x(1 - x)^2$ and choosing $a = 0$ in the usual form for the mean-value theorem:

$$y'(\bar{x}) = \frac{y(x) - y(a)}{x - a} = \frac{y(x)}{x} = (1 - x)^2$$

Thus $y'(\bar{x}) \geq 0$, and as

x goes from 0 to 1, $\bar{x}$ goes from 0 to $\frac{1}{3}$.
x goes from 1 to 2, $\bar{x}$ goes from $\frac{1}{3}$ to 0.
x goes beyond 2, $\bar{x}$ jumps over the interval where $y' < 0$ to $\bar{x} > 1$.

Hence $\bar{x} = \bar{x}(x)$ is not a continuous function of x.

We next examine the special case of *Hermite interpolation* where both the function and the derivative are given at each of the $n + 1$ sample points. To find the error we note that $y(x) - P(x)$ has a double zero at each x_i, and we set

$$y(x) - P(x) = [(x - x_1)(x - x_2) \cdots (x - x_{n+1})]^2 K(x)$$

We proceed as above to deduce that

$$y(x) = P(x) + \frac{(x - x_1)^2(x - x_2)^2 \cdots (x - x_{n+1})^2}{(2n + 2)!} y^{(2n+2)}(\bar{x}) \qquad (8.6\text{-}2)$$

The details are left to the reader.

The form of the error term that we have been using corresponds to the error term in a Taylor series known as the Lagrange form of the error.

EXERCISES

8.6-1. Find the error term in case the function and the first two derivatives are assigned at each sample point. (See Exercise 8.3-1.)

8.6-2. Show that the error term in the case where at x_i the function and the first $m_i - 1$ derivatives are assigned is

$$(x - x_1)^{m_1}(x - x_2)^{m_2} \cdots (x - x_{n+1})^{m_{n+1}} \frac{y^{(m)}(x)}{m!} \qquad (8.6\text{-}3)$$

where $m = m_1 + m_2 + \cdots + m_{n+1}$. (*Note:* This is not the most general case; at any point where a derivative is given we also require that all lower-order derivatives be also given. In Sec. 8.2 we did not do so, but so far as the author knows, no corresponding error term has been published.)

8.7 DIFFICULTY OF POLYNOMIAL APPPOXIMATION

It is customary, as in the previous section, to express the error of a polynomial approximation in terms of a suitable derivative of the function being approximated. It is usually thought that for "most" reasonable functions such expressions for the error become small for sufficiently large n, but this is not so.

Suppose that we restrict our discussion to analytic functions, that is, functions having a convergent Taylor series

$$y(x) = \sum_{n=0}^{\infty} \frac{(x - x_0)^n}{n!} y^{(n)}(x_0)$$

at all points x_0 in our range of interest. If, further, the function is an integral (entire) function, that is, converges everywhere in the finite part of the complex plane, as sin x, exp x, polynomials in x, etc., then it is indeed possible that all the higher derivatives are small. But if the function has a singularity in the finite part of the complex plane, as tan x, log x, rational functions in x, etc., then the Taylor series must have a finite radius of convergence R, and this in turn means that for an infinite number of values of n

$$\frac{(R + \epsilon)^n}{n!} |y^{(n)}(x_0)| \geq 1 \qquad (\epsilon > 0)$$

$$|y^{(n)}(x_0)| \geq \frac{n!}{(R + \epsilon)^n}$$

In words, the upper bound on the nth derivative grows as $n!$. As an example, consider

$$y = \ln x$$

$$y' = \frac{1}{x}$$

$$y'' = \frac{-1}{x^2}$$

$$y''' = \frac{2!}{x^3}$$

$$\cdots\cdots$$

$$y^{(n)} = \frac{(-1)^{n-1}(n - 1)!}{x^n}$$

Thus, even though the curve $y = \ln x$ looks smooth near some value x, nevertheless as n gets large the derivatives at this point become very large in size and tend to behave as $n!$.

This is the general case; for "most" functions some of the higher-order derivatives tend to grow as $n!$. It is only for certain integral functions that all the derivatives can remain bounded.[1] Even for polynomials there is a tendency for the derivatives to grow in size until the nth one which is $a_0 n!$, after which they suddenly all become zero. Of course, many of the higher derivatives of a function may be small while some of

[1] The converse is *not* true; the derivatives of an integral function need not be bounded. For example, if $y = xe^x$, then $y^{(n)}(0) = n + 1$.

the others are large. For example, an even function, say $f(x^2)$, has all its odd-order derivatives equal to zero at the origin, but if the function is not an integral function, then there are also an infinite number of the even-order derivatives which tend to behave as $n!$.

It would be helpful if we had to deal only with integral functions which can have nicely bounded derivatives, but the facts seem to be that, if the function is an integral function, it is quite likely that the whole problem can be solved analytically, whereas if it is necessary to use numerical methods the function is likely to be rather poorly behaved.

Weierstrass's theorem, which states, loosely speaking, that a continuous function in a closed interval can be uniformly approximated by a polynomial, is often cited [20, p. 19] as a justification for using polynomial approximation. However, the method that we have been using of exact matching at the sample points is *not* the way in which the Weierstrass polynomial is defined; hence the theorem, while possibly suggestive, does not apply. Indeed, the simple function [39, pp. 35–39]

$$y(x) = \frac{1}{1 + x^2}$$

is well known to be a bad example, since as a selected set of equally spaced points is increased in number the approximating polynomial (in the sense of exact matching) diverges from the function between some of the sample points. Thus, even for equally spaced sample points, we cannot rely on a polynomial to be a good approximation if exact matching at the sample points is the criterion used to select the polynomial. The explanation of this phenomenon is, of course, that the derivatives grow too rapidly.

As an example of this effect, consider the Riemann zeta function in Table 8.7-1. The top line of differences does not tend to become small very rapidly, which is owing to the obvious singularity at $x = 1$.

It should be noted, however, that, if a table is given at one-half the spacing of another table of the same function, then the first differences are one-half as large, the second, one-fourth as large, etc., as those of the second table. This suggests an empirical rule: If the differences of a table approach zero rapidly, then probably too small a spacing was used and the table was overcomputed, whereas if the differences do not become small, then a finer spacing should be considered.

A final, and perhaps much more basic, objection to polynomial approximation is that it rarely has any physical implications that will lead to useful insights.

On the other hand, the theory is simple, well developed, requires the minimum of computation, and is useful. Experience shows that polynomials often do a good job, although the error term may be either unobtainable or pessimistic. The values of the difference table can be highly suggestive of the values of the derivative, and they indicate the smallness of the contribution due to taking more terms in the Newton interpolation polynomial. However, the use of the differences as a guide to the values of the derivatives is sometimes dangerous; thus, for integer

TABLE 8.7-1. RIEMANN ZETA FUNCTION, $\zeta(x) = \sum_{k=1}^{\infty} \frac{1}{k^x}$

x	$\zeta(x)$	$\Delta\zeta$	$\Delta^2\zeta$	$\Delta^3\zeta$	$\Delta^4\zeta$	$\Delta^5\zeta$
2	1.64493					
		−0.44287				
3	1.20206		0.32313			
		−0.11974		−0.24878		
4	1.08232		7435		0.20023	
		−0.04539		−4855		−0.16688
5	1.03693		2580		3335	
		−0.01959		−1520		−2403
6	1.01734		1060		932	
		−0.00899		−588		−596
7	1.00835		472		336	
		−0.00427		−252		−199
8	1.00408		220		137	
		−0.00207		−115		−75
9	1.00201		105		62	
		−0.00102		−53		−35
10	1.00099		52		27	
		−0.00050		−26		−16
11	1.00049		26		11	
		−0.00024		−15		−0
12	1.00025		11		11	
		−0.00013		−4		−11
13	1.00012		7		0	
		−0.00006		−4		+3
14	1.00006		3		3	
		−0.00003		−1		−4
15	1.00003		2		−1	
		−0.00001		−2		
16	1.00002		0			
		−0.00001				
17	1.00001					

x, the values of $\sin \pi x$ are all zero, so that the difference table is also all zeros, suggesting zero error in the approximation

$$\sin \pi x = 0$$

which is hardly true. The fallacy is obvious here, but in some situations it might be overlooked.

EXERCISE

8.7-1. Compute the nth derivative of

$$\frac{1}{1+x^2} = \frac{1}{2i}\left(\frac{1}{x-i} - \frac{1}{x+i}\right) \qquad (i = \sqrt{-1})$$

8.8 ON SELECTING SAMPLE POINTS

The problem of which sample points to select occurs when we try to interpolate. If we had complete freedom in choosing, we would select the value x for which we were going to interpolate, thus making the problem trivial. But it often happens that we have an extensive set of values (x_i,y_i) which we might use, none of which coincides with the desired value x. In the absence of knowledge of the size of the derivative occurring in the error term, we can only pick our samples to minimize the factor [see Eq. (8.6-3) of Exercise 8.6-2]

$$(x - x_1)^{m_1}(x - x_2)^{m_2} \cdots (x - x_n)^{m_n}$$

in front of the derivative term.

Common sense and the minimization of this factor coincide, namely, use information close to the place at which we want to interpolate. In using Newton's interpolation formula

$$y(x) = y(x_1) + (x - x_1)([x_2,x_1] + (x - x_2)\{[x_3,x_2,x_1] \cdots\})$$

there is a tendency to try to keep x near x_1, but this should be resisted; it is the balance of *all* the factors in the error term that is desired. Thus we pick our x_i values grouped on both sides, if possible, of the desired x value.

Another question (whose answer cannot be given completely here) sometimes occurs. Suppose that you wish to interpolate values throughout an entire interval. Where should you select the samples to minimize the maximum error? The answer is given by the theory of the Chebyshev polynomials of Chap. 19.

CHAPTER 9

Polynomial Interpolation—Equally Spaced Data

9.1 NEWTON'S FORMULA FOR INTERPOLATION

Very often our information about a function (in the form of samples) is given at a set of equally spaced values of x. This simplifies much of the notation and computation as well as the ideas involved.

For equally spaced data it is customary to use the notation of Part I:

$$\Delta y_n = y_{n+1} - y_n$$

This is the familiar notation of the difference calculus except that we have fixed $\Delta x = h$ at all times:

$$\Delta x_n = x_{n+1} - x_n = h$$

We also have

$$\Delta^2 y_n = y_{n+2} - 2y_{n+1} + y_n$$

etc.

These differences are related to the divided differences as follows:

$$[x_2,x_1] = \frac{y_2 - y_1}{x_2 - x_1} = \frac{\Delta y_1}{h}$$

$$[x_3,x_2,x_1] = \frac{[x_3,x_2] - [x_2,x_1]}{x_3 - x_1} = \frac{\Delta y_2/h - \Delta y_1/h}{2h}$$

$$= \frac{\Delta^2 y_1}{2!h^2}$$

and, in general,

$$[x_1,x_2, \ . \ . \ . \ ,x_n] = \frac{\Delta^{n-1} y_1}{(n-1)!h^{n-1}}$$

The differences are approximations to derivatives at the middle of the range of the samples used:

$$\Delta y_1 \sim h \frac{dy(x + h/2)}{dx}$$

$$\Delta^2 y_1 \sim h^2 \frac{d^2y(x + h)}{dx^2}$$

$$. \ . \ . \ . \ . \ . \ . \ . \ . \ . \ . \ . \ .$$

$$\Delta^n y_1 \sim h^n \frac{d^ny(x + nh/2)}{dx^n}$$

It is these relationships that enable the finite difference calculus to approximate expressions in the differential calculus.

Newton's formula in this new notation for equally spaced sample points is [see Eq. (1.5-4) and Sec. 8.4]

$$y = y_0 + (x - x_0)\frac{\Delta y_0}{h} + (x - x_0)(x - x_0 - h)\frac{\Delta^2 y_0}{2h^2} + (x - x_0)(x - x_0 - h)(x - x_0 - 2h)\frac{\Delta^3 y_0}{3!h^3} + \cdots$$

If we suppose that $x_0 = 0$, we get

$$y = y_0 + x\frac{\Delta y_0}{h} + x(x - h)\frac{\Delta^2 y_0}{2h^2} + x(x - h)(x - 2h)\frac{\Delta^3 y_0}{3!h^3} + \cdots$$

and if we further assume $h = 1$ we have [compare with Eq. (1.5-4)]

$$y = y_0 + x\,\Delta y_0 + x(x - 1)\frac{\Delta^2 y_0}{2!} + x(x - 1)(x - 2)\frac{\Delta^3 y_0}{3!} + \cdots$$

EXERCISE

9.1-1. Write the error term for Newton's formula when the data are given at $-n$, $-(n - 1)$, . . . , 0, 1, 2, . . . , n and a $2n$th difference is used in the formula. Estimate the maximum of the coefficient of the derivative.

9.2 INTERPOLATION IN TABLES

One of the major uses of equally spaced interpolation formulas is for interpolation in tables of equally spaced data. A typical example is the error integral whose difference table is given in Table 9.2-1. The differences are well behaved in the sense that they first tend toward zero, being fairly smooth in the third difference column, but beginning with the fourth difference, and clearly in the fifth difference, we find that roundoff noise is dominating. For this table we would not expect, then, to go beyond fourth differences in any interpolation formula.

In principle, Newton's formula for interpolation gives the answer, and except for roundoff effects the answer is exactly the same as could be found by any other interpolating polynomial using the same sample points. Nevertheless, in practice, many other formulas are used, and in the interest of general education, so that the reader can understand the methods given in the prefaces of most tables, we shall discuss some of those which are most popular.

TABLE 9.2-1. TABLE OF THE ERROR INTEGRAL $\frac{1}{\sqrt{2\pi}}\int_0^x e^{-\theta^2/2}\,d\theta$

t	$f(t)$	Δ	Δ^2	Δ^3	Δ^4	Δ^5
0.00	0.0000					
		987				
0.25	0.0987		−59			
		928		−50		
0.50	0.1915		−109		+19	
		819		−31		+4
0.75	0.2734		−140		+23	
		679		−8		−10
1.00	0.3413		−148		+13	
		531		+5		+4
1.25	0.3944		−143		17	
		388		+22		−11
1.50	0.4332		−121		6	
		267		+28		−10
1.75	0.4599		−93		−4	
		174		+24		+4
2.00	0.4773		−69		0	
		105		+24		−7
2.25	0.4878		−45		−7	
		60		+17		+3
2.50	0.4938		−28		−4	
		32		+13		−4
2.75	0.4970		−15		−8	
		17		+5		+10
3.00	0.4987		−10		+2	
		7		+7		−9
3.25	0.4994		−3		−7	
		4		+0		+9
3.50	0.4998		−3		+2	
		1		+2		
3.75	0.4999		−1			
		1				
4.00	0.5000					

9.3 THE LOZENGE DIAGRAM

The lozenge diagram is a device for showing that a large number of formulas which appear to be different are really all the same. We defined the binomial coefficients

$$C(u + k, n) = \frac{(u + k)(u + k - 1)(u + k - 2) \cdots (u + k - n + 1)}{n!}$$

in Sec. 1.3. There are n factors in the numerator and n in the denominator. Viewed as a function of u, $C(u + k, n)$ is a polynomial of degree n.

$y(-3)$	$C(u+3,1)$	$\Delta^2 y(-4)$	$C(u+4,3)$	$\Delta^4 y(-5)$
1	$\Delta y(-3)$	$C(u+3,2)$	$\Delta^3 y(-4)$	$C(u+4,4)$
$y(-2)$	$C(u+2,1)$	$\Delta^2 y(-3)$	$C(u+3,3)$	$\Delta^4 y(-4)$
1	$\Delta y(-2)$	$C(u+2,2)$	$\Delta^3 y(-3)$	$C(u+3,4)$
$y(-1)$	$C(u+1,1)$	$\Delta^2 y(-2)$	$C(u+2,3)$	$\Delta^4 y(-3)$
1	$\Delta y(-1)$	$C(u+1,2)$	$\Delta^3 y(-2)$	$C(u+2,4)$
$y(0)$	$C(u,1)$	$\Delta^2 y(-1)$	$C(u+1,3)$	$\Delta^4 y(-2)$
1	$\Delta y(0)$	$C(u,2)$	$\Delta^3 y(-1)$	$C(u+1,4)$
$y(1)$	$C(u-1,1)$	$\Delta^2 y(0)$	$C(u,3)$	$\Delta^4 y(-1)$
1	$\Delta y(1)$	$C(u-1,2)$	$\Delta^3 y(0)$	$C(u,4)$
$y(2)$	$C(u-2,1)$	$\Delta^2 y(1)$	$C(u-1,3)$	$\Delta^4 y(0)$
1	$\Delta y(2)$	$C(u-2,2)$	$\Delta^3 y(1)$	$C(u-1,4)$
$y(3)$	$C(u-3,1)$	$\Delta^2 y(2)$	$C(u-2,3)$	$\Delta^4 y(1)$
1	$\Delta y(3)$	$C(u-3,2)$	$\Delta^3 y(2)$	$C(u-2,4)$
$y(4)$				

FIG. 9.3-1. Lozenge diagram.

Figure 9.3-1 shows the lozenge diagram. A line starting at a point on the left edge and following some path across the page defines an interpolation formula if the following rules are used:

1*a*. For a *left to right* step, *add*.
1*b*. For a *right to left* step, *subtract*.

2*a*. If the *slope* of the step is *positive*, use the product of the difference crossed times the factor immediately *below*.

2*b*. If the *slope* of the step is *negative*, use the product of the difference crossed times the factor immediately *above*.

3*a*. If the step is *horizontal* and passes through a *difference*, use the product of the difference times the *average* of the factors *above and below*.

3*b*. If the step is *horizontal* and passes through a *factor*, use the product of the factor times the *average* of the differences *above and below*.

As an example of rules 1*a* and 2*b*, consider starting at $y(0)$ and going down and to the right. We get, term by term,

$$
\begin{aligned}
y(u) &= y(0) + C(u,1)\,\Delta y(0) + C(u,2)\,\Delta^2 y(0) + C(u,3)\,\Delta^3 y(0) + \cdots \\
&= y(0) + u\,\Delta y(0) + \frac{u(u-1)}{2}\,\Delta^2 y(0) \\
&\qquad + \frac{u(u-1)(u-2)}{3!}\,\Delta^3 y(0) + \cdots
\end{aligned}
$$

which is Newton's formula.

Had we gone up and to the right, we would have used rules 1*a* and 2*a* to get Newton's backward formula:

$$\begin{aligned} y(u) &= y(0) + C(u,1)\,\Delta y(-1) + C(u+1,\,2)\,\Delta^2 y(-2) \\ &\qquad + C(u+2,\,3)\,\Delta^3 y(-3) + \cdots \\ &= y(0) + u\,\Delta y(-1) + \frac{(u+1)u}{2}\,\Delta^2 y(-2) \\ &\qquad + \frac{(u+2)(u+1)u}{3!}\,\Delta^3 y(-3) + \cdots \end{aligned} \tag{9.3-1}$$

To get Stirling's formula, we start at $y(0)$ and go horizontally to the right, using rules 3*a* and 3*b*:

$$\begin{aligned} y(u) &= y(0) + u\,\frac{\Delta y_0 + \Delta y_{-1}}{2} + \frac{C(u+1,\,2) + C(u,2)}{2}\,\Delta^2 y_{-1} \\ &\qquad + C(u+1,\,3)\,\frac{\Delta^3 y_{-2} + \Delta^3 y_{-1}}{2} + \cdots \\ &= y_0 + u\,\frac{\Delta y_0 + \Delta y_{-1}}{2} + \frac{u^2}{2}\,\Delta^2 y_{-1} + \frac{u(u^2-1)}{3!}\,\frac{\Delta^3 y_{-2} + \Delta^3 y_{-1}}{2} + \cdots \end{aligned} \tag{9.3-2}$$

If we start midway between $y(0)$ and $y(1)$, we get Bessel's formula:

$$\begin{aligned} y(u) &= 1\,\frac{y_0 + y_1}{2} + \frac{C(u,1) + C(u-1,\,1)}{2}\,\Delta y_0 \\ &\qquad + C(u,2)\,\frac{\Delta^2 y_{-1} + \Delta^2 y_0}{2} + \cdots \\ &= \frac{y_0 + y_1}{2} + \left(u - \frac{1}{2}\right)\Delta y_0 + \frac{u(u-1)}{2}\,\frac{\Delta^2 y_{-1} + \Delta^2 y_0}{2} + \cdots \end{aligned} \tag{9.3-3}$$

If we zigzag properly, we can get Gauss' formula for interpolation:

$$y(u) = y_0 + u\,\Delta y_0 + \frac{u(u-1)}{2}\,\Delta^2 y(-1) + \frac{u(u^2-1)}{3!}\,\Delta^3 y(-1) + \cdots \tag{9.3-4}$$

All sorts of paths can be chosen, and each will give some formula. What we need to show is that these are all valid interpolation formulas. Such a proof requires showing the following:

1. At least one valid formula results, and since we have found Newton's interpolation formula from the diagram this step is completed.
2. The contribution around any closed path is zero; hence we may deform a path into any other path that we wish.
3. If two formulas end at the same place they are the same, which is necessary to prove since the entry points into the lozenge diagram need not be the same for different formulas.

To prove part 2, we take a single complete lozenge (Fig. 9.3-2).

Path 1: $C(u - s, n)\,\Delta^n y(s - 1) + C(u - s + 1, n + 1)\,\Delta^{n+1} y(s - 1)$
Path 2: $C(u - s, n)\,\Delta^n y(s) + C(u - s, n + 1)\,\Delta^{n+1} y(s - 1)$
Path 3: $C(u - s, n)\,\dfrac{\Delta^n y(s - 1) + \Delta^n y(s)}{2}$

$$+ \frac{C(u - s + 1, n + 1) + C(u - s, n + 1)}{2}\,\Delta^{n+1} y(s - 1)$$

Recalling that going from right to left produces negative terms (rule 1b), we have only to show that the three paths are the same.

$$\begin{aligned}\text{Path 1} - \text{path 2} &= C(u - s, n)[\Delta^n y(s - 1) - \Delta^n y(s)] \\ &\qquad + [C(u - s + 1, n + 1) - C(u - s, n + 1)]\,\Delta^{n+1} y(s - 1) \\ &= C(u - s, n)[-\Delta^{n+1} y(s - 1)] \\ &\qquad + C(u - s, n)\,\frac{u - s + 1 - (u - s - n)}{n + 1}\,\Delta^{n+1} y(s - 1) \\ &= C(u - s, n)[-\Delta^{n+1} y(s - 1) + \Delta^{n+1} y(s - 1)] = 0\end{aligned}$$

Also, path 3 equals the average of paths 1 and 2. Hence step 2 of the proof is complete.

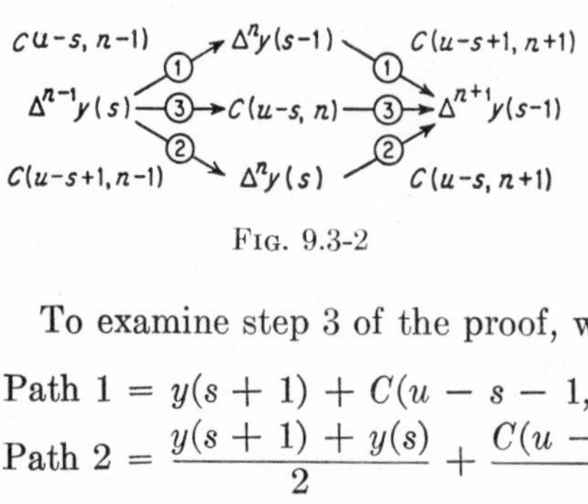

FIG. 9.3-2

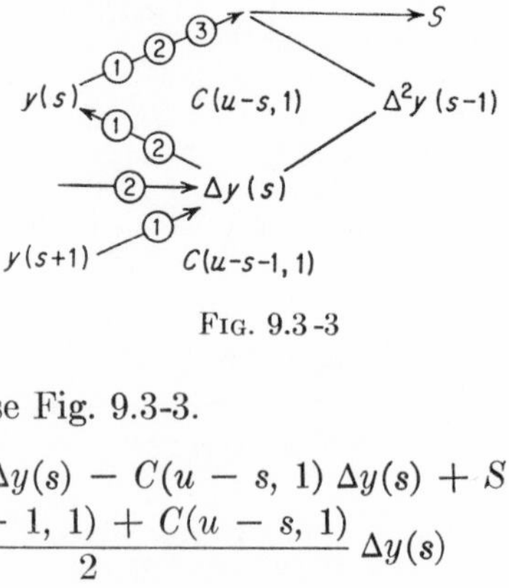

FIG. 9.3-3

To examine step 3 of the proof, we use Fig. 9.3-3.

$$\text{Path 1} = y(s + 1) + C(u - s - 1, 1)\,\Delta y(s) - C(u - s, 1)\,\Delta y(s) + S$$

$$\text{Path 2} = \frac{y(s + 1) + y(s)}{2} + \frac{C(u - s - 1, 1) + C(u - s, 1)}{2}\,\Delta y(s) - C(u - s, 1)\,\Delta y(s) + S$$

$$\text{Path 3} = y(s) + S$$

where S is the rest of the formula. Now using

$$y(s + 1) - y(s) = \Delta y(s)$$

we obtain

$$\begin{aligned}\text{Path 1} &= y(s) + \Delta y(s) + (u - s)\,\Delta y(s) - \Delta y(s) - (u - s)\,\Delta y(s) + S \\ &= y(s) + S = \text{path 3}\end{aligned}$$

Similarly, path 2 may be reduced to path 3. Thus we conclude that the interpolation formula depends on the ending values and not on the path used to reach there.

EXERCISE

9.3-1. Verify formulas (9.3-2) to (9.3-4).

9.4 REMARKS ON THESE FORMULAS

How do these different formulas compare with each other and with the Lagrange formula found earlier? The value obtained in an interpolation depends on the polynomial used, and the polynomial depends on the sample points used. The error term has the form (see Sec. 8.6)

$$\frac{(x - x_1)(x - x_2) \cdots (x - x_{n+1})y^{(n+1)}(\bar{x})}{(n + 1)!}$$

The coefficient of the derivative is minimized when x is in the middle of the range of samples. Thus there is a tendency to use an even number of samples when the interpolation point is in the middle of an interval, and an odd number when it is near a sample point.

The methods derived by the lozenge diagram explicitly exhibit the differences, and these differences give some idea of the accuracy but require more arithmetic than does the Lagrange method of interpolating. The Lagrange method, however, does not directly use the differences so that no indication of the accuracy is available. This is one of the fundamental problems in numerical computation—the use of differences gives some clues to the accuracy of the method being used, while the Lagrange approach minimizes the amount of arithmetic. As a result, the difference methods tend to be used in exploratory work, and the Lagrange methods in well-understood, routine work.

TABLE 9.4-1. LAGRANGE CUBIC INTERPOLATION
$f(x) = A_{-1}(x)f(-1) + A_0(x)f(0) + A_1(x)f(1) + A_2(x)f(2)$

x	A_{-1}	A_0	A_1	A_2	
0	0	1	0	0	1.0
0.1	−0.0285	0.9405	0.1045	−0.0165	0.9
0.2	−0.0480	0.8640	0.2160	−0.0320	0.8
0.3	−0.0595	0.7735	0.3315	−0.0455	0.7
0.4	−0.0640	0.6720	0.4480	−0.0560	0.6
0.5	−0.0625	0.5625	0.5625	−0.0625	0.5
	A_2	A_1	A_0	A_{-1}	x

The Lagrange coefficients for equally spaced data have been tabulated extensively [32]. Some coefficients for cubic interpolation are given in Table 9.4-1. Note that the symmetry allows us to read the table from the top, using the left-hand entries, or from the bottom, using the right-hand entries.

EXERCISES

9.4-1. Using these Lagrange interpolating coefficients, compute the error integral at 1.40 from Table 9.2-1.

9.4-2. Using Table 9.4-1, compute the entry in the table for $x = 0.35$.

9.5 MISCELLANEOUS INTERPOLATION FORMULAS

A number of other formulas which are not directly obtained from the lozenge diagram are occasionally useful. They usually rest on the fact that any particular difference may be eliminated by using

$$\Delta^n y(s+1) - \Delta^n y(s) = \Delta^{n+1} y(s)$$

The cost of this elimination is the inclusion of an extra difference of some other order.

For example, in Bessel's interpolation formula we may eliminate all odd-order differences; this gives

$$y(u) = (1-u)y(0) + uy(1) + \frac{(2-u)(1-u)(-u)}{3!} \Delta^2 y(-1) + \frac{(u+1)u(u-1)}{3!} \Delta^2 y(0) + \cdots$$

which is known as Everett's formula and is quite popular since the table maker need only publish the function and even-order differences.

Similarly we may start with almost any formula and eliminate any-order differences that we please (at the cost of several differences of the orders that we leave in the formula). We could, if we wish, eliminate, say, both the second- and third-order differences, thus using the function, the first differences, and the fourth and higher differences. Carried to the extreme, this leads to the Lagrange form which uses no differences but many function values. The purpose of eliminating the differences is mainly to save type and space in the printing of the table, and this is done at the cost of extra work on the part of the user. The proper balance depends on circumstances and cannot be given once for all.

A second device that is often used is called "throwback." This idea, due mainly to Comrie, uses the fact that the coefficients of the successive differences tend to be proportional to each other in various interpolation formulas, such as Everett's, and consequently, if suitable amounts of the higher differences are combined with those of the lower when the table is printed, then much of the effect of the higher-order differences in the interpolation is automatically achieved by using the lower-order formula.

Thus, in Bessel's formula, the ratio of the coefficients of the Δ^4 to the coefficient of Δ^2 is

$$\frac{B^{\text{IV}}}{B^{\text{II}}} = \frac{(u+1)(u-2)}{12} \qquad (0 \le u \le 1)$$

which varies from $-1/6$ to $-3/16$. Hence if we put c units of Δ^4 with Δ^2 to form the modified second difference column, we make an error of

$$(B^{\text{IV}} - cB^{\text{II}})\,\Delta^4$$

The number c is often taken as -0.184 as an average compromise.

We again remind the reader that table making is a highly skilled art and refer the reader to Fox [10] as well as to the standard texts of Kopal [20] and Hildebrand [14].

EXERCISES

9.5-1. Find a formula which uses only the function and third and sixth differences by eliminating the others from Bessel's interpolation formula.

9.5-2. Same as Exercise 9.5-1, but use Stirling's formula.

CHAPTER 10

A Uniform Method for Finding Formulas

10.1 INTRODUCTION

The purpose of this chapter is to present a uniform method for finding formulas. In terms of the four basic questions:

1. What samples?
2. What class of functions?
3. What criterion of goodness of fit?
4. What accuracy?

we restrict ourselves in Chaps. 10 to 16 to the class of polynomials and to the criterion of exact matching. The advantage of this presentation is that not only does a uniform method of derivation make learning easier but, more importantly, it leads to putting the process of deriving the appropriate formula onto the computer itself.

Superficial thinking about the mechanization of numerical analysis suggests that all the formulas should be labeled in some appropriate way and placed on a storage tape of a machine. The user has only to know what formula he wants and then call it in by name whenever he wants it. I suggest that this is not the proper way to approach the problem. Past experience with the idea of an extensive library of subroutines, covering a large, ill-defined field of human thought, has shown that it is often not a satisfactory approach; too many cases fall between those supplied on the library tape.

The problem may be viewed as one of information retrieval; we wish to find a particular piece of information, namely, the details of, say, a seventh-order interpolation formula. Rather than label it with the name given by historical accident, we shall use the answers to the first three questions as our label.

Instead of information retrieval, we propose to use information regeneration; rather than search a tape for a formula that may or may not be on it, we shall describe a program that will *derive* any of a broad class of formulas. This is not completely novel in computer circles; we normally recompute the values of the elementary transcendental functions rather than look them up in a table. Recent work in the field of theorem proving suggests that the proposed approach of re-deriving a formula whenever we need it is not such an impossible task as it might at first seem. It is partly a question of economics as to which of the two approaches, information retrieval or information regeneration, is best or what blend of the two should be used. When the number of formulas that are to be available is large, the storage costs and the search time can be very high, whereas when the formula has a long, complex derivation the regeneration time is expensive. For the limited class that we are going to discuss, it appears that the regeneration approach is quite feasible.

This proposed approach has several assets. First, it is a powerful stimulus for further research. Second, it means that the user can select the formula to fit the situation and not try to find some already derived formula that does the job approximately. Third, it makes the learning much simpler, since only one method is used. Lastly, it is in the spirit of modern computing—in the division of labor between the human being and the computer, we want to put onto the computer as much as we can of those parts of the work that it can do best.

In this chapter we shall concentrate on the formal methods for deriving formulas. In Chap. 11 we shall examine methods for estimating the error made when using a formula. In Chap. 12 we shall examine and compare various specific formulas for numerically computing definite integrals. The reader who is looking for a specific integration formula should first look elsewhere in the book. If unsuccessful he should return to this point and find out how to derive it. In Chaps. 13 and 15 we shall develop extensions to the technique for finding formulas.

10.2 SOME TYPICAL INTEGRATION FORMULAS

We examine a few typical examples before plunging into the study of a systematic method for deriving formulas. It so happens that the method will handle much more than integration formulas, but it is preferable at this point to think in terms of numerical integration. It might be thought that numerical differentiation is simpler than integration, but this is not so, and we postpone the problem of estimating derivatives from noisy data until much later.

We pointed out in Sec. 7.3 that the problem of numerically integrating

$$\int_0^1 f(x)\,dx$$

is equivalent to estimating the average value of $f(x)$ in the interval $0 \leq x \leq 1$ from n samples. This is fundamentally a statistical problem in sequential sampling and design of experiments, and statisticians are beginning to think about such problems. However, since their results so far are meager,[1] we shall fall back on the classical methods which specify in advance both the positions x_i where the samples are to be taken and the weights w_i to attach to these samples.

We shall not use the analytic substitution process of constructing an interpolating polynomial passing exactly through the sample points and then integrating the polynomial between the limits of integration; rather, we shall find the coefficients of the formula directly. We shall not use the Taylor-series method but instead we shall make the proposed formula exact for $1, x, x^2, \ldots$ as far as we can go. In Sec. 7.2, it was shown that these three methods are equivalent, and the third method is the easiest to use in practice.

The simplest nontrivial formula for computing an integral uses one sample point, namely,

$$\int_0^1 f(x)\,dx = w_1 f(x_1)$$

where w_1 and x_1 are to be determined. With two parameters available, we can make it exact for $f(x) = 1$ and x. Using these two functions, we get

$$\int_0^1 1 \cdot dx = 1 = w_1 \qquad \text{and} \qquad \int_0^1 x \cdot dx = \tfrac{1}{2} = w_1 x_1$$

from which

$$w_1 = 1 \qquad x_1 = \tfrac{1}{2}$$

Thus we have the formula that we expected:

$$\int_0^1 f(x)\,dx = f(\tfrac{1}{2}) \tag{10.2-1}$$

It is reasonable to estimate the error of this formula by using the next higher power of x, in this case x^2, in the equation:

$$\int_0^1 x^2\,dx = \tfrac{1}{3} = \tfrac{1}{4} + E_2 \qquad \text{or} \qquad E_2 = \tfrac{1}{12}$$

As we shall show in the next chapter, the error, or remainder term as it is often called, is given by

$$\frac{h^3 E_2 f''(\theta)}{2!} = \frac{h^3 f''(\theta)}{24} = \frac{f''(\theta)}{24} \tag{10.2-2}$$

where θ is some value in the interval $0 < \theta < 1$, and h is the interval size ($h = 1$ in this case).

[1] But see J. H. Halton, Thesis, Princeton University, Princeton, N.J.

The standard trapezoid formula uses the two samples at the ends of the interval

$$\int_0^1 f(x)\,dx = w_0 f(0) + w_1 f(1)$$

Again we use $f(x) = 1$ and x:

$$1 = w_0 + w_1$$
$$\tfrac{1}{2} = w_1$$

From these equations we get

$$\int_0^1 f(x)\,dx = \tfrac{1}{2}[f(0) + f(1)] \tag{10.2-3}$$

To estimate the error, we again use x^2:

$$\tfrac{1}{3} = \tfrac{1}{2} + E_2 \qquad \text{or} \qquad E_2 = -\tfrac{1}{6} \tag{10.2-4}$$

which is *twice* the size of the error of the single-sample formula (10.2-2).

We could have placed our two samples in arbitrary positions

$$\int_0^1 f(x)\,dx = w_1 f(x_1) + w_2 f(x_2) \tag{10.2-5}$$

With four parameters available, we use 1, x, x^2 and x^3:

$$\begin{aligned} 1 &= w_1 + w_2 \\ \tfrac{1}{2} &= w_1 x_1 + w_2 x_2 \\ \tfrac{1}{3} &= w_1 x_1^2 + w_2 x_2^2 \\ \tfrac{1}{4} &= w_1 x_1^3 + w_2 x_2^3 \end{aligned} \tag{10.2-6}$$

These equations can be solved, but it is much easier to choose a different coordinate system and take advantage of the symmetry; thus we choose a coordinate system whose origin is located at the middle of the range. We can also rescale the size without altering the condition that the formula be exact for 1, x, x^2, x^3. We therefore recast the problem as

$$\int_{-1}^1 f(x)\,dx = w_1 f(x_1) + w_2 f(x_2) \tag{10.2-7}$$

which leads to the equations

$$\begin{aligned} 2 &= w_1 + w_2 \\ 0 &= w_1 x_1 + w_2 x_2 \\ \tfrac{2}{3} &= w_1 x_1^2 + w_2 x_2^2 \\ 0 &= w_1 x_1^3 + w_2 x_2^3 \end{aligned}$$

We now eliminate $w_2 x_2$ from the second and fourth equations and get

$$0 = w_1 x_1^3 - w_1 x_1 x_2^2$$

or

$$x_1^2 = x_2^2 \qquad (w_1 x_1 \neq 0)^1$$
$$x_1 = -x_2$$

[1] The choice $w_1 x_1 = 0$ does not lead to a solution of the equations.

(The plus sign would give only one sample point.) The second of the four equations now gives

$$w_1 = w_2$$

and the first equation gives

$$w_1 = w_2 = 1$$

while finally the third equation gives

$$2/3 = 2x_1^2$$

or

$$x_1 = \frac{1}{\sqrt{3}}$$

The formula is, therefore,

$$\int_{-1}^{1} f(x)\,dx = f\left(\frac{-1}{\sqrt{3}}\right) + f\left(\frac{1}{\sqrt{3}}\right)$$

If we set $y = (1 + x)/2$, we have for the original integral (10.2-5)

$$\int_0^1 f(y)\,dy = \frac{1}{2}\left[f\left(\frac{1 - 1/\sqrt{3}}{2}\right) + f\left(\frac{1 + 1/\sqrt{3}}{2}\right)\right] \tag{10.2-8}$$
$$= \tfrac{1}{2}[f(0.2113\cdots) + f(0.7887\cdots)]$$

This type of formula in which we do not assign the places at which the samples of the function are to be taken is known as a *gaussian formula*, and we shall later study it more in detail.

As another example, suppose that we know both the function and its derivative at the two end points. We can write

$$\int_0^1 f(x)\,dx = w_0 f(0) + w_1 f(1) + w_2 f'(0) + w_3 f'(1)$$

and use $1, x, x^2, x^3$ to get the equations

$$\begin{aligned} 1 &= w_0 + w_1 \\ 1/2 &= w_1 + w_2 + w_3 \\ 1/3 &= w_1 + 2w_3 \\ 1/4 &= w_1 + 3w_3 \end{aligned}$$

It is easy to solve these to obtain, in turn,

$$w_3 = -1/12 \qquad w_1 = 1/2 \qquad w_0 = 1/2 \qquad w_2 = 1/12$$

and we have

$$\int_0^1 f(x)\,dx = \tfrac{1}{2}[f(0) + f(1)] + \tfrac{1}{12}[f'(0) - f'(1)] \tag{10.2-9}$$

To estimate the error, we try x^4:

$$\begin{aligned} 1/5 &= w_1 + 4w_3 + E_4 \\ E_4 &= 1/30 \end{aligned}$$

Using the results of the next chapter, we have the error

$$\frac{1}{30}\frac{h^5}{4!}f^{(4)}(\theta) = \frac{1}{720}f^{(4)}(\theta)$$

As a final example, consider the integral

$$\int_{-1}^{1} f(x) \sin \frac{\pi}{2} x \, dx = w_{-1}f(-1) + w_0 f(0) + w_1 f(1) \qquad (10.2\text{-}10)$$

The method in no way depends on the properties of sin x and will equally well apply to integrals of the form

$$\int_a^b K(x)f(x) \, dx$$

although the w_i will depend on the $K(x)$ chosen.

With three parameters, we try 1, x, x^2:

$$\begin{aligned} 0 &= w_{-1} + w_0 + w_1 \\ \frac{8}{\pi^2} &= -w_{-1} \qquad + w_1 \\ 0 &= w_{-1} \qquad + w_1 \end{aligned}$$

These are easily solved for

$$w_1 = -w_{-1} = \frac{4}{\pi^2} \qquad w_0 = 0$$

which gives, using (10.2-10),

$$\int_{-1}^{1} f(x) \sin \frac{\pi}{2} x \, dx = \frac{4}{\pi^2} [f(1) - f(-1)] \qquad (10.2\text{-}11)$$

EXERCISES

10.2-1. Find the formula

$$\int_{-\pi}^{\pi} f(x) \sin x \, dx = \left(1 - \frac{8}{\pi^2}\right) [f(\pi) - f(-\pi)] + \frac{16}{\pi^2}\left[f\left(\frac{\pi}{2}\right) - f\left(-\frac{\pi}{2}\right)\right]$$

10.2-2. Find the formula

$$\int_0^{2h} f(x) \, dx = \frac{h}{15} [7f(0) + 16f(h) + 7f(2h)] + \frac{h^2}{15} [f'(0) - f'(2h)] + \frac{h^7 f^{(6)}(\theta)}{4,725}$$

10.2-3. Find the formula

$$\int_{-1}^{1} f(x)\,dx = \frac{1}{4}\left[3f\left(-\frac{2}{3}\right) + 2f(0) + 3f\left(\frac{2}{3}\right)\right] + \frac{7}{1{,}620} f^{(4)}(\theta)$$

10.3 ASSIGNED SAMPLE POINTS

We begin our examination of a uniform method for finding formulas by considering the simple case where the sample points are assigned in advance and only function values are used.

Let $L(f)$ be some linear operator (see Sec. 7.2) acting on $f(x)$; again, typically, we think of $L(f)$ as being integration. Then in analogy to the method used for showing how to find an interpolation formula in Sec. 8.2, we can simply write the determinant

$$\begin{vmatrix} L(f) & f_1 & f_2 & \cdots & f_n \\ m_0 & 1 & 1 & \cdots & 1 \\ m_1 & x_1 & x_2 & \cdots & x_n \\ m_2 & x_1^2 & x_2^2 & \cdots & x_n^2 \\ \cdots & \cdots & \cdots & \cdots & \cdots \\ m_{n-1} & x_1^{n-1} & x_2^{n-1} & \cdots & x_n^{n-1} \end{vmatrix} = 0$$

where m_k is the kth moment of the operator L, which is the result of applying the operator L to x^k,

$$m_k = L(x^k)$$

We shall not use this method but rather shall use the methods of the previous section. These methods are formally equivalent but practically different.

Each of the examples in the previous section had a similar set of *defining equations* of the general form

$$\begin{aligned} m_0 &= w_1 + w_2 + w_3 + \cdots + w_n \\ m_1 &= w_1x_1 + w_2x_2 + w_3x_3 + \cdots + w_nx_n \\ m_2 &= w_1x_1^2 + w_2x_2^2 + w_3x_3^2 + \cdots + w_nx_n^2 \\ &\cdots\cdots\cdots\cdots\cdots\cdots \\ m_{n-1} &= w_1x_1^{n-1} + w_2x_2^{n-1} + w_3x_3^{n-1} + \cdots + w_nx_n^{n-1} \end{aligned} \qquad (10.3\text{-}1)$$

The matrix of the unknowns, $X = (x_i^k)$, has a determinant (called a Vandermonde determinant) which is not zero if $x_i \neq x_j$. (See Sec. 8.2.)

We propose to examine the analytical inversion of the matrix X. We could, of course, invert each case numerically as we came to it, but if we did, then questions of accuracy would arise that would be hard to answer; indeed, in special cases the matrices are often hard to invert. Given the analytic form of the inverse matrix, it is relatively easy to estimate the accuracy of the computation.

To find the inverse matrix, we first introduce the *fundamental polynomials* (compare with Sec. 8.3)

$$\begin{aligned}\pi_i(x) &= (x - x_1)(x - x_2) \cdots (x - x_{i-1})(x - x_{i+1}) \cdots (x - x_n) \\ &= \sum_{k=0}^{n-1} C_{i,k}x^k \qquad (i = 1, 2, \ldots, n) \qquad (10.3\text{-}2)\end{aligned}$$

The ith polynomial omits the ith factor $(x - x_i)$. We note the important fact that

$$\pi_i(x_j) = 0 \qquad (i \neq j) \qquad \text{and} \qquad \pi_i(x_i) \neq 0$$

This fact suggests the means for finding the inverse of X. Since

$$X^{-1}X = I$$

the top row of X^{-1} can be taken as

$$\frac{C_{1,k}}{\pi_1(x_1)}$$

because this row times the jth column of X gives

$$\frac{\Sigma C_{1,k}x_j^k}{\pi_1(x_1)} = \frac{\pi_1(x_j)}{\pi_1(x_1)} = \begin{cases} 1 & j = 1 \\ 0 & j \neq 1 \end{cases}$$

In general, the mth row of X^{-1} can be written

$$\frac{C_{m,k}}{\pi_m(x_m)}$$

The $C_{m,k}$ are symmetric functions of the appropriate sample points and are easily found by multiplications and additions; they are polynomials in the sample points x_i.

The special case of equally spaced sample points when the sample points are symmetrically arranged about the origin is worth tabulation. These inverse matrices are labeled S_n and are tabulated below. The numbers in the parentheses following the S are the locations of the sample points; thus $S_4(-\frac{3}{2}, -\frac{1}{2}, \frac{1}{2}, \frac{3}{2})$ means that the samples are located at $-\frac{3}{2}$, $-\frac{1}{2}$, $\frac{1}{2}$, $\frac{3}{2}$. When the matrix is multiplied on the right by the column vector of the moments, the resulting column vector gives the weights.

Universal Matrices

$$S_2 = S_2(-\tfrac{1}{2},\tfrac{1}{2}) = \begin{pmatrix} \tfrac{1}{2} & -1 \\ \tfrac{1}{2} & 1 \end{pmatrix} = \tfrac{1}{2}\begin{pmatrix} 1 & -2 \\ 1 & 2 \end{pmatrix}$$

$$S_3 = S_3(-1,0,1) = \begin{pmatrix} 0 & -\tfrac{1}{2} & \tfrac{1}{2} \\ 1 & 0 & -1 \\ 0 & \tfrac{1}{2} & \tfrac{1}{2} \end{pmatrix} = \tfrac{1}{2}\begin{pmatrix} 0 & -1 & 1 \\ 2 & 0 & -2 \\ 0 & 1 & 1 \end{pmatrix}$$

$$S_4 = S_4(-\tfrac{3}{2},-\tfrac{1}{2},\tfrac{1}{2},\tfrac{3}{2}) = \tfrac{1}{48}\begin{pmatrix} -3 & 2 & 12 & -8 \\ 27 & -54 & -12 & 24 \\ 27 & 54 & -12 & -24 \\ -3 & -2 & 12 & 8 \end{pmatrix}$$

$$S_5 = S_5(-2,-1,0,1,2) = \tfrac{1}{24}\begin{pmatrix} 0 & 2 & -1 & -2 & 1 \\ 0 & -16 & 16 & 4 & -4 \\ 24 & 0 & -30 & 0 & 6 \\ 0 & 16 & 16 & -4 & -4 \\ 0 & -2 & -1 & 2 & 1 \end{pmatrix}$$

$$S_6 = S_6(-\tfrac{5}{2},-\tfrac{3}{2},\ \ldots\ ,\tfrac{5}{2})$$

$$= \frac{1}{3{,}840}\begin{pmatrix} 45 & -18 & -200 & 80 & 80 & -32 \\ -375 & 250 & 1{,}560 & -1{,}040 & -240 & 160 \\ 2{,}250 & -4{,}500 & -1{,}360 & 2{,}720 & 160 & -320 \\ 2{,}250 & 4{,}500 & -1{,}360 & -2{,}720 & 160 & 320 \\ -375 & -250 & 1{,}560 & 1{,}040 & -240 & -160 \\ 45 & 18 & -200 & -80 & 80 & 32 \end{pmatrix}$$

$$S_7 = \tfrac{1}{720}\begin{pmatrix} 0 & -12 & 4 & 15 & -5 & -3 & 1 \\ 0 & 108 & -54 & -120 & 60 & 12 & -6 \\ 0 & -540 & 540 & 195 & -195 & -15 & 15 \\ 720 & 0 & -980 & 0 & 280 & 0 & -20 \\ 0 & 540 & 540 & -195 & -195 & 15 & 15 \\ 0 & -108 & -54 & 120 & 60 & -12 & -6 \\ 0 & 12 & 4 & -15 & -5 & 3 & 1 \end{pmatrix}$$

10.4 SOME EXAMPLES OF FORMULAS

Let us examine the kinds of formulas that we can find from these matrices. Consider $S_3 = S_3(-1,0,1)$.

$$S_3 = \begin{pmatrix} 0 & -\tfrac{1}{2} & \tfrac{1}{2} \\ 1 & 0 & -1 \\ 0 & \tfrac{1}{2} & \tfrac{1}{2} \end{pmatrix}$$

Suppose that we wish to derive Simpson's formula for the integral

$$\int_{-1}^{1} f(x)\,dx = w_{-1}f(-1) + w_0 f(0) + w_1 f(1)$$

We need to find the moments m_0, m_1, m_2, which are defined by

$$m_k = \int_{-1}^{1} x^k \, dx = \frac{1 + (-1)^k}{k + 1}$$

Regarding these as a column vector m, we multiply it on the left by S_3 to get the weights w_i in the form of a column vector w:

$$S_3 m = \begin{pmatrix} 0 & -\frac{1}{2} & \frac{1}{2} \\ 1 & 0 & -1 \\ 0 & \frac{1}{2} & \frac{1}{2} \end{pmatrix} \begin{pmatrix} 2 \\ 0 \\ \frac{2}{3} \end{pmatrix} = \begin{pmatrix} \frac{1}{3} \\ \frac{4}{3} \\ \frac{1}{3} \end{pmatrix} = w \tag{10.4-1}$$

which is, of course, the correct answer [Eq. (7.2-4)].

Next consider finding the "half Simpson" formula for

$$\int_{-1}^{0} f(x) \, dx = a_{-1} f(-1) + a_0 f(0) + a_1 f(1)$$

The moments are

$$m_0 = 1$$
$$m_1 = -\tfrac{1}{2}$$
$$m_2 = \tfrac{1}{3}$$

Hence

$$\begin{pmatrix} 0 & -\frac{1}{2} & \frac{1}{2} \\ 1 & 0 & -1 \\ 0 & \frac{1}{2} & \frac{1}{2} \end{pmatrix} \begin{pmatrix} 1 \\ -\frac{1}{2} \\ \frac{1}{3} \end{pmatrix} = \begin{pmatrix} \frac{5}{12} \\ \frac{8}{12} \\ -\frac{1}{12} \end{pmatrix} = \frac{1}{12} \begin{pmatrix} 5 \\ 8 \\ -1 \end{pmatrix} \tag{10.4-2}$$

Suppose that we had tried to find a formula of the form

$$\left.\frac{dy}{dx}\right|_{x=0} = w_{-1} f(-1) + w_0 f(0) + w_1 f(1)$$

The moment vector is

$$\begin{pmatrix} 0 \\ 1 \\ 0 \end{pmatrix}$$

and we have

$$\begin{pmatrix} 0 & -\frac{1}{2} & \frac{1}{2} \\ 1 & 0 & -1 \\ 0 & \frac{1}{2} & \frac{1}{2} \end{pmatrix} \begin{pmatrix} 0 \\ 1 \\ 0 \end{pmatrix} = \begin{pmatrix} -\frac{1}{2} \\ 0 \\ \frac{1}{2} \end{pmatrix} \tag{10.4-3}$$

The last example of Sec. 10.2 [Eq. (10.2-10)] was a formula for

$$\int_{-1}^{1} f(x) \sin \frac{\pi}{2} x \, dx$$

whose moments are 0, $8/\pi^2$, and 0. Hence we have

$$\begin{pmatrix} 0 & -\frac{1}{2} & \frac{1}{2} \\ 1 & 0 & -1 \\ 0 & \frac{1}{2} & \frac{1}{2} \end{pmatrix} \begin{pmatrix} 0 \\ \frac{8}{\pi^2} \\ 0 \end{pmatrix} = \begin{pmatrix} -\frac{4}{\pi^2} \\ 0 \\ \frac{4}{\pi^2} \end{pmatrix}$$

which is (10.2-11).

Suppose that we had wanted to interpolate at a point x, using these same three sample points. This time the moment vector depends on the position x at which we are interpolating. We get

$$\begin{pmatrix} 0 & -\frac{1}{2} & \frac{1}{2} \\ 1 & 0 & -1 \\ 0 & \frac{1}{2} & \frac{1}{2} \end{pmatrix} \begin{pmatrix} 1 \\ x \\ x^2 \end{pmatrix} = \begin{pmatrix} \frac{x^2 - x}{2} \\ 1 - x^2 \\ \frac{x^2 + x}{2} \end{pmatrix} \tag{10.4-4}$$

or, more usually,

$$f(x) = \frac{x^2 - x}{2} f(-1) + (1 - x^2) f(0) + \frac{x^2 + x}{2} f(1)$$

which is clearly the correct answer.

This last example suggests that we can find formulas for indefinite integrals as well as for definite integrals. With a little more thought, it becomes clear that integrals such as

$$\int_{-1}^{0} f(x)\, dx \qquad \int_{0}^{5} f(x)\, dx \qquad \int_{0}^{x} K(x) f(x)\, dx$$

can all be found in this same form, provided that we can find the necessary moments.

EXERCISES

10.4-1. Derive the result of Exercise 10.2-1.

10.4-2. Find a quartic interpolating polynomial corresponding to (10.4-4).

10.4-3. Find formulas for $y'(0)$, $y''(0)$, $y'''(0)$, $y^{\mathrm{IV}}(0)$, using S_5.

10.5 FUNCTION VALUE AND DERIVATIVE AT ASSIGNED POINTS

It frequently happens that either we have both the function values and the derivatives at the sample points or, having found the function values, it is only slightly more labor to find the derivatives. While it is possible to use any number of derivatives, we shall confine our discussion to a single derivative; the more complex cases are easy to do once the basic idea of the derivation is understood. For the sake of standardization, we shall place all the function values first and follow them by the derivative values, as in (10.2-9). As a general rule, "Never ignore any information that you happen to have," but at times it may be too expensive to use all that you have.

The right-hand side of the formula now appears as

$$w_1 f(x_1) + w_2 f(x_2) + \cdots + w_n f(x_n) + w_{n+1} f'(x_1) + \cdots + w_{2n} f'(x_n)$$

Thus the matrix to be inverted is

$$\begin{pmatrix} 1 & 1 & \cdots & 1 & 0 & 0 & \cdots & 0 \\ x_1 & x_2 & \cdots & x_n & 1 & 1 & \cdots & 1 \\ x_1^2 & x_2^2 & \cdots & x_n^2 & 2x_1 & 2x_2 & \cdots & 2x_n \\ \cdots & \cdots & \cdots & \cdots & \cdots & \cdots & \cdots & \cdots \\ x_1^{2n-1} & x_2^{2n-1} & \cdots & x_n^{2n-2} & (2n-1)x_1^{2n-2} & (2n-1)x_2^{2n-2} & \cdots & (2n-1)x_n^{2n-2} \end{pmatrix}$$

This time we wish to find the coefficients of a polynomial having *all* the properties

$$\begin{aligned} P(x_i) &= 0 \\ P'(x_i) &= 0 \end{aligned} \qquad (i = 1, 2, \ldots, n)$$

except one. (Compare with Sec. 8.3 and the Hermite interpolation.) The excluded condition depends on the row; as in Sec. 10.3, the first row *excludes*

	$P(x_1) = 0$
The second	$P(x_2) = 0$
The third	$P(x_3) = 0$
. . . . . .	. . . . . .
The nth	$P(x_n) = 0$
The $(n + 1)$st	$P'(x_1) = 0$
. . . .	. . . .
The $2n$th	$P'(x_n) = 0$

These polynomials are easy to construct. For the $(n + k)$th row we use

$$P_{n+k}(x) = (x - x_k) \prod_{j=1}^{n}{}' (x - x_j)^2$$

(the prime means omit $j = k$). For the kth row $(k \leq n)$ we try an arbitrary linear factor $ax + b$:

$$P_k(x) = (ax + b) \prod_{j=1}^{n}{}' (x - x_j)^2$$

Now $P'_k(x_k) = 0$ can be solved to determine the ratio of a to b (which is all that we need to know). Occasionally the polynomial that we want is of lower degree; this is indicated by the inability to determine the ratio.

As an example, let us construct the inverse matrix for two points $x = 0$ and $x = 1$. The polynomial for the first row is of the form

$$P_1(x) = (ax + b)(x - 1)^2$$
$$P_1'(0) = 0 = a - 2b$$

and so we set $b = 1$, $a = 2$, for convenience, as it makes $P_1(0) = 1$. This gives $P_1(x) = 1 + 0x - 3x^2 + 2x^3$, and the top row of the inverse matrix is, therefore, 1, 0, −3, 2, since $P_1(0) = 1$.

For the second row we obtain the polynomial $-2x^3 + 3x^2$ so that the row in the matrix is 0, 0, 3, −2. The third row has the polynomial $x - 2x^2 + x^3$, while the fourth row has $-x^2 + x^3$ as its polynomial. Thus we have the inverse matrix [since $P_2(1) = 1$, $P_3'(0) = 1$, $P_4'(1) = 1$].

$$\begin{pmatrix} 1 & 0 & -3 & 2 \\ 0 & 0 & 3 & -2 \\ 0 & 1 & -2 & 1 \\ 0 & 0 & -1 & 1 \end{pmatrix}$$

When we apply this to (10.2-9), which had moments equal to 1, ½, ⅓, ¼, we get exactly the same answer.

EXERCISES

10.5-1. Show that the inverse matrix for the function and first derivative values at $x = -1, 0, 1$ is

$$\begin{pmatrix} 0 & 0 & \frac{4}{4} & -\frac{5}{4} & -\frac{2}{4} & \frac{3}{4} \\ 1 & 0 & -2 & 0 & 1 & 0 \\ 0 & 0 & \frac{4}{4} & \frac{5}{4} & -\frac{2}{4} & -\frac{3}{4} \\ 0 & 0 & \frac{1}{4} & -\frac{1}{4} & -\frac{1}{4} & \frac{1}{4} \\ 0 & 1 & 0 & -2 & 0 & 1 \\ 0 & 0 & -\frac{1}{4} & -\frac{1}{4} & \frac{1}{4} & \frac{1}{4} \end{pmatrix}$$

10.5-2. Apply Exercise 10.5-1 to the integral in Exercise 10.2-2.

10.6 NO ASSIGNED SAMPLE POINTS; GAUSSIAN QUADRATURE

In Sec. 10.2 we examined a pair of formulas, (10.2-1) and (10.2-8), in which the positions of all the sample points as well as the weights were taken as parameters. The general case of this form would be to use n sample points x_i

$$\int_a^b K(x)f(x)\,dx = \sum_{i=1}^{n} w_i f(x_i) \qquad K(x) > 0 \tag{10.6-1}$$

which has $2n$ parameters and can be made exact for $1, x, \ldots, x^{2n-1}$.

We get the *defining equations*

$$
\begin{aligned}
m_0 &= w_1 + w_2 + w_3 + \cdots + w_n \\
m_1 &= w_1x_1 + w_2x_2 + w_3x_3 + \cdots + w_nx_n \\
m_2 &= w_1x_1^2 + w_2x_2^2 + w_3x_3^2 + \cdots + w_nx_n^2 \\
&\cdots\cdots\cdots\cdots\cdots\cdots \\
m_{2n-1} &= w_1x_1^{2n-1} + \cdots + w_nx_n^{2n-1}
\end{aligned}
\tag{10.6-2}
$$

There is a well-known method for solving this particular system of nonlinear equations. We first define the *sample polynomial*

$$\pi(x) = \prod_{i=1}^{n} (x - x_i) = \sum_{k=0}^{n} C_k x^k \tag{10.6-3}$$

Note that

$$\pi(x_i) = 0 \qquad (i = 1, 2, \ldots, n)$$

We multiply the top equation by C_0, the next by C_1, . . . , the nth by $C_n = 1$, and add to get

$$\sum_{k=0}^{n} C_k m_k = \sum_{i=1}^{n} w_i \pi(x_i) = 0$$

We now shift each coefficient down one equation and repeat the process to obtain

$$\sum_{k=0}^{n} C_k m_{k+1} = \sum_{i=1}^{n} w_i x_i \pi(x_i) = 0$$

If we do this a total of n times, we have

$$\sum_{k=0}^{n} C_k m_{k+j} = 0 \qquad (j = 0, 1, \ldots, n-1) \tag{10.6-4}$$

The determinant of the C_k is called "persymmetric" by Muir [30, p. 419].

If $|m_{k+j}| \neq 0$, we can solve for the C_k (using $C_n = 1$). If the determinant is zero, then one of two things can happen. First, the equations can be inconsistent. An example is given by

$$
\begin{aligned}
\int_{-1}^{1} xf(x)\,dx &= w_1 f(x_1) \\
m_0 = 0 &= w_1 \cdot 1 \\
m_1 = \tfrac{2}{3} &= w_1 \cdot x_1 \qquad \text{false}
\end{aligned}
$$

Thus there is no such formula. Second, the rank may be $n - 1$, and we add one more equation to the set of defining equations (10.6-2) and derive another equation (10.6-4) involving the m_{k+j} and the C_k. We repeat this process until we have n linearly independent equations for the

C_k, which may or may not be consistent. Each additional equation indicates the increased accuracy of the formula.

From the C_k, which are the coefficients of the sample polynomial (10.6-3)

$$\pi(x) = x^n + C_{n-1}x^{n-1} + \cdots + C_0 = 0$$

we can find the zeros of $\pi(x)$ which are the sample points x_i.

We are now reduced to the previous case of known sample points and can use the first n of the defining equations (10.6-2) to find the weights w_i.

As an example, consider Eq. (10.2-6) which we avoided because of the threat of messy algebra. We had (10.2-5) and (10.2-6):

$$\int_0^1 f(x)\,dx = w_1 f(x_1) + w_2 f(x_2)$$

$$\begin{array}{rcll|c|c} 1 &=& w_1 & + w_2 & C_0 & \\ \frac{1}{2} &=& w_1x_1 & + w_2x_2 & C_1 & C_0 \\ \frac{1}{3} &=& w_1x_1^2 & + w_2x_2^2 & 1 & C_1 \\ \frac{1}{4} &=& w_1x_1^3 & + w_2x_2^3 & & 1 \end{array}$$

$$\pi(x) = (x - x_1)(x - x_2) = x^2 + C_1x + C_0$$

The multipliers to form the new equations are shown in the two columns on the right. These give

$$C_0 + (\tfrac{1}{2})C_1 + \tfrac{1}{3} = 0$$

$$(\tfrac{1}{2})C_0 + (\tfrac{1}{3})C_1 + \tfrac{1}{4} = 0$$

or

$$C_0 = \tfrac{1}{6} \qquad C_1 = -1$$

Hence

$$\pi(x) = x^2 - x + \tfrac{1}{6}$$

and

$$x_1 = \frac{1 - \sqrt{1 - \frac{2}{3}}}{2} = \frac{1 - \sqrt{\frac{1}{3}}}{2}$$

$$x_2 = \frac{1 + \sqrt{1 - \frac{2}{3}}}{2} = \frac{1 + \sqrt{\frac{1}{3}}}{2}$$

as before [Eq. (10.2-8)].

10.7 SOME SAMPLE POINTS ASSIGNED

It often happens that we want to use one or both of the end points as sample points. Let us consider using one end point only. We, of course, lose one of the $2n$ parameters. The defining equations (10.3-1) are

$$m_k = w_1a^k + \sum_{i=2}^{n} w_ix_i^k \qquad (k = 0, 1, \ldots, 2n - 2)$$

where a is the given sample point.

We partially eliminate a by multiplying each equation by a and subtracting it from the equation immediately below it.

$$\bar{m}_k = m_{k+1} - am_k = \sum_{i=2}^{n} w_i(x_i - a)x_i^k \qquad (k = 0, 1, \ldots, 2n - 3)$$

We now form the sample polynomial, using *only* the unknown sample points

$$\pi(x) = \prod_{i=2}^{n} (x - x_i) = \sum_{k=0}^{n-1} C_k x^k$$

and repeat the elimination process used in the previous section:

$$\sum_{k=0}^{n-1} C_k \bar{m}_{k+j} = \sum_{i=2}^{n} w_i(x_i - a)x_i^j \pi(x_i) = 0 \qquad (j = 0, 1, \ldots, n - 2)$$

We can then solve for the C_k and the unknown x_i as before.

EXERCISES

10.7-1. Discuss the case where both end points are to be used in the formula.

10.7-2. Discuss the general problem of eliminating any number of assigned sample points.

10.8 ANALYTICAL REMARKS

In this section we consider what will happen when we try to carry out the operations that we have been discussing. Thus this section is not part of the routine for finding the formula but is an examination of what we shall find when we try to compute a formula.

There is no question that we can go from the defining equations (10.6-2) to Eqs. (10.6-4):

$$\sum_{k=0}^{n} C_k m_{k+j} = 0$$

If the rank of the matrix (m_{k+j}) is not high enough to allow us to solve for the C_k, then we add another defining equation and try again until we either raise the rank to the required amount, obtain inconsistent equations indicating that there is no such formula, or become tired. Each time we add another defining equation, we get a formula that is exact for one more power of x, and it is unlikely, except in the most trivial cases, that the process will not stop very quickly.

The next difficulty comes when we try to find the roots of the sample polynomial. We now show that in many cases the roots are all real, distinct, and lie in the interval of integration. Consider first the case where none of the sample points is assigned and $K(x) \geq 0$. We observe that, since the formula is exact for all powers of x up through $2n - 1$, we have [$\pi(x)$ as in Eq. (10.6-3)]

$$\int_a^b K(x)\pi(x)x^k\,dx = \sum_{i=1}^{n} w_i x_i^k \pi(x_i) = 0 \qquad (k = 0, \ldots, n-1) \quad (10.8\text{-}1)$$

[since the $\pi(x_i)$ which occur on the right-hand side are all zero]. Now suppose that $\pi(x)$ has less than n distinct zeros in the range of integration. Let

$$x_1, \quad x_2, \quad \ldots, \quad x_m \qquad (m < n)$$

be the odd multiplicity zeros in the interval. The polynomial

$$p(x) = \prod_{i=1}^{m} (x - x_i)$$

is of degree $m \leq n - 1$; hence we have, by (10.8-1)

$$\int_a^b K(x)\pi(x)p(x)\,dx = 0$$

But in the interval of integration the integrand is of constant sign and cannot be identically zero, and so we are led to a contradiction. Thus there are n real, distinct zeros of $\pi(x)$ in the interval, and we are assured that the real roots for which we are looking actually exist.

In the case where one end point a is assigned, we have $\pi(x)$ of degree $n - 1$. We form

$$\int_a^b K(x)(x - a)\pi(x)x^k\,dx = 0 \qquad (k = 0, 1, \ldots, n-2)$$

and assume that $\pi(x)$ has only $m < n - 1$ distinct roots of odd degree in the interval. Using $p(x)$ as before, we obtain

$$\int_a^b K(x)(x - a)\pi(x)p(x)\,dx = 0$$

This leads to the same contradiction.

We leave it as an exercise to show that the same type of argument applies to the case where both end points are assigned.

We can also show that the weights w_i at the unassigned sample points are all positive. This is an interesting result because it shows that the formula will tend to resist noise in the data as there will be no cancellation due to the weights being of opposite signs (although there can be cancellation due to the function being of opposite signs at various sample points). We define, for the gaussian case of no assigned sample points [compare with (10.3-2)]

$$\pi_k(x) = (x - x_1)(x - x_2) \cdot \cdot \cdot (x - x_{k-1})(x - x_{k+1}) \cdot \cdot \cdot (x - x_n)$$

Since $\pi_k^2(x)$ is of degree $2n - 2$, we have

$$\int_a^b K(x)\pi_k^2(x)\,dx = \sum_{i=1}^{n} w_i\pi_k^2(x_i) = w_k\pi_k^2(x_k)$$

Hence $w_k > 0$, provided that $K(x) \geq 0$. A similar argument applies to the cases where one or both end points are used. When an assigned sample point falls in the interval, the result is not in general true, although special cases can be found where it is.

We cannot show that the weights are positive in the case of all assigned sample points, because, in fact, they can be negative.

If we attempt to extend the gaussian-type formula where the sample points are unassigned to include the use of the function and the first derivative at the same points, then we find (at least in the few cases that have been investigated) that the zeros of the sample polynomial are complex numbers.

EXERCISE

10.8-1. Examine the case where both end points are assigned. Show that the zeros are all real, distinct, and in the interval and that the weights are positive.

10.9 LINEAR RESTRAINTS ON THE WEIGHTS

It sometimes happens that we either have given, or would like to put, linear restraints on the weights. Such restraints are likely to involve the weights associated with assigned sample points; they are not likely to involve the weights associated with unassigned sample points, since it is hard to know in what order they will be labeled—and to label them in order of size makes for very difficult mathematical problems. The main exception is when the same restraint is applied to all the unknown weights.

The best-known example of restraints involving the weights is Chebyshev's integration, where

$$w_1 = w_2 = \cdot \cdot \cdot = w_n = w$$

and all the sample points are unassigned. This gives us $n + 1$ parameters and consequently the $n + 1$ defining equations

$$m_k = w \sum_{i=1}^{n} x_i^k \qquad (k = 0, 1, \ldots, n)$$

Using $k = 0$, we find

$$w = \frac{m_0}{n}$$

The remaining equations become

$$\frac{nm_k}{m_0} = \bar{m}_k = \sum_{i=1}^{n} x_i^k$$

The Newton identities enable us to express the sums of powers of the roots in terms of the coefficients of the sample polynomial C_k. We now have the sample polynomial and can find the zeros x_i.

It is known that, when $K(x) = 1$, the sample points are all real, distinct, and in the interval for $n = 1, 2, \ldots, 7, 9$ and only for those n.

A. Ralston has published[1] an interesting example of a mixed type of integration in which two sample points and one linear restraint are assigned. Specifically, the formula

$$\int_{-1}^{1} f(x)\,dx = \sum_{i=1}^{n} w_i f(x_i)$$

has the conditions

$$x_1 = -1$$
$$x_n = 1$$
$$w_1 = -w_n$$

We have, therefore, $2n - 3$ parameters and can make the formula exact for $1, x, \ldots, x^{2n-4}$. The defining equations are

$$2 = \sum_{i=2}^{n-1} w_i$$

$$0 = -2w_1 + \sum w_i x_i$$

$$2/3 = \sum w_i x_i^2$$

$$0 = -2w_1 + \sum w_i x_i^3$$

$$\cdots\cdots\cdots\cdots$$

$$\frac{2}{2n-3} = \sum w_i x_i^{2n-4}$$

In order to follow the subsequent algebra more easily, and at the same

[1] A. Ralston, *J. Assoc. Computing Machinery*, vol. 6, pp. 384–394, July, 1959.

time not oversimplify it, we take the case $n = 5$ which gives $2n - 3 = 7$ equations.

The sample polynomial is

$$\pi(x) = (x - x_2)(x - x_3)(x - x_4) = x^3 + C_2x^2 + C_1x + C_0$$

The usual elimination process leads to $n - 1$ (in this case, four) equations:

$$\begin{aligned}
C_0 + (1/3)C_2 &= -w_1C_1 - w_1 \\
(1/3)C_1 + 1/5 &= -w_1C_0 - w_1C_2 \\
(1/3)C_0 + (1/5)C_2 &= -w_1C_1 - w_1 \\
(1/5)C_1 + 1/7 &= -w_1C_0 - w_1C_2
\end{aligned}$$

It is easy to see that we can form $n - 3$ $(= 2)$ equations which do not involve the quadratic terms w_iC_i $(i = 0, 1, 2)$,

$$\begin{aligned}
(2/3)C_0 + (2/15)C_2 &= 0 \\
(2/15)C_1 + (2/35) &= 0
\end{aligned}$$

We have, therefore, $n - 3$ linear, nonhomogeneous equations in the unknowns $C_0, C_1, \ldots, C_{n-3}$, or one more unknown than equations. Solving the equations in terms of, say, C_0, we have

$$\begin{aligned}
C_1 &= -3/7 \\
C_2 &= -5C_0
\end{aligned}$$

These are put in the first two quadratic equations of the set to give

$$\begin{aligned}
(-1/3)C_0 &= -2/7 w_1 \\
1/35 &= +2C_0w_1
\end{aligned}$$

Thus in the general case, as well as in the specific case of $n = 5$, when we eliminate C_0 we come to a quadratic in w_1. In the case $n = 5$,

$$w_1 = \pm\sqrt{1/60}$$

The plus or minus sign has the effect of reversing the formula from left to right or right to left.

Knowing w_1, we can easily find the C_i and the sample polynomial

$$\pi(x) = x^3 - 5\sqrt{(3/245)}\,x^2 - (3/7)x + \sqrt{(3/245)}$$

Reasoning similar to that of Sec. 10.8 shows that the zeros are real distinct, and in the interval of integration.

EXERCISE

10.9-1. Show that the Ralston case of $n = 6$ has the sample polynomial

$$\pi(x) = x^4 + \frac{4}{(3\sqrt{6})}x^3 - \frac{2}{3}x^2 - \frac{4}{(7\sqrt{6})}x + \frac{1}{21}$$

10.10 THE GREGORY FORMULA

The two cases that we have examined which had linear restraints on the coefficients both kept most sample points unassigned. The present section examines the case of equally spaced assigned sample points with many restraints. We also introduce a new concept, that of having a formula involving an arbitrary number of sample points and yet keeping the same general pattern of weights. To be specific, we shall examine formulas whose weights have the following general patterns:

$$a \quad b \quad b \quad b \quad b \quad \cdots \quad b \quad b \quad b \quad a \tag{10.10-1}$$

$$a \quad b \quad c \quad c \quad c \quad \cdots \quad c \quad c \quad b \quad a \tag{10.10-2}$$

$$a \quad b \quad c \quad d \quad d \quad \cdots \quad d \quad c \quad b \quad a \tag{10.10-3}$$

etc. Since each formula has one more parameter than the one above it, we can expect to find a family of increasingly accurate formulas.

Formula (10.10-1), for example, suggests that the integral using $n + 1$ sample points

$$\int_0^{nh} f(x)\,dx = af(0) + bf(h) + bf(2h) + \cdots + bf((n-1)h) + af(nh)$$

can be made exact for both 1 and x. This leads to two equations [using Eqs. (3.1-3)]

$$nh = b(n-1) + 2a$$

$$\frac{n^2h^2}{2} = bh\,\frac{n(n-1)}{2} + ahn$$

which are exactly the same equation. If we then try x^2, we get

$$\frac{n^3h^3}{3} = bh^2\,\frac{n(n-1)(2n-1)}{6} + ah^2n^2$$

The solution of the two equations is

$$a = \frac{n}{2(n+1)}\,h = \frac{h}{2} - \frac{h}{2(n+1)}$$

$$b = \frac{n^2}{n^2-1}\,h = h + \frac{h}{n^2-1}$$

Thus the formula is exact for quadratic polynomials.

The coefficients depend on n, the number of intervals, and this is often a nuisance. We could have required the first equation that we derived to be *identically true for all* n and not tried to make the formula exact for x^2. This approach gives

$$b = h$$

$$a = \frac{h}{2}$$

which is the usual composite trapezoid rule.

The patterns of the weights were chosen to be symmetric, and, as we shall show, this implies that, if one of these formulas is exact for all powers of x up to some even power, it is exact for the next higher odd power. This effect was seen in the above example when both 1 and x gave the same equation. To see the truth of this remark in general, we observe that the transformation $\bar{x} = x - n/2$ translates the origin to the middle of the range of integration. With this origin, all odd powers are automatically exact since both sides, because of symmetry, give zero. When the odd power is translated back to the original interval of integration, all the lower powers appear. If the formula is exact for these lower powers, both sides of the equation will cancel again, leaving the formula true for the odd power alone. We need, therefore, consider only even powers of x, since the symmetry of the problem will take care of the odd powers automatically.

We pursue the problem of finding formulas whose patterns of weights are independent of the number of samples taken, provided, of course, that enough sample points are taken so that all the parameters appear. These formulas could be found one at a time independently of each other, but the following observation reduces the problem to that of modifying each formula in turn to get the next one in the family. The third pattern (10.10-3), to take a specific case, can be written

$$\int_0^{nh} f(x)\,dx = \text{(trapezoid rule)} + A_1[\Delta f(0) - \Delta f(n-1)] + A_2[\Delta^2 f(0) + \Delta^2 f(n-2)]$$

The addition of the difference terms will not affect the accuracy of the formula for 1 and x. It is also clear that there exist constants A_1 and A_2 so that this will reduce to the pattern (10.10-3).

In the general case, the form

$$\int_0^{nh} f(x)\,dx = \text{(trapezoid rule)} + A_1[\Delta f(0) - \Delta f(n-1)] + A_2[\Delta^2 f(0) + \Delta^2 f(n-2)] + A_3[\Delta^3 f(0) - \Delta^3 f(n-3)] + A_4[\Delta^4 f(0) + \Delta^4 f(n-4)] + \cdots$$

provides an approach in which each coefficient can apparently be found in turn. Actually, because of the symmetry which produces the extra exactness of an odd power of x, pairs of coefficients are determined at the same time. Thus to determine A_1 and A_2, we impose the condition that the formula be exact for x^2. Using the results of Sec. 3.1, we have (for $h = 1$)

$$\frac{n^3}{3} = \frac{n(n-1)(2n-1)}{6} + \frac{1}{2}n^2 + A_1[1 - (2n-1)] + A_2[2+2]$$

Equating like powers of n on both sides, we obtain

$$\begin{aligned} n^3&: & \tfrac{1}{3} &= \tfrac{1}{3} \\ n^2&: & 0 &= -\tfrac{3}{6} + \tfrac{1}{2} \\ n&: & 0 &= +\tfrac{1}{6} - 2A_1 \\ 1&: & 0 &= +2A_1 + 4A_2 \end{aligned}$$

The last two equations give

$$A_1 = +\tfrac{1}{12} \qquad A_2 = -\tfrac{1}{24}$$

Using x^4 we can get, in a like manner,

$$A_3 = \tfrac{19}{720} \qquad A_4 = -\tfrac{3}{160}$$

and using x^6 we have

$$A_5 = \frac{863}{60{,}480} \qquad A_6 = \frac{-275}{24{,}192}$$

We have, therefore, Gregory's formula

$$\begin{aligned} \frac{1}{h}\int_0^{nh} f(x)\,dx = {} & (\tfrac{1}{2}f_0 + f_1 + f_2 + \cdots + f_{n-1} + \tfrac{1}{2}f_n) \\ & + \tfrac{1}{12}(\Delta f_0 - \Delta f_{n-1}) - \tfrac{1}{24}(\Delta^2 f_0 + \Delta^2 f_{n-2}) + \tfrac{19}{720}(\Delta^3 f_0 - \Delta^3 f_{n-3}) \\ & - \tfrac{3}{160}(\Delta^4 f_0 + \Delta^4 f_{n-4}) + \cdots \end{aligned}$$

The Lagrange form of Gregory's formula in which the linear operator is written out in terms of the function values is often more useful and is the form with which we started in this section. In this form the coefficients are given in the following table, where the left-hand column gives the highest-order coefficient A_k. By symmetry, we need give only one end of the formula.

A_0:	$\frac{1}{2}$	1	1	1	1	. . .
A_1:	$\frac{5}{12}$	$\frac{13}{12}$	1	1	1	. . .
A_2:	$\frac{9}{24}$	$\frac{28}{24}$	$\frac{23}{24}$	1	1	. . .
A_3:	$\frac{251}{720}$	$\frac{897}{720}$	$\frac{633}{720}$	$\frac{739}{720}$	1	. . .

A corresponding formula which uses derivatives in place of differences is known as the Euler-Maclaurin expansion and is sometimes useful but requires the coding of the derivatives when used on a machine [see Eq. (4.8-2)].

EXERCISES

10.10-1. Using the methods of this section, develop the corresponding formulas for

$$\begin{aligned} \int_0^n f(x)\,dx = {} & af(\tfrac{1}{2}) + bf(\tfrac{3}{2}) + cf(\tfrac{5}{2}) + cf(\tfrac{7}{2}) + \cdots + cf(n - \tfrac{5}{2}) \\ & + bf(n - \tfrac{3}{2}) + af(n - \tfrac{1}{2}) \end{aligned}$$

10.10-2. Using the same samples as in Exercise 10.10-1, carry out the work to find the coefficient of the first through fourth difference terms.

10.11 SUMMARY

The main purpose of this chapter has been to show how to derive various kinds of formulas, and we have not been concerned to any great extent with the actual results obtained. One of the reasons for adopting this attitude is that there are many more formulas than can be put in a

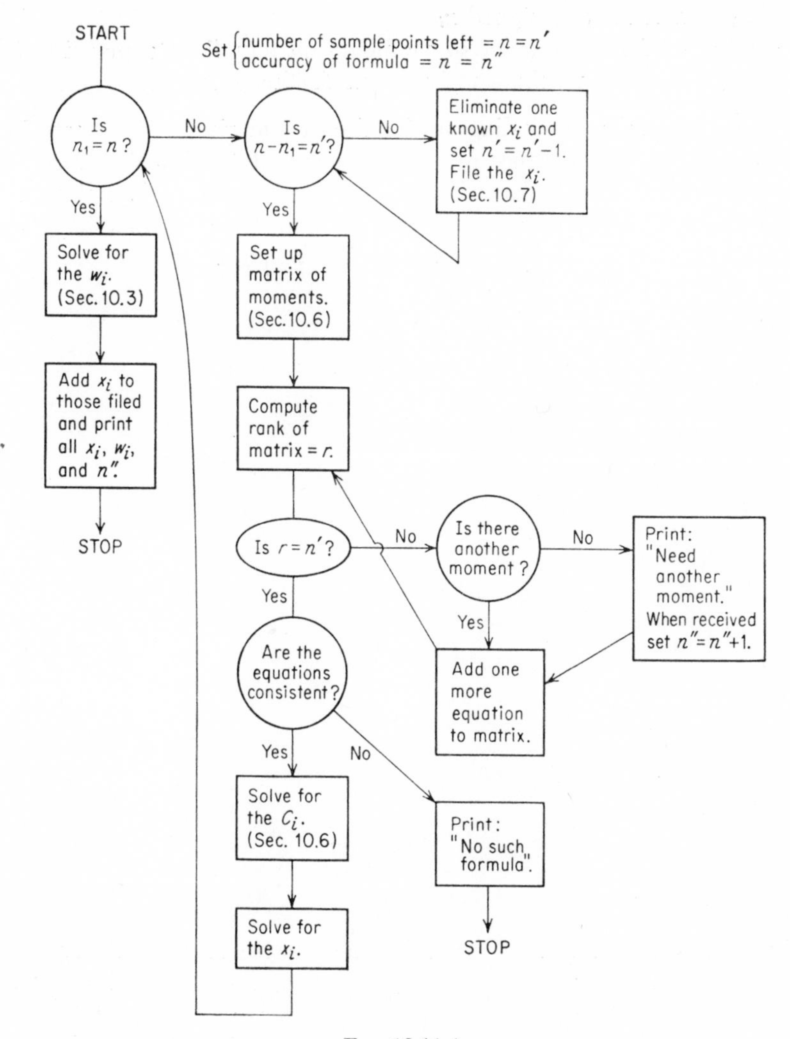

FIG. 10.11-1

single book, and the investigation of all the results would be such a mass of detail as to discourage even the most ardent lover of numerical analysis. When the methods of finding a formula are understood, then it is easy to find the specific formula needed at a particular time.

We also pointed out the possibility of the machine actually doing much of the work of the derivations. To further this aim, we present a flow diagram (Fig. 10.11-1) of the derivation scheme for formulas having no linear restraints on the weights.

Let n = number of sample points x_i

n_1 = number of assigned x_i

$2n - n_1$ = minimum number of moments m_k supplied to machine (on occasion machine will demand further moments)

The corresponding flow diagram for the case of given linear restraints is not included here because of the vagueness of the kinds of formulas that one might want. As remarked before, the linear restraints will probably be applied to assigned sample points or, as in Chebyshev integration, uniformly on all the unassigned weights.

CHAPTER 11

On Finding the Error Term of a Formula

11.1 NEED FOR AN ERROR TERM

Having described a somewhat uniform method for finding formulas, we now describe a uniform method for finding the corresponding error terms. As in the case of finding formulas, the uniform method is not always the shortest or most elegant, but its virtues seem to outweigh the liabilities of occasionally doing more work.

In this chapter, we concentrate on finding an actual expression for the error of the formula while ignoring the roundoff effects. The error in the formula is often called the *truncation error* because in the Taylor-series type of derivation (see Sec. 7.2) the error comes from the truncation of the series. As in the case of polynomial interpolation, the form of the error will be written in terms of some high-order derivative (or derivatives). This is the traditional form, but it has all the faults discussed in Sec. 8.7.

In this chapter we are concerned with finding the error term, and we postpone the discussion of the actual results until the next chapter where some formulas and their error terms are brought together. The need for an error term should, however, be evident. We would like to know the error before choosing one formula over another, although, of course, the size of the truncation error is not the sole criterion for selection.

11.2 THE ORDER OF THE ERROR TERM

We again imagine that we are examining a formula for estimating a definite integral, although linear operators such as indefinite integrals, interpolation, etc., are actually included. We introduce the notation

$$E_k = E_k(x)$$

to represent the error in the formula when we substitute x^k in place of $f(x)$ in the formula. In our method of derivation we have been setting (except in the Gregory formula) $E_k = 0$ for $k = 0, 1, \ldots, m - 1$, where m is the number of free parameters in the formula. We expect that $E_m \neq 0$ but are prepared to find that it is zero, since we know that in some cases, such as Simpson's formula, the accuracy is higher than the number of free parameters.

We need to test E_m to see if it is exactly zero. In case all the sample points are given, E_m is rationally expressible in terms of the given moments and given sample points. But in the Gauss quadrature case, to take a specific example, we had to solve an algebraic equation for the sample points x_j. The sample polynomial that we had to solve to get the unknown sample points gives a relation which enables us to reduce powers of the x_j above n (the number of unknown sample points) to powers below n. The reduction involves the coefficients C_k of the sample polynomial. Now the powers of x_j below n are multiplied by their weights w_j and hence are expressible in terms of the moments m_k. As a result, we again have a rational expression in the moments.[1] The test to see if $E_m = 0$ is thus a rational operation on the given moments and the given sample points, and if we are prepared to do enough algebra on rational expressions, we can settle the question of whether or not $E_m = 0$ without getting involved with the roundoff troubles that arise in solving the sample polynomial for the unknown sample points.

If we find that one value gives zero, we try the next until we find an m such that $E_m \neq 0$, or else we become tired and quit; the latter is not likely except in trivial cases, since one or two stages of extra accuracy are all that can be expected.

EXERCISES

11.2-1. Determine m for Eq. (10.2-8), using the method of Sec. 11.2.

11.2-2. Determine m for Simpson's formula (10.4-1).

11.2-3. Determine m for Exercise 10.2-1.

11.3 THE INFLUENCE FUNCTION $G(s)$

Let us first be clear about the class of formulas that we are going to examine. We shall use the operation of integration for purposes of discussion, although, as we know, other linear operators are also admissible. On the right-hand side of the equation we have used values of the function and its derivatives, but the maximum derivative allowed will always be less than the order of accuracy of the formula, which in the previous

[1] The moments m_k need not be rational, of course, and they can present problems (in principle) of exact cancellation or not, but in practice this is not usually troublesome.

section we have labeled $m - 1$. Thus we consider formulas of the form

$$\int_a^b f(x)\,dx = \sum_{k=1}^{n} a_k f(x_k) + \sum_{k=1}^{n} b_k f'(x_k) + \cdots + \sum_{k=1}^{n} m_k f^{(m-2)}(x_k) + R \quad (11.3\text{-}1)$$

where some of the a_k, b_k, . . . , m_k may be zero. It should be noted that the sample points x_k may lie inside or outside the range of integration. This formula is supposed to be exact for $1, x, \ldots, x^{m-1}$, and $E_m \neq 0$, with m having been determined as in the preceding section.

To obtain an expression for the remainder R, we start with the Taylor series

$$f(x) = f(A) + (x - A)f'(A) + \frac{(x-A)^2}{2!} f''(A) + \cdots + \frac{(x-A)^{m-1}}{(m-1)!} f^{(m-1)}(A) + \frac{1}{(m-1)!} \int_A^x f^{(m)}(s)(x-s)^{m-1}\,ds \quad (11.3\text{-}2)$$

This formula may be found by integrating the last term by parts m times; thus in many respects this formula is merely an identity. We choose A as the minimum of a and all the x_k.

We now substitute expression (11.3-2) into formula (11.3-1) on both sides, that is, for $f(x)$ and for the $f'(x_k)$, $f''(x_k)$, . . . , $f^{(m-2)}(x_k)$. The error in the formula (11.3-1) is the difference $R(f)$ between the two sides of the equation. Now the sum of all the terms of the Taylor expansion *except* the integral term is a polynomial in x of degree $m - 1$ or less, and since we have determined the coefficients of our formula (11.3-1) so that it is exact for $1, x, \ldots, x^{m-1}$ [that is, $E_k = 0$ $(k = 0, 1, \ldots, m-1)$], the polynomial exactly cancels out leaving only the integral term to be substituted into the formula. We have, therefore, the remainder (or error) when we substitute $f(x)$ in the formula

$$\begin{aligned} R = R(f) = &\int_a^b \frac{1}{(m-1)!} \int_A^x f^{(m)}(s)(x-s)^{m-1}\,ds\,dx \\ &- \sum \frac{a_k}{(m-1)!} \int_A^{x_k} f^{(m)}(s)(x_k - s)^{m-1}\,ds \\ &- \sum \frac{b_k}{(m-1)!} \frac{d}{dx_k} \int_A^{x_k} f^{(m)}(s)(x_k - s)^{m-1}\,ds \\ &- \cdots - \sum \frac{m_k}{(m-1)!} \frac{d^{m-2}}{dx_k^{m-2}} \int_A^{x_k} f^{(m)}(s)(x_k - s)^{m-1}\,ds \end{aligned} \quad (11.3\text{-}3)$$

In order to eliminate the notational trouble at the upper limits x_k in the integrals, we introduce a notational trick. Define for $j > 0$

$$(x - s)_+^j = \begin{cases} 0 & \text{if } x - s \leq 0 \\ (x - s)^j & \text{if } x - s \geq 0 \end{cases} \tag{11.3-4}$$

If $j = 0$, then

$$(x - s)_+^0 = \begin{cases} 0 & \text{if } x - s < 0 \\ 1 & \text{if } x - s > 0 \end{cases}$$

but $j = 0$ cannot occur in our case since the maximum derivative allowed is $m - 2$.

By examining these terms as x approaches s it is easy to see that

$$\begin{aligned} \frac{d}{dx}(x - s)_+^{m-1} &= (m - 1)(x - s)_+^{m-2} \\ \frac{d^2}{dx^2}(x - s)_+^{m-1} &= (m - 1)(m - 2)(x - s)_+^{m-3} \end{aligned} \tag{11.3-5}$$

etc., so long as the order of the derivative is less than $m - 1$ (which we are supposing in our formula). It is also easy to see that

$$\int_a^b (x - s)_+^{m-1}\, dx = \frac{(b - s)_+^m - (a - s)_+^m}{m} \tag{11.3-6}$$

Using this new notation, we may increase the upper limit of integration in s to B in Eq. (11.3-3), where B is taken to be the maximum of b and all the x_k. This increase in the upper limit does not affect the value of the integral, since each term is zero throughout the increased range.

Using these observations, interchanging the order of integration with the other operations in each term, and carrying out the differentiations using (13.3-5) and the integration using (13.3-6), we get for (11.3-3)

$$\begin{aligned} R = R(f) = \frac{1}{(m - 1)!} \int_A^B f^{(m)}(s) \Bigg[& \frac{(b - s)_+^m - (a - s)_+^m}{m} \\ & - \sum_{k=1}^n a_k (x_k - s)_+^{m-1} - \sum_{k=1}^n b_k (m - 1)(x_k - s)_+^{m-2} - \cdots \\ & - \sum_{k=1}^n m_k (m - 1)(m - 2) \cdots 3 \cdot 2 (x_k - s)_+ \Bigg] ds \end{aligned} \tag{11.3-7}$$

The quantity $1/(m - 1)!$ times the expression in brackets depends on s and the particular formula being examined but *does not depend* on the function $f(s)$. Thus we write

$$R = R(f) = \int_A^B f^{(m)}(s) G(s)\, ds \tag{11.3-8}$$

where $G(s)$ is called the "influence function." The particular function $f(x) = x^m$ enables us to compute the integral of $G(s)$ from (11.3-7):

$$R(f) = R(x^m) = m! \int_A^B G(s)\, ds = E_m \neq 0 \qquad (11.3\text{-}9)$$

We therefore know the value of the integral of $G(s)$ as well as an expression for $(m - 1)!G(s)$ as given in the brackets. Note that this is the $G(s)$ for (11.3-1) and does not apply to other types of formulas.

An examination of $G(s)$ shows that it is piecewise a polynomial of degree at most m, and the pieces are bounded by the sample points x_k and the two values of the integration range a and b.

11.4 WHEN $G(s)$ HAS A CONSTANT SIGN

In most of the important cases it turns out that the influence function $G(s)$ has a constant sign. When this is so, the equation for the remainder [Eq. (11.3-8)]

$$R = R(f) = \int_A^B f^{(m)}(s) G(s)\, ds$$

satisfies the conditions for the mean-value theorem, and we can write, using (11.3-9),

$$R = f^{(m)}(\theta) \int_A^B G(s)\, ds = \frac{E_m}{m!} f^{(m)}(\theta) \qquad (A \le \theta \le B)$$

where E_m is the error in the formula when x^m is substituted for $f(x)$ and is known to be not zero by the way in which m was chosen.

The problem to be considered now is "How can we decide whether or not $G(s)$ is of constant sign?" As a first approach to this problem, consider the error of the trapezoid rule

$$R(f) = \int_0^1 f(x)\, dx - \frac{f(1) + f(0)}{2}$$

which is exact for $f(x) = 1$ and for x. Thus $m = 2$. Then we get from (11.3-7)

$$1!G(s) = \frac{(1 - s)_+^2 - (-s)_+^2}{2} - \frac{(1 - s)_+ + (-s)_+}{2}$$

Suppose that we graph this piecewise polynomial function. For $s \ge 1$ it is clear that $G(s) \equiv 0$, since each term equals zero. For $s < 0$ we likewise, but with a bit more algebra, find $G(s) \equiv 0$. For $0 < s < 1$ we have no breaks in the polynomial, and

$$G(s) = \frac{(1 - s)^2 - (1 - s)}{2} = \frac{-s + s^2}{2} = -\frac{s(1 - s)}{2} < 0$$

which is clearly of constant sign. See Fig. 11.4-1.

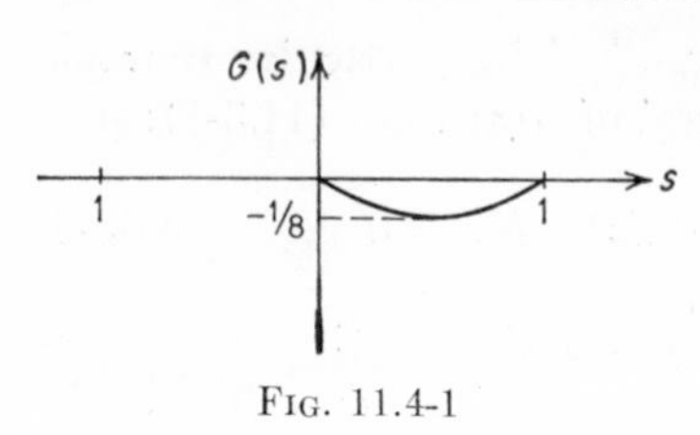

FIG. 11.4-1

As a second example to get our bearings on the problem of finding whether $G(s)$ has a constant sign or not, consider Simpson's formula

$$\int_{-1}^{1} f(x)\,dx = \frac{f(-1) + 4f(0) + f(1)}{3}$$

A brief examination shows that $m = 4$; hence, using (11.3-7),

$$3!G(s) = \frac{(1-s)_+^4 - (-1-s)_+^4}{4} - \frac{(-1-s)_+^3 + 4(-s)_+^3 + (1-s)_+^3}{3}$$

We have four regions to consider, $s > 1$, $1 > s > 0$, $0 > s > -1$, and $-1 > s$.

$s > 1$:

$$3!G(s) \equiv 0 \qquad \text{since all } (\ \ )_+ \equiv 0$$

$1 > s > 0$:

$$3!G(s) = \frac{(1-s)^4}{4} - \frac{(1-s)^3}{3} = (1-s)^3\left(\frac{-1-3s}{12}\right) < 0$$

$0 > s > -1$:

$$3!G(s) = \frac{(1-s)^4}{4} - \frac{(1-s)^3}{3} - \frac{4(-s)^3}{3} = \frac{-1 + 6x^2 + 8x^3 + 3x^4}{12}$$

$$= (1+s)^3\left(\frac{-1+3s}{12}\right) < 0$$

$-1 > s$:

$$3!G(s) \equiv 0 \qquad \text{(See Fig. 11.4-2)}$$

These two examples illustrate two points. First, outside the range (A,B), $G(s) \equiv 0$. Second, if the formula is symmetric, then the $G(s)$

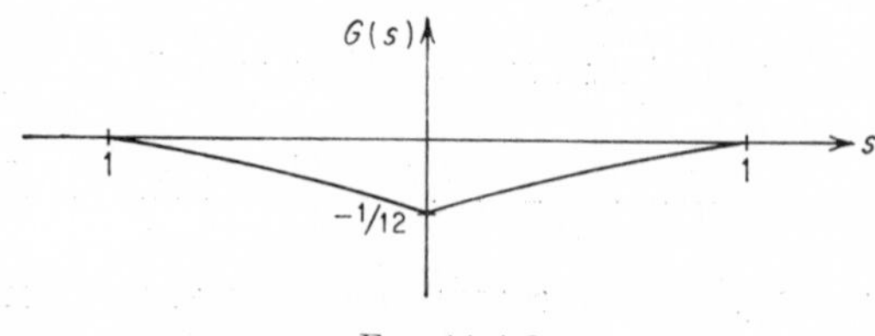

FIG. 11.4-2

is also symmetric. A proof of this follows from the observation that, if the formula is symmetric, then we must get the same error whether we integrate $f(x)$ or $f(-x)$, and, since we have the free choice of the function $f(x)$ that we are using, any lack of symmetry in $G(s)$ could be exploited to give different answers.

Inside the range (A,B) the breaks in the piecewise polynomial occur at the sample points x_k and the ends of the integration range, a and b. Inside each subinterval the polynomial is of degree m at most (for single integrals, but for repeated integration it could go higher).

The problem is to determine if these pieces of polynomials have zeros in their ranges. The direct attack of finding the zeros of the polynomial pieces and then checking to see if those zeros fall in the proper subinterval is likely to be expensive of machine time. On the other hand, the simple process of computing $G(s)$ at a fairly tight mesh of points and plotting the result may not carry conviction if the spacing is too large. The algebraic method that we have used in the two examples is likely to be difficult to program on a machine, although it is often adequate for hand verification. Thus there appears to be no entirely satisfactory way to check if $G(s)$ changes sign.

In any case, it should be realized that $G(s)$ depends only on the formula and hence the verification need be done only once for any particular formula.

EXERCISES

11.4-1. Check that the formula of Exercise 10.2-1

$$\int_{-\pi}^{\pi} f(x) \sin x \, dx$$

has a $G(s)$ of constant sign even though the kernel $K(x) = \sin x$ changes sign.

11.4-2. Check the formula of Exercise 10.2-2.

11.5 WHEN $G(s)$ IS NOT OF CONSTANT SIGN

When $G(s)$ is not of constant sign there are three things to observe. First, there are then functions $f(x)$ for which there is no θ such that the mean-value theorem is true. To show this, suppose that $G(s)$ is positive except for a small interval in which it is negative and that the integral of $G(s)$ is positive. Consider a function $f(x)$ such that $f^{(m)}(x)$ is positive and continuous but is small outside this small interval and very large inside. Thus the integral

$$\int_A^B f^{(m)}(s)G(s)\,ds \tag{11.5-1}$$

is negative, but $f^{(m)}(\theta)$ and the integral of $G(s)$ are both positive. Hence there can be no value θ. Similar arguments can be made for other $G(s)$.

The second point to observe is that when $G(s)$ changes sign we can still get a bound on the error. The bound follows readily by taking the absolute values of $G(s)$:

$$\int_A^B f^{(m)}(s)|G(s)|\, ds = f^{(m)}(\theta) \int_A^B |G(s)|\, ds$$

Hence
$$\left| \int_A^B f^{(m)}(s)G(s)\, ds \right| \le \max\,[|f^{(m)}(x)|] \int_A^B |G(x)|\, ds$$

where the max is over all x $(A \le x \le B)$. The computation of the integral

$$\int_A^B |G(s)|\, ds$$

can be done analytically or estimated numerically since, again, it need be done only once for any particular formula.

The third observation is that when the sign of $G(s)$ changes then we can still find an error term. As an illustration, consider the function

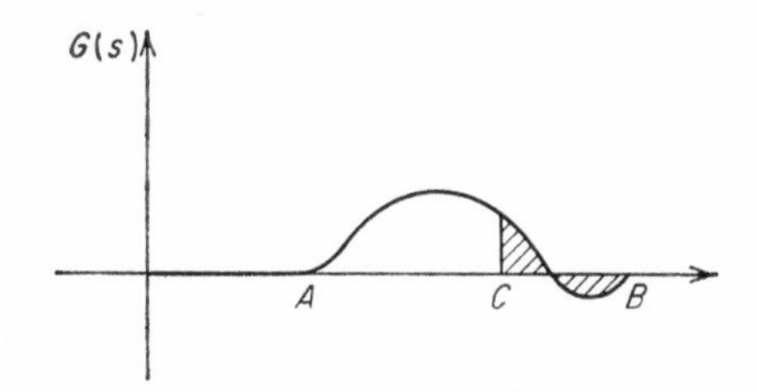

FIG. 11.5-1. A particular $G(s)$.

$G(s)$ in Fig. 11.5-1. The point $x = C$ has been chosen so that the two shaded areas are equal in size, that is,

$$\int_C^B G(s)\, ds = 0$$

We had [Eqs. (11.3-8) or (11.5-1)]

$$R = R(f) = \int_A^B f^{(m)}(s)G(s)\, ds$$
$$= \int_A^C + \int_C^B$$

We integrate the second term by parts and set

$$\int_C^s G(s)\, ds = H(s) \qquad [\textit{Note: } H(B) = H(C) = 0]$$

Then

$$R = \int_A^C f^{(m)}(s)G(s)\, ds + f^{(m)}(s)H(s) \Big|_C^B - \int_C^B f^{(m+1)}(s)H(s)\, ds$$
$$= \int_A^C f^{(m)}(s)G(s)\, ds - \int_C^B f^{(m+1)}(s)H(s)\, ds$$

We can now apply the mean-value theorem to each piece.

$$R = f^{(m)}(\theta_1) \int_A^C G(s)\, ds - f^{(m+1)}(\theta_2) \int_C^B H(s)\, ds$$

If $G(s)$ is as shown in Fig. 11.5-2, we could simply set

$$H(s) = \int_A^s G(s)\, ds$$

and integrate (11.5-1) by parts:

$$\begin{aligned} R(f) &= f^{(m)}(s)H(s) \Big|_A^B - \int_A^B f^{(m+1)}(s)H(s)\, ds \\ &= f^{(m)}(B)H(B) - f^{(m+1)}(\theta) \int_A^B H(s)\, ds \end{aligned}$$

We shall not analyze the general case since the results are not used in

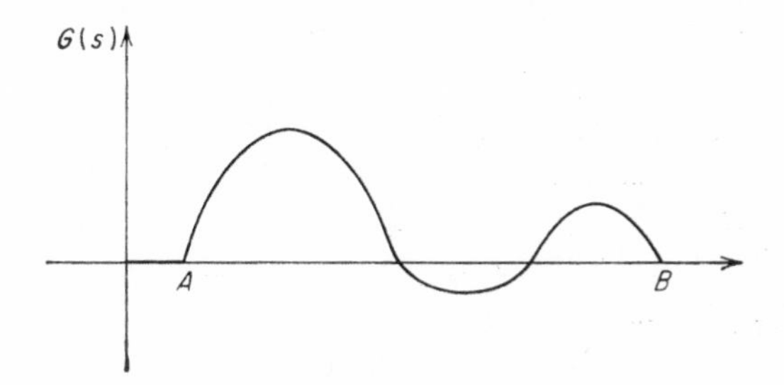

FIG. 11.5-2. Another particular $G(s)$.

the text but shall merely observe that such forms can be useful at times.

11.6 THE FLAW IN THE TAYLOR-SERIES APPROACH

Since there exists an error term from the Taylor series of the form[1]

$$\begin{aligned} f(a + x) = f(a) + (x - a)f'(a) + \frac{(x-a)^2}{2!} f''(a) + \cdots \\ + \frac{(x-a)^{m-1}}{(m-1)!} f^{(m-1)}(a) + \frac{(x-a)^m f^{(m)}(a + \theta x)}{m!} \end{aligned}$$

it is natural to ask why we cannot use the Taylor-series approach (second method of Sec. 7.2) to obtain the error of the formula. Let us suppose that we do use the Taylor-series approach. The coefficients of the formula have been determined to make the powers of x cancel on both sides up to the mth power, which is assumed not to cancel. Let us fix

[1] This is sometimes called the Lagrange form of the error, and sometimes the whole expression is called the extended mean-value theorem.

the x values in our mind. We therefore get a series of terms in $f^{(m)}(a + \theta_i x_i)$ for various θ_i,

$$\alpha_1 f^{(m)}(a + \theta_1 x_1) + \alpha_2 f^{(m)}(a + \theta_2 x_2) + \cdots + \alpha_k f^{(m)}(a + \theta_k x_k)$$

where, of course, the θ_i depend on x_i, $\theta_i = \theta_i(x_i)$. We now want to replace this expression by

$$(\alpha_1 + \alpha_2 + \cdots + \alpha_k) f^{(m)}(a + \theta x)$$

for some θ (and we expect many of the terms α_i to cancel). Under what circumstances does there exist such a θ?

We know from the influence function approach that, if $G(s)$ is of constant sign, then there is such a θ, and if $G(s)$ is not of constant sign, then there are some functions $f(x)$ for which there is no such θ. In this indirect manner we can answer the question. There seems to be no direct way of finding the answer, but if one were found, then it might provide an alternative, and perhaps better, approach to the theory of the error term.

In the case of the gaussian-type formula with some unassigned sample points, there are special techniques[1] for determining the error term more conveniently than through the $G(s)$ function, and this suggests that there is some hope of finding a different uniform approach.

[1] See Hildebrand [14] or Kopal [20].

CHAPTER 12

Formulas for Definite Integrals

12.1 INTRODUCTION

There is a significant difference between computing a definite integral

$$\int_a^b f(x)\,dx$$

and computing an indefinite integral, which may be written in the form

$$\int_a^x f(x)\,dx$$

The result of the first computation is a single number, while the result of the second is a table of numbers. This chapter covers definite integrals, while the next discusses indefinite integrals.

Another distinction should be made. A definite integral may be computed by a single formula, or it may be divided into a number of pieces and the formula applied to each piece. An example of the latter is the widely used *Simpson's composite formula*

$$\int_a^{a+2nh} f(x)\,dx = \frac{h}{3}\,[f(a) + 4f(a+h) + 2f(a+2h) + 4f(a+3h) + \cdots + f(a+2nh)] \qquad (12.1\text{-}1)$$

which is made up from the repeated use of

$$\int_a^{a+2h} f(x)\,dx = \frac{h}{3}\,[f(a) + 4f(a+h) + f(a+2h)]$$

for each double interval.

The advantage usually cited in favor of the single formula is that for the same number of samples the formula has a higher-order error term. But if there are nearby singularities of the integrand in the complex plane, then the high-order error term is quite likely to be large. The point is sometimes raised that when $h \to 0$ the error term of a high-accuracy formula approaches zero more rapidly than does the error term for a low-accuracy formula. Unfortunately, the place where this advantage sets in is generally not known, even if the integral is computed at two different spacings, say the second at one-half the spacing of the first. Thus in practice the composite formula is often used in preference to the single formula.

Another basis for choosing one formula over another is the effect that it produces on the roundoff of the sample integrand values. This effect is measured by the sum of the squares of the weights of the samples. If the formula is to give correct results when $f(x)$ is a constant, then the sum of the weights w_i is a constant. For minimum noise we want to minimize

$$\sum_{i=1}^{n} w_i^2$$

subject to the restraint that the sum of the w_i is a constant. The best choice is well known to be that all the w_i be equal. Any divergence from equality causes some amplification of the roundoff over the minimum possible. However, this effect is not as large as might be expected. For example, in Simpson's composite formula (12.1-1) the fluctuations of 2 to 1 in the weights produce an average square effect of

$$\frac{(4/3)^2 + (2/3)^2}{2} = \frac{10}{9} = 1.111$$

as compared with 1 when the weights are all equal. When some of the weights are negative, then the noise amplification can be much greater, as in the Newton-Cotes formulas (Sec. 12.2) of high order.

The problem of numerical integration, as was noted in Sec. 7.3, is one of taking samples of the integrand from which to estimate the average height of the integrand. The methods so far developed in this book approximate the integrand, or a factor of the integrand, by a polynomial, and using the principle of analytic substitution, the integration is performed on the polynomial approximation. Thus we consider the integrand $f(x)$ as the product of two factors, $K(x)$ and $g(x)$. We require three properties for this split into two factors. First, the variation from case to case of integration is contained in the factor $g(x)$; $K(x)$ can, of course, be simply the factor 1. Second, the moments m_k

$$m_k = \int_a^b K(x)x^k\,dx \tag{12.1-2}$$

can be found analytically. Third, the function $g(x)$ can be approximated adequately by a polynomial of degree n.

Under these circumstances, the machine can compute a formula which will provide an estimate of the integral. The methods of Chap. 10 show that the desired formula exists and can be found by a computing machine if we assign the sample points, provided that when we assign any derivative at a sample point we also assign all lower-order derivatives. When some or all of the sample points are unassigned and are chosen by the method of integration, then there is a risk that some of the sample points may turn out to be complex numbers. This, however, need not be a fatal objection to the use of the formula, although it certainly provides some discouragement.

As an example of the above approach, suppose that we want to compute the integral

$$I = -\int_0^1 \frac{\log x}{1 + e^x}\, dx \tag{12.1-3}$$

The factor $\log x$ has a singularity at $x = 0$ and hence the complete integrand is not readily approximated in the interval by a polynomial. The choice of $K(x) = \log x$ leads to the moments

$$m_k = -\int_0^1 x^k \log x\, dx = \frac{1}{(k+1)^2} \tag{12.1-4}$$

This leaves the problem of approximating $1/(1 + e^x)$ by a polynomial, which is not hard to do, since the approximating may be done by exact matching of samples as we have been doing in Part II, or $1/(1 + e^x)$ can be expanded in a Maclaurin series and integrated term by term.

Unfortunately, there is no explicit description which states when a function can be readily approximated by a polynomial. The error terms that we have given are generally not easily found in practice, and we must rely on other clues for breaking up $f(x)$ into the two factors $K(x)$ and $g(x)$. The presence of a singularity is certainly one clear indication that it should be put in the factor $K(x)$. Other clues are that a polynomial of degree n does not have $n + 1$ zeros and therefore does not approximate rapidly oscillating function very well. Still another clue is that a polynomial does not remain bounded in the neighborhood of infinity. To some extent, these considerations determine the choice of $K(x)$, along with the considerations that the variability from case to case should be removed from $K(x)$ and that the moments of $K(x)$ can be found analytically.

Having discussed the general considerations, let us turn to the examination of some specific formulas. One purpose of listing specific formulas is to provide a background for designing new formulas. The other reason is to provide a supply of formulas for small problems requiring only a few cases of integration.

EXERCISES

12.1-1. Using the universal matrices of Sec. 10.3, find a five-point formula for

$$-\int_0^1 f(x) \log x \, dx$$

[*Note:* Rewrite the integral so that it has a range $(-2,2)$ and then compute the moments, using (12.1-4).]

12.1-2. Find the coefficients for

$$\int_{-1}^{1} \frac{f(x)}{\sqrt{1-x^2}} dx = a_{-1}f_{-1} + a_{-1/2}f_{-1/2} + a_0 f_0 + a_{1/2}f_{1/2} + a_1 f_1$$

12.1-3. Discuss Exercise 10.2-3 as a composite formula. Compare with Simpson's.

12.1-4. If $\Sigma w_i = C$, prove that $\Sigma w_i^2 = \min$ if, and only if, all w_i are equal.

12.2 NEWTON-COTES FORMULAS

Logically, the simplest formulas for integration are those which use a set of equally spaced samples and arise from integrating the polynomial that exactly matches the samples. The samples[1] $0, h, 2h, \ldots, nh$ lead to the Newton-Cotes formulas

$$\int_0^{nh} f(x)\, dx = w_0 f(0) + w_1 f(h) + w_2 f(2h) + \cdots + w_n f(nh)$$

The first case of the Newton-Cotes formula is the well-known trapezoid rule

$$\int_0^h f(x)\, dx = \frac{h}{2}[f(0) + f(h)] - \frac{1}{12} h^3 f''(\theta) \qquad (12.2\text{-}1)$$

The second is Simpson's formula

$$\int_0^{2h} f(x)\, dx = \frac{h}{3}[f(0) + 4f(h) + f(2h)] - \frac{1}{90} h^5 f^{\text{IV}}(\theta) \qquad (12.2\text{-}2)$$

The third is the three-eighths rule

$$\int_0^{3h} f(x)\, dx = \frac{3h}{8}[f(0) + 3f(h) + 3f(2h) + f(3h)] - \frac{3}{80} h^5 f^{\text{IV}}(\theta) \qquad (12.2\text{-}3)$$

Table 12.2-1 gives the coefficients for the form

$$\int_0^{nh} f(x)\, dx = Ah[B_0 f(0) + B_1 f(h) + B_2 f(2h) + \cdots + B_n f(nh)] + K_n$$

where n is the number of intervals and *not* the number of sample points. The coefficients are symmetric, and hence we need give only part of the table of coefficients.

[1] Note that we are using $n + 1$ samples, rather than n as we have done previously.

TABLE 12.2-1

n	A	B_0	B_1	B_2	B_3	B_4	B_5	K_n
1	$1/2$	1	1					$-(1/12)h^3f''(\theta)$
2	$1/3$	1	4	1				$-(1/90)h^5f^{IV}(\theta)$
3	$3/8$	1	3	3	1			$-(3/80)h^5f^{IV}(\theta)$
4	$2/45$	7	32	12	32	7		$-(8/945)h^7f^{VI}(\theta)$
5	$5/288$	19	75	50	50	75	19	$-(275/12{,}096)h^7f^{VI}(\theta)$
6	$1/140$	41	216	27	272	27	216	$-(9/1{,}400)h^9f^{VIII}(\theta)$
7	7/17,280	751	3,577	1,323	2,989	2,989	1,323	$-(8{,}183/518{,}400)h^9f^{VIII}(\theta)$
8	4/14,175	989	5,888	−928	10,496	−4,540	10,496	
9	9/89,600	2,857	15,741	1,080	19,344	5,778	5,778	
10	5/299,376	16,067	106,300	−48,525	272,400	−260,550	427,368	

Some of the coefficients become negative for $n = 8$. For $n = 9$ they are all positive, but for $n \geq 10$ there are negative coefficients. This tends to produce poor roundoff properties, and as a result the higher-order Newton-Cotes formulas are seldom used. The order of the error terms jumps by 2 in going from an odd number to the next even number, which tends to favor the even-order formulas.

Notice the appearance of the powers of h in the error term. When we use other than unit spacing the powers of h automatically appear in the defining equations (10.3-1). We may either compute with the h values present, or else pretend we have unit spacing and supply the proper power of h later (for equally spaced samples); the latter is usually preferable.

If these formulas are to be compared for the same interval, then it must be remembered that $h =$ (range of integration)/n and the corresponding adjustments made in the error terms.

The Newton-Cotes formulas can be derived in many ways. Perhaps the simplest way to find the actual coefficients is to observe that the Gregory formula, when written to include all the differences that can be computed from the sample points, is exact for polynomials of the maximum degree and hence when written in the Lagrange form must be the same as if it were derived directly. Thus for the case $n = 4$ we have

$$\int_0^4 f(x)\,dx = \tfrac{1}{2}f_0 + f_1 + f_2 + f_3 + \tfrac{1}{2}f_4 + \tfrac{1}{12}(\Delta f_0 - \Delta f_3)$$
$$- \tfrac{1}{24}(\Delta^2 f_0 + \Delta^2 f_2) + \tfrac{19}{720}(\Delta^3 f_0 - \Delta^3 f_1)$$
$$- \tfrac{3}{160}(\Delta^4 f_0 + \Delta^4 f_0) = \tfrac{14}{45}f_0 + \tfrac{64}{45}f_1 + \tfrac{24}{45}f_2 + \tfrac{64}{45}f_3 + \tfrac{14}{45}f_4$$

EXERCISES

12.2-1. Apply the $n = 2, 4, 6, 8$ Newton-Cotes cases to $\int_0^1 e^{-x}\,dx$. Compare with the correct answer.

12.2-2. Discuss the noise amplification of the Newton-Cotes formulas.

12.2-3. Derive the Newton-Cotes formula from the Gregory formula for $n = 6$.

12.3 THE USE OF THE GREGORY FORMULA

We have just seen that, by suitably selecting terms in the Gregory formula, we can get the coefficients of the Newton-Cotes formulas, and this explains why the Gregory formula is so useful in practice; it is both flexible and accurate.

Composite formulas, such as the Simpson composite formula

$$\int_0^{2nh} f(x)\,dx = \frac{h}{3}(f_0 + 4f_1 + 2f_2 + 4f_3 + \cdots + 4f_{2n-1} + f_{2n})$$

are often used to avoid the negative coefficients (which produce high noise amplification) and the high-order derivative in the error term (which can be very large). The appropriately terminated Gregory formula (in this case ending with the Δ^2 terms) gives the same-order error term and better roundoff protection, since most of the coefficients will be the same size and the sum of the squares of the coefficients will therefore be less.

Thus for equally spaced data the Gregory formula

$$\begin{aligned}\frac{1}{h}\int_0^{nh} f(x)\,dx = {} & \tfrac{1}{2}f_0 + f_1 + f_2 + \cdots + f_{n-1} + \tfrac{1}{2}f_n + \tfrac{1}{12}(\Delta f_0 - \Delta f_{n-1}) \\ & - \tfrac{1}{24}(\Delta^2 f_0 + \Delta^2 f_{n-2}) + \tfrac{19}{720}(\Delta^3 f_0 - \Delta^3 f_{n-3}) \\ & - \tfrac{3}{160}(\Delta^4 f_0 + \Delta^4 f_{n-4}) + \cdots\end{aligned}$$

is probably the most useful, flexible formula and is widely used.

When computing an integral such as

$$\int_{-\infty}^{\infty} e^{-x^2}\,dx$$

the end corrections to the sum of the equally spaced, equally weighted, integrand values (the differences that appear at each end of Gregory's formula) do not contribute anything. This result has been noted both theoretically and experimentally, provided that the spacing is not too large. Hartree gives a table (see Table 12.3-1) for results obtained when computing

$$I = \int_{\infty}^{\infty} e^{-x^2}\,dx = h\left(\tfrac{1}{2} + \sum_{k=1}^{\infty} e^{-h^2k^2}\right)$$

The remarkable accuracy for $h = \frac{1}{2}$ is unfortunately accompanied by a lack of knowledge of when it occurs or when h is too large; the error analysis of Gregory's formula is a difficult topic.

TABLE 12.3-1. COMPUTATION OF $\int_0^\infty e^{-x^2}\,dx$*

h	I	
0.5	0.88622 69254 5	correct to 11 decimals
0.6	69254 8	
0.7	69285	
0.8	0.88622 72808	
0.9	23 598	
1.0	32 0	
1.1	0.88674	

* D. R. Hartree, "Numerical Analysis," 2d ed., p. 116, Oxford University Press, London, 1958. See also R. A. Fisher, *Phil. Trans. Roy. Soc. London, A222, 1922,* for general h and error theory.

EXERCISE

12.3-1. Make a flow diagram for the Gregory formula which involves the machine choice of when to stop using more differences. Discuss your reasoning for terminating as you do.

12.4 OPEN FORMULAS

The Newton-Cotes formulas are sometimes called *closed* because they include the end points of the interval as contrasted with the following *open* formulas which do not make use of the end values:

$$\int_0^{nh} f(x)\,dx = w_1 f(h) + w_2 f(2h) + \cdots + w_{n-1} f((n-1)h)$$

For purposes of listing the coefficients (see Table 12.4-1) we adopt the form

$$\int_0^{nh} f(x)\,dx = Ah[B_1 f(h) + B_2 f(2h) + \cdots + B_{n-1} f((n-1)h)] + K_n$$

TABLE 12.4-1

n	A	B_1	B_2	B_3	B_4	K_n
3	$3/2$	1	1			$(3h^3/4)f''(\theta)$
4	$4/3$	2	-1	2		$(14/45)h^5 f^{IV}(\theta)$
5	$5/24$	11	1	1	11	$(95/144)h^5 f^{IV}(\theta)$
6	$3/10$	11	-14	26	-14	$(41/140)h^7 f^{VI}(\theta)$

These formulas are used occasionally in the numerical integration of differential equations. The user should beware of using an open-type formula when he cannot compute the value of the integrand at the end of an interval, since the difficulty of computing probably means that there is a singularity there; hence the assumption of polynomial approximation to the integrand is not valid.

Between the open and closed types of formulas there are the mid-point formulas (known to Maclaurin) which use samples at the mid-points of the n intervals.

$$\int_0^{nh} f(x)\,dx = w_1 f\left(\frac{h}{2}\right) + w_2 f\left(\frac{3h}{2}\right) + \cdots + w_n f\left(\frac{(2n-1)h}{2}\right)$$

EXERCISE

12.4-1. Derive the Maclaurin-type formulas in Lagrange form from Exercise 10.10-1.

12.5 GAUSSIAN QUADRATURE

The name gaussian quadrature is associated with situations in which all the sample points are unassigned. There are three widely used cases: the range $(-1,1)$ with weight factor 1, sometimes called Gauss-Legendre; the range $(0,\infty)$ with weight factor e^{-x}, called Gauss-Laguerre; and the range $(-\infty,\infty)$ with weight factor e^{-x^2}, called Gauss-Hermite. In each case the second name is attached because the sample points turn out to be the zeros of the corresponding orthogonal polynomial of degree n. In the Legendre ($n = 1$ to 16) and Laguerre ($n = 1$ to 15) cases, the zeros and weights have been tabulated in a National Bureau of Standards publication [31].

We have shown that the sample points are all distinct, real, and lie in the interval of integration, and the weights are all positive. The error term is proportional to the $2n$th derivative,

$$E = \frac{f^{(2n)}(\theta)}{(2n)!} \int_a^b \pi^2(x)\,dt$$

where $\pi(x) = (x - x_1)(x - x_2) \cdots (x - x_n)$.

Many other special cases have been investigated in which the kernel $K(x)$ has singularities at one or both ends of the range of integration. Of special interest is the Gauss-Chebyshev case

$$\int_{-1}^{1} \frac{f(x)}{\sqrt{1-x^2}}\,dx = \sum_{i=1}^{n} w_i f(x_i)$$

The sample points turn out to be

$$x_j = \cos\frac{2j-1}{2n}\pi$$

while the weights turn out to be all equal:

$$w_i = \frac{\pi}{n}$$

Perhaps the best references for many other special cases are Kopal [20] and Mineur [28].

The gaussian methods of integration are remarkably effective in estimating the value of an integral from a few samples, provided that the function (excluding the kernel) can be accurately approximated by a polynomial. This is true of most worked examples in textbooks; much depends on the closeness of singularities to the range of integration.

The feeling is widespread that, because the gaussian-quadrature methods do so well, they are sensitive to slight errors. A little thought shows that if the location of the sample points is slightly in error then the estimate of the integral is changed by this error times the derivative of the integrand at the sample point multiplied by the corresponding weight w_i. The same is, of course, true for any method of integration. The fact that the weights are all positive means that no unnecessary serious cancellation occurs. Thus the gaussian methods are generally very effective (when care is given to the possibility of nearby complex zeros of the part approximated by the polynomial).

EXERCISE

12.5-1. Show that

$$\int_{-1}^{1} \frac{f(x)}{\sqrt{1 - x^2}}\,dx = \frac{\pi}{3}\left[f\left(-\frac{\sqrt{3}}{2}\right) + f(0) + f\left(\frac{\sqrt{3}}{2}\right)\right] + \frac{\pi f^{(6)}(\theta)}{23{,}040}$$

is exact for fifth-degree polynomials.

12.6 MISCELLANEOUS GAUSS-TYPE INTEGRATION FORMULAS

The effectiveness of the Gauss approach to integration and the mathematics involved in the method have combined to produce many special cases. The purpose of this section is to list a few of them.

Radau investigated the matter of assigning some values for the sample points and leaving some free. The main case of interest is when both end points are assigned. The reason for this is that, when a composite formula is made from a single formula, the common end points need be computed only once. Kopal [20] has tables for this case for $n = 3(1)11$.

Ralston (see Sec. 10.9) investigated the interesting case when the weight at one end is the negative of the weight at the other. In the composite formula the weight cancels out, except for the two extreme ends. Thus the accuracy is improved by the available free weight parameter, the cost of computing is only slightly raised, and the longer the range of integration the relatively less is the extra cost.

In all these formulas the error term is usually of the form

$$\frac{f^{(m)}(\theta)}{m!} \int_a^b K(x)\pi^2(x) \text{ (product of factors at assigned sample points) } dx$$

EXERCISE

12.6-1. Discuss the use of the formula in Exercise 10.2-2 as a composite formula. Compare it with Simpson's composite formula.

12.7 SUMMATION OF SERIES

Gregory's formula expresses an integral as the sum of a sequence of equally spaced, equally weighted, integrand values *plus* corrections involving the end values only. If the range of the integral goes to infinity, then the sum is an infinite series.

We may reverse the situation and express the sum of an infinite series as an infinite integral plus correction terms at the starting end (those at infinity vanish because of the convergence of the series). Thus, to take a simple example,

$$\begin{aligned}\sum_{x=1}^{\infty} \frac{1}{x^2} &= \int_1^{\infty} \frac{dx}{x^2} + \frac{1}{2}\left(\frac{1}{1^2}\right) - \frac{1}{12}\left(\frac{1}{2^2} - \frac{1}{1^2}\right) + \frac{1}{24}\left(\frac{1}{3^2} - \frac{2}{2^2} + \frac{1}{1^2}\right) \cdots \\ &= 1 + 1/2 + 1/16 + 11/432 \cdots \\ &= 1.588 \cdots \qquad \text{(correct ans.} = 1.64 \cdots)\end{aligned}$$

It is often advantageous to add the first few (say 10) terms separately and apply the integral to the rest.

It may be that the integral arising from the series cannot be computed analytically. We may then either integrate it numerically or make a transformation of variables and then estimate the integral by Gregory's formula. In passing from series to integral to series, two sets of end corrections occur. The gain is that, by a proper transformation, the new series to be summed converges much more rapidly.

12.8 CHANGE-OF-SCALE EFFECTS

In two of the gaussian formulas

$$\int_{-\infty}^{\infty} e^{-x^2} f(x)\, dx$$

$$\int_0^{\infty} e^{-x} f(x)\, dx$$

a single degree of freedom in scaling occurs. Using the second integral as an example, we write it as

$$I = \int_0^\infty e^{-ax}[e^{-(1-a)x}f(x)]\,dx$$

If we set $ax = \theta$ $(a > 0)$,

$$I = \frac{1}{a}\int_0^\infty e^{-\theta}F(\theta)\,d\theta$$

where
$$F(\theta) = e^{-[(1-a)/a]\theta}f\left(\frac{\theta}{a}\right)$$

then we may apply the same integration formula to this new form as to the original integral. Which value of a to choose depends on how well the corresponding $F(\theta)$ is approximated by a polynomial of the assigned degree. Loosely speaking, if most of the terms

$$w_iF(x_i)$$

were small, one would be suspicious that computation effort was being wasted for that choice of the parameter a.

This change of scale also raises the question of change of variable. Again using

$$\int_0^\infty e^{-x}f(x)\,dx$$

we set $x = -\ln\theta$. We get

$$\int_0^1 f(-\ln\theta)\,d\theta$$

The question arises "Which function is more readily approximated by a polynomial

$$f(x) \qquad \text{or} \qquad f(-\ln\theta)$$

remembering their respective ranges of approximation, $(0 \leq x \leq \infty)$ or $(0 \leq \theta \leq 1)$?"

One closing remark on the subject: One can usually change variables to eliminate singularities. In an earlier example (Sec. 1.9), we had

$$g(y) = \frac{d}{dy}\int_0^y \frac{f(x)}{\sqrt{y-x}}\,dx$$

The substitution $x = y\sin^2\theta$ led to

$$g = \frac{d}{dy}\int_0^{\pi/2} 2\sqrt{y}\,f(y\sin\theta)\sin\theta\,d\theta$$

which removed the singularity.

12.9 INTEGRALS WITH A PARAMETER

It is often necessary to evaluate an integral of the form

$$F(\lambda) = \int_0^\infty f(x,\lambda)\,dx$$

for a number of parameter values λ.

Experience shows that frequently, by differentiating with respect to the parameter, a differential equation for $F(\lambda)$ can be found, and this equation can be solved in the sense that it is reduced to an integral of the form

$$F(\lambda) = A(\lambda) + \int_0^\lambda g(\lambda)\,d\lambda$$

In this form, each step forward in λ gives another solution $F(\lambda)$, and it is generally a more efficient way to approach the problem.

Another rather isolated observation is that frequently an integral with a rapidly oscillating integrand can be converted to another with no oscillation but a rapid decay instead.

These few remarks may be said to be on the topic of how not to compute an integral. The actual practice of these suggestions falls in the field of classical manipulative analysis, which is seldom taught these days and falls outside the scope of this book.[1]

[1] See M. Abramowitz, On the Practical Evaluation of Integrals, *J. Soc. Ind. Appl. Math.*, vol. 2, pp. 20–35, 1954.

CHAPTER 13

Indefinite Integrals

13.1 OUTLINE OF CHAPTER AND NOTATION

The main purpose of this chapter is to introduce, after a brief survey, two new concepts. The first concept is that of *instability* which can occur when values of a function that have already been computed are used to compute the next value. This repeated use of the function values may be compared to "feedback" and produces instability in the same manner. Feedback is the use of part of the output at a particular time as input at a slightly later time. Sometimes this time delay can be neglected and sometimes it cannot. Perhaps the simplest physical example is the feedback amplifier which is widely used in control circuits (see Fig. 13.1-1). The output is the sum of the two inputs on the left times the gain of the amplifier, which is taken as -10^9. We have

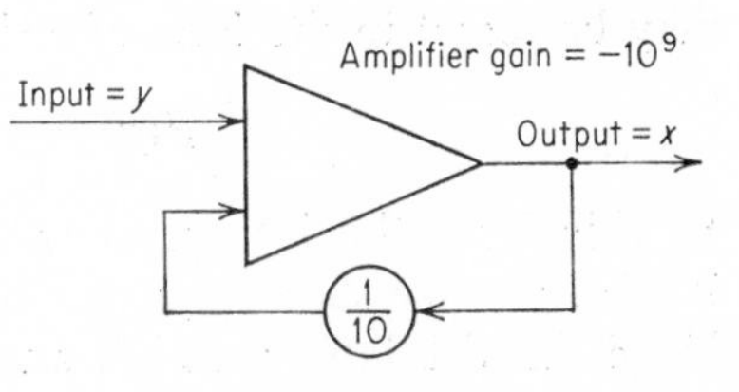

Fig. 13.1-1

$$(y + \tfrac{1}{10}x)(-10^9) = x$$

or

$$x = -\frac{10^{10}y}{10^9 + 10} = -\frac{10y}{1 + 10^{-8}} \sim -10y$$

Thus the output is -10 times the input, quite independent of small changes in the characteristics of the amplifier; that is, it is not sensitive to the exact value of the gain.

Under some circumstances, if there is enough delay in returning the signal from the output to the input, the system oscillates—"sings" is the word often used. Such situations arise frequently in public auditoriums which have a microphone (y) and an amplifier system output (x). The output from the loudspeaker is delayed by the time that it takes the sound to get from the speaker to the microphone, and the result is heard by the audience unless care is taken to control it.

The theory of feedback systems is well developed and cannot be treated in detail here. It should be clear that, when we use an output result as an input to the same process at a later stage, we run the risk of producing an unwanted oscillation, whose period and behavior depend on the particular process.

The second new concept is an extension of the technique that we have been using to find formulas. In the past we have used all the parameters that we had in order to make the formula exact for as many powers of x as we could, or, in other words, we have concentrated on making the truncation error of as high an order as possible. In order to cope with instability and other effects, such as roundoff propagation, we shall find it necessary to leave some of the coefficients of the general formula undetermined at the time we are finding the error term and later select values for them which will produce suitable formulas.

The problem of calculating an indefinite integral

$$y(x) = \int_a^x f(x)\,dx \tag{13.1-1}$$

may be recast in the form of a simple differential equation

$$y'(x) = f(x) \qquad y(a) = 0 \tag{13.1-2}$$

In this form the integrand values are labeled y' while the answer is labeled y. We shall adopt the differential equation approach because it blends in with later notation when the more general differential equation

$$y'(x) = f(x,y) \qquad y(a) = b \tag{13.1-3}$$

is examined. (See Chap. 15.) This approach tends to limit the class of formulas that we find. We shall also make a slight change in our notation. Up to this time we have been using $y(x)$ and y_x to mean the same thing. We shall now use the subscript notation to count; that is, we shall number 0, 1, 2, . . . the points at which we compute the answer. But for reasons of clarity it is desirable to keep the spacing of the problem at h and to let the origin fall where it may. Thus $y(a + nh)$ and y_n are the same quantity. No confusion should arise once the convention is understood, and the notation becomes a great deal simpler.

13.2 SOME SIMPLE FORMULAS FOR INDEFINITE INTEGRALS

Probably the simplest formula for approximating one step of an indefinite integral is

$$y_{n+1} = y_n + hy'[a + (n + \tfrac{1}{2})h] = y_n + hy'_{n+\frac{1}{2}} \tag{13.2-1}$$

where
$$y' = f(x)$$
or
$$y = \int_a^x y'\,dx$$

This is clearly the mid-point formula (10.2-1), and it has one-half the error of the usual trapezoid rule (10.2-3)

$$y_{n+1} = y_n + \frac{h}{2}(y'_{n+1} + y'_n)$$

The mid-point formula also has better roundoff properties.

A more accurate formula, which may be found by using the universal matrices of Sec. 10.3, is

$$y_{n+1} = y_n + \frac{h}{24}(-y'_{n-1} + 13y'_n + 13y'_{n+1} - y'_{n+2}) \tag{13.2-2}$$

where the error term is

$$\tfrac{11}{720}h^5y^{(5)}(\theta) \simeq 0.0153h^5y^{(5)}(\theta) \tag{13.2-3}$$

If this formula is used[1] to compute an integral, then two extra points which lie outside the range of integration have to be computed (one at each end). If the values of the integrand cannot be found at such points because of singularities in the formula, then the assumption that the integrand can be approximated accurately by a polynomial is probably false. An alternative formula can be used for the ends of the interval, namely,

$$y_1 = y_0 + \frac{h}{24}(9y'_0 + 19y'_1 - 5y'_2 + y'_3) \tag{13.2-4}$$

Simpson's formula (10.4-1) is widely used for computing indefinite integrals. In our present notation this takes the form

$$y_{n+1} = y_{n-1} + \frac{h}{3}(y'_{n-1} + 4y'_n + y'_{n+1}) \tag{13.2-5}$$

with an error term

$$-\frac{h^5y^{(5)}(\theta)}{90} \simeq -0.0111h^5y^{(5)}(\theta) \tag{13.2-6}$$

which is slightly less than (13.2-3).

There is one often overlooked trouble that occurs when Simpson's formula is used in the conventional manner and, although the trouble is not serious, it can be annoying. Simpson's formula carries the value of the integral two steps forward, and the special half Simpson's formula (10.4-2) is used to start the chain for the odd-numbered sample points. The result of this jumping two steps ahead is that the accumulated errors at the odd- and even-numbered points are rather independent of each

[1] We of course actually compute $(24/h)y_n$ and multiply by $h/24$ when we print out values of y_n. In this way, while we do the same amount of computing, the roundoff effects do not build up so rapidly, since the final multiplication by $h/24$ does not propagate further.

other, especially as the weights attached to the computed integrand values are different. All this tends to produce an oscillation. While this oscillation is due to errors being committed and hence gives some measure of the accuracy of the results, it can be very annoying at times. The oscillation can be avoided if the even-numbered point chain only is computed and the half Simpson is used to produce each odd-numbered point.

EXERCISES

13.2-1. Derive (13.2-2) and (13.2-4) with their error terms.

13.2-2. Compare Simpson's-formula method with the correspondingly accurate formula based on the Gregory formula (Sec. 10.10).

13.3 A GENERAL APPROACH

Many special formulas could be investigated one at a time, but we shall turn to a general approach and examine a whole class at once. In estimating the next value, y_{n+1}, of the integral, we could use old values of the integral, $y_n, y_{n-1}, \ldots$, as well as both new values of the integrand $y'_{n+1} \cdots$ and old values $y'_n, y'_{n-1}, \ldots$. We shall arbitrarily limit the present study to the form

$$y_{n+1} = a_0 y_n + a_1 y_{n-1} + a_2 y_{n-2} + h(b_{-1} y'_{n+1} + b_0 y'_n + b_1 y'_{n-1} + b_2 y'_{n-2}) + E_5 \frac{h^5 y^{(5)}(\theta)}{5!} \qquad (13.3\text{-}1)$$

which uses three old values of the integral together with one new and three old values of the integrand. The question immediately arises "Why not try the more reasonable approach of using integrand values symmetrically placed about the value being computed?" Such an approach would probably give more accurate formulas, but we shall not do so for reasons that will appear when we investigate the general differential equation (13.1-3) in Chap. 15.

With seven parameters in the general form, we shall impose only five conditions, namely, that it be exact for $y = 1, x, \ldots, x^4$, or what is the same thing, exact for $y = 1, (x - n), \ldots, (x - n)^4$ and shall save two parameters for other purposes. Earlier we made the formula exact when the *integrand* was $1, x, x^2, \ldots$; now we make the *answer* exact for $1, x, x^2, \ldots$. Since the integrals of powers of x are powers one higher in x, we need only account for why we have added the condition of exactness for $y = 1$. The consequences of not doing so are so drastic that no further argument is given here. Using these five conditions and taking a_1 and a_2 as parameters, we get from the defining equations

$$\begin{aligned} a_0 &= 1 - a_1 - a_2 & b_0 &= \tfrac{1}{24}(19 + 13a_1 + 8a_2) \\ a_1 &= a_1 & b_1 &= \tfrac{1}{24}(-5 + 13a_1 + 32a_2) \\ a_2 &= a_2 & b_2 &= \tfrac{1}{24}(1 - a_1 + 8a_2) \\ b_{-1} &= \tfrac{1}{24}(9 - a_1) & E_5 &= \tfrac{1}{6}(-19 + 11a_1 - 8a_2) \end{aligned} \qquad (13.3\text{-}2)$$

We have computed E_5 (and the other coefficients) as if we had unit spacing, and then supplied the proper powers of h. Working one example with spacing h will show the reader why we have found this process (which really changes the meaning of E_5 slightly) is preferable in practice. No serious confusion should arise from the change in the meaning of E_5.

EXERCISE

13.3-1. Derive Eqs. (13.3-2).

13.4 TRUNCATION ERROR

We have written the general form (13.3-1) as if the influence function $G(s)$ (see Sec. 11.3) had a constant sign (Sec. 11.4). It is necessary to examine this assumption and find which values of a_1, a_2 make this true.

Using Chap. 11 as our guide, we first notice that, except for special values of a_1, a_2, $m = 5$ (see Sec. 11.2). The influence function of Sec. 11.3 has an $A = -2h$, so that (11.3-2) becomes

$$\begin{aligned} y(x) &= y(-2h) + (x + 2h)y'(-2h) + \frac{(x+2h)^2}{2!}y''(-2h) \\ &\quad + \frac{(x+2h)^3}{3!}y'''(-2h) + \frac{(x+2h)^4}{4!}y^{(4)}(-2h) \\ &\quad + \frac{1}{4!}\int_{-2h}^{x} y^{(5)}(s)(x - s)^4\,ds \end{aligned} \qquad (13.4\text{-}1)$$

If this is substituted in the general form (13.3-1), we finally come to (11.3-8) with $B = h$,

$$R(y) = \int_{-2h}^{h} y^{(5)}(s)G(s)\,ds$$

where $G(s)$ is given by Eq. (13.4-2). (Note that no integration is needed in constructing it.)

$$\begin{aligned} G(s) = \frac{1}{4!}\{&(h - s)_+^4 - a_0(-s)_+^4 - a_1(-h - s)_+^4 - a_2(-2h - s)_+^4 \\ &- 4h[b_{-1}(h - s)_+^3 + b_0(-s)_+^3 + b_1(-h - s)_+^3 + b_2(-2h - s)_+^3]\} \end{aligned} \qquad (13.4\text{-}2)$$

Now if $G(s)$ is of constant sign (Sec. 11.4), then we have

$$R(x^5) = 5! \int_{-2h}^{h} G(s)\, ds = E_5 h^5$$

where E_5 is the result of substituting $y(x) = x^5$ for unit spacing.

The problem is to find those values of a_1, a_2 such that $G(s)$ has a constant sign. Thus we need to find the zeros of $G(s)$ for each pair a_1,a_2. Since this is difficult to do, we resort to a common mathematical trick and invert the problem: We set $G(s) = 0$ and examine the resulting values of a_1, a_2. Using (13.3-2), we can write $G(s)$ as a *linear function* in a_1, a_2,

$$G(s) = G_0(s) + a_1 G_1(s) + a_2 G_2(s)$$

where

$$G_0(s) = \frac{1}{4!}\left[(h - s)_+^4 - (-s)_+^4 - \frac{3}{2}h(h - s)_+^3 - \frac{19}{6}h(-s)_+^3 \right.$$
$$\left. + \frac{5}{6}h(-h - s)_+^3 - \frac{h}{6}(-2h - s)_+^3\right]$$

$$G_1(s) = \frac{1}{4!}\left[(-s)_+^4 - (-h - s)_+^4 + \frac{h}{6}(h - s)_+^3 - \frac{13}{6}h(-s)_+^3 \right.$$
$$\left. - \frac{13}{6}h(-h - s)_+^3 + \frac{h}{6}(-2h - s)_+^3\right]$$

$$G_2(s) = \frac{1}{4!}\left[(-s)_+^4 - (-2h - s)_+^4 - \frac{4}{3}h(-s)_+^3 - \frac{16}{3}(-h - s)_+^3 \right.$$
$$\left. - \frac{4}{3}h(-2h - s)_+^3\right]$$

There are three intervals of s in which we must examine $G(s)$, $h \geq s \geq 0$, $0 \geq s \geq -h$, and $-h \geq s \geq -2h$. In the first interval

$$(h - s)^3\left[(h - s) - \frac{3}{2}h + a_1\frac{h}{6}\right] = 0$$

leads to

$$a_1 = 3 + \frac{6s}{h}$$

which describes a family of vertical lines in the (a_1,a_2) plane. The line moves from $a_1 = 9$ to $a_1 = 3$ as s goes from h to 0 (Fig. 13.4-1).

In the second interval we have

$$\left[(h - s)^4 - s^4 - \frac{3}{2}h(h - s)^3 + \frac{19}{6}hs^3\right]$$
$$+ a_1\left[s^4 + \frac{h}{6}(h - s)^3 + \frac{13}{6}hs^3\right] + a_2\left[s^4 + \frac{4h}{3}s^3\right] = 0$$

This describes a line which continues to move to the left but gradually tilts until, at $s = -h$,

$$a_2 = \frac{a_1 - 1}{2}$$

This line passes through the point (1,0) with a slope of 1:2.

In the third interval we have the line rotating about the point (1,0) and ending up as

$$a_2 = \frac{a_1 - 1}{4}$$

Above and to the left of these lines (see Fig. 13.4-1), we have a $G(s)$ which

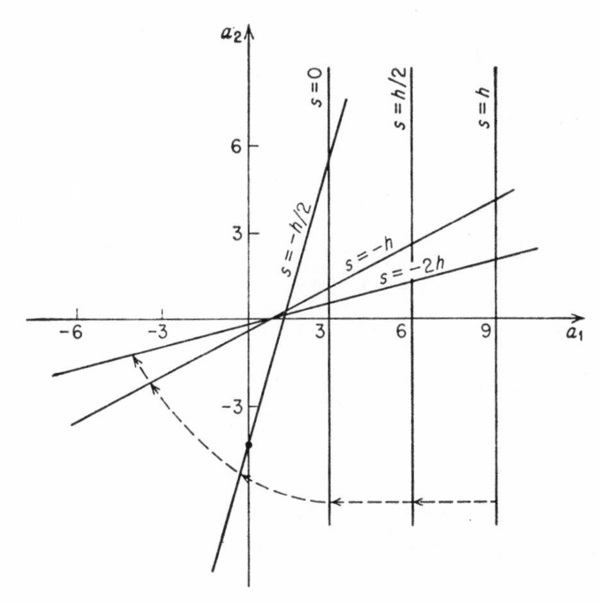

FIG. 13.4-1. Region of constant sign for $G(s)$. [The lines are those along which $G(s) = 0$ for the value s marked on the line.]

has no zeros[1] and hence is of constant sign. Thus in this region we have the error term of the form

$$\frac{E_5 h^5}{5!} y^{(5)}(\theta)$$

Outside this region we do not have such an error term. In view of future work, we shall not investigate the matter further.

In the region the error is measured by E_5, from (13.3-2). The "equi-

[1] There is also a region in the lower right for which $G(s)$ does not change sign, but both the size of the error term and roundoff effects keep us from using such formulas, and we shall ignore it in the future.

error" lines are shown as light lines in Fig. 13.4-2. These lines show that, other things being the same, we should try to stay down toward the lower right where the error term is small. The slope of the lines, 11:8,

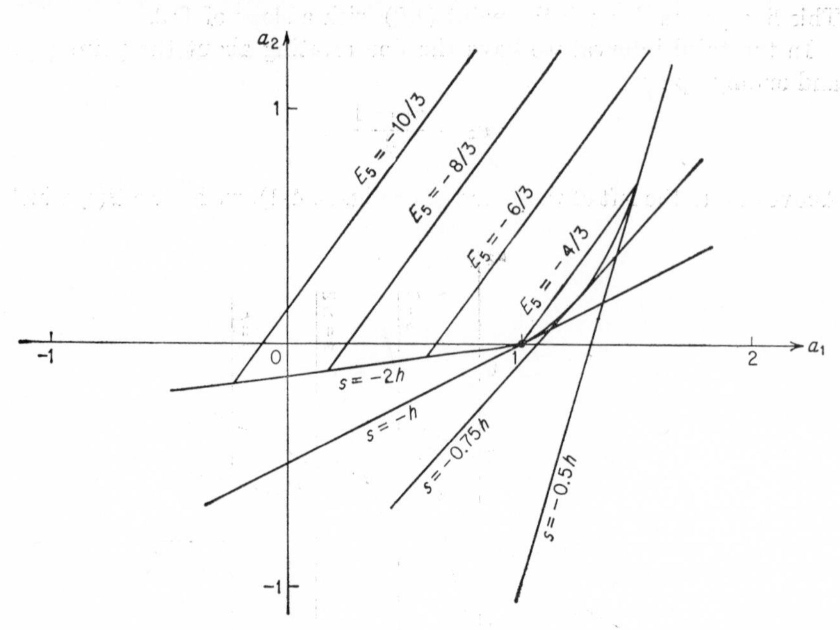

FIG. 13.4-2. Equi-truncation error curves. (The value of E along the line gives the truncation error on that line.)

indicates the exchange ratio between the two parameters; if we increase a_2 by 11 units and a_1 by 8, then the error is the same.

13.5 STABILITY

We first examine the formulas by considering the trivial case

$$y' = 0 \tag{13.5-1}$$

If this case does not behave favorably, then the more general case of

$$y' = f(x) \tag{13.5-2}$$

will also give trouble.

If we choose a point such as $a_1 = -1$, $a_2 = 0$ in the (a_1,a_2) plane, then the formula becomes

$$y_{n+1} = 2y_n - y_{n-1}$$

Now consider what happens to the solution $y = 0$ when a small error ϵ

(which we can imagine as coming from roundoff) is put in the solution at y_0. It is easy to compute Table 13.5-1.

TABLE 13.5-1

$$
\begin{aligned}
&\vdots \\
y_{-2} &= 0 \\
y_{-1} &= 0 \\
y_0 &= \epsilon \\
y_1 &= 2\epsilon \\
y_2 &= 3\epsilon \\
y_3 &= 4\epsilon \\
&\vdots
\end{aligned}
$$

Thus the effect of the small error grows steadily.

As another example, choose $a_1 = 1$, $a_2 = 1$. The equation is

$$y_{n+1} = -y_n + y_{n-1} + y_{n-2}$$

and we get Table 13.5-2.

TABLE 13.5-2

$$
\begin{aligned}
&\vdots \\
y - 2 &= 0 \\
y - 1 &= 0 \\
y_0 &= \epsilon \\
y_1 &= -\epsilon \\
y_2 &= 2\epsilon \\
y_3 &= -2\epsilon \\
y_4 &= 3\epsilon \\
y_5 &= -3\epsilon \\
&\vdots
\end{aligned}
$$

This time the error oscillates as well as grows.

These two examples indicate that we need to examine the phenomena further; clearly some points (a_1,a_2) are very poor choices.

The general form (13.3-1) that we are investigating is a linear difference equation with constant coefficients, and we may apply the methods of Chap. 5. Since the values of y' are the integrand values which we compute independently of the y values, we consider the difference equation as being in the form

$$y_{n+1} - a_0 y_n - a_1 y_{n-1} - a_2 y_{n-2} = F_n \tag{13.5-3}$$

We first solve the homogeneous equation

$$y_{n+1} - (1 - a_1 - a_2)y_n - a_1y_{n-1} - a_2y_{n-2} = 0 \qquad (13.5\text{-}4)$$

To solve this, we set $y_n = \rho^n$ to get the characteristic equation

$$\rho^3 - (1 - a_1 - a_2)\rho^2 - a_1\rho - a_2 = 0$$

or

$$(\rho - 1)[\rho^2 + (a_1 + a_2)\rho + a_2] = 0 \qquad (13.5\text{-}5)$$

Let the zeros of the quadratic factor be ρ_1 and ρ_2. The solution of (13.5-4) is

$$y_n = C_1(\rho_1)^n + C_2(\rho_2)^n + C_3 \qquad (13.5\text{-}6)$$

provided that no two roots are equal. If two roots are equal, as they happen to be in the two examples above, then (13.5-6) needs to be changed. If either of the roots in (13.5-6) is greater than 1 in absolute value, then the solution will grow similarly to a geometric progression as n increases (provided that the corresponding coefficient is not zero).

We therefore wish to explore the region of the (a_1,a_2) plane for which $|\rho_i| < 1$. The easiest way to do this is to determine the boundaries of the region. We first examine the boundary along which a root becomes equal to 1. To do this, we set $\rho = 1$ in the quadratic factor. This defines the line

$$1 + a_1 + 2a_2 = 0$$
$$a_2 = -\frac{1 + a_1}{2} \qquad (13.5\text{-}7)$$

Next we examine where $\rho = -1$. To do this, we set $\rho = -1$ in the quadratic factor. This defines the line

$$a_1 = 1 \qquad (13.5\text{-}8)$$

Next we examine where the roots become complex. This curve is defined by

$$(a_1 + a_2)^2 - 4a_2 = 0$$

or

$$a_1 = \pm 2\sqrt{a_2} - a_2 \qquad (13.5\text{-}9)$$

which is shown in Fig. 13.5-1. Inside this parabola the roots are complex conjugates of each other and both have the modulus

$$|\rho| = \sqrt{a_2}$$

Thus the region in which $|\rho_i| < 1$ is bounded by the line

$$a_2 = 1 \qquad (13.5\text{-}10)$$

which just cuts off the parabola (13.5-9) where it is tangent to the other lines (13.5-7) and (13.5-8). These three straight lines, then, define the triangle in Fig. 13.5-1 within which the characteristic roots must lie in order for the method of integration to have an error that does not grow as

CHAPTER 17

Least Squares: Theory

17.1 INTRODUCTION

In Sec. 7.1 we introduced the four basic questions:

1. What samples shall we use?
2. What class of approximating functions shall we use?
3. What criterion of goodness of fit shall we use?
4. What accuracy do we want?

Up to this point we have used a wide variety of different methods for choosing our samples, but we have used polynomials and the criterion of exact matching of the polynomial to the samples. We now begin the examination of other criteria for selecting the particular polynomial from the general class of all polynomials of degree n.

The actual choice of the criterion to be used in a particular problem depends on the background of the particular equations and cannot be given precisely in a textbook on computing. We shall, however, try to develop some feeling for a few of the available criteria by discussing them at some length.

The criterion of exact matching is open to criticism. When we were using it, we were aware that our knowledge of the function at the selected points was contaminated by roundoff noise. As long as the noise level is low, and in many situations with a modern 8- to 12-decimal-place computing machine this is so, then exact matching is a reasonable procedure.

When the noise level is high, as can happen in some computations—and almost always happens when the data come from physical measurements—then it is necessary to examine more closely the wisdom of trying to find an approximating function which exactly coincides with the "noisy" samples that we have. The theory of *least-squares* approximation is probably the most widely used approach to "noisy" situations.

17.2 THE PRINCIPLE OF LEAST SQUARES

Suppose that we try to measure some quantity and we make n measurements,

$$x_i = x + \epsilon_i \qquad (i = 1, 2, \ldots, n)$$

where the ϵ_i are the errors (or noise) of the measurements and x is the "true value," whatever that may mean.

The principle of least squares states that the best estimate $\hat{x}$ of x is that number which minimizes the sum of the squares of the deviations of the data from their estimate,[1]

$$f(\hat{x}) = \sum_{i=1}^{n} \hat{\epsilon}_i^2 = \sum_{i=1}^{n} (x_i - \hat{x})^2$$

In the final analysis, the usefulness of this principle rests on the questions of how well the model fits experience and how easy we find it to use in practice.

It is worth noting that this principle is equivalent to the assumption that the average,

$$x_a = \frac{1}{n} \sum_{i=1}^{n} x_i$$

is the best estimate. To prove this equivalence, we first need to show that the least-squares principle leads to the average. We observe that

$$f(\hat{x}) = \sum_i (x_i - \hat{x})^2$$

can be regarded as a function of $\hat{x}$ to be minimized by the usual calculus method,

$$\frac{df}{d\hat{x}} = -2 \sum_i (x_i - \hat{x}) = 0$$

$$\sum_i x_i - \sum_i \hat{x} = 0$$

$$\sum_i x_i - n\hat{x} = 0$$

$$\hat{x} = \frac{1}{n} \sum_i x_i = x_a$$

[1] The Gauss-Markoff theorem asserts that if av $(\epsilon_i) = 0$, cov $(\epsilon_i \epsilon_j) = 0$, then this estimate has the smallest variance of all linear estimates.

Thus $\hat{x} = x_a$ minimizes $\Sigma(x_i - \hat{x})^2$. We also note that

$$\frac{d^2f}{d\hat{x}^2} = +2n > 0$$

hence we have obtained a minimum.

We can also show this by forming

$$f(x_a) = \Sigma(x_i - x_a)^2$$

Then

$$\begin{aligned} f(x_a) &= \Sigma x_i^2 - 2x_a\Sigma x_i + \Sigma x_a^2 \\ &= \Sigma x_i^2 - 2x_a n x_a + n x_a^2 \\ &= \Sigma x_i^2 - n x_a^2 \end{aligned}$$

If we choose any value other than x_a, say x_b, then

$$\begin{aligned} f(x_b) &= \Sigma(x_i - x_b)^2 \\ &= \Sigma x_i^2 - 2n x_a x_b + n x_b^2 \end{aligned}$$

We then have

$$\begin{aligned} f(x_b) - f(x_a) &= n(x_a^2 - 2x_a x_b + x_b^2) \\ &= n(x_a - x_b)^2 \geq 0 \end{aligned}$$

Thus the sum of the squares of the $\hat{\epsilon}_i$ is minimum *only* when the estimated value is taken as x_a. We have therefore proved that "the principle of least squares" and "the choice of the average" are equivalent.

EXERCISE

17.2-1. Find the least-squares estimate of the numbers 2, 3, 2, 1, 2, 3 both ways.
Ans. $\hat{x} = 13/6$

17.3 OTHER CHOICES

If we select the average of n measurements as the best choice, then we are led to the least-squares principle. But other choices are possible and often more appropriate.

Suppose that, instead of choosing to minimize the sum of the squares, we choose to minimize the sum of the deviations in absolute value

$$\sum_i |x_i - \hat{x}| = f(\hat{x})$$

This leads to the choice of the *median* (middle) value x_m of the x_i (if there are an even number of x_i we are led to choose the average of the middle two). The proof is straightforward. Supposing that there are an odd number, $2k + 1$, of x_i, we choose the middle one in size as x_m (which is the median). Then any shift of the choice from the median to x, say upward, will increase the k terms $|x_i - x|$ that have x_i below x_m and decrease the k terms $|x_i - x|$ that have x_i above x_m, each by the same amount, but it will also increase the term $|x_m - x|$, thus increasing the entire sum of the deviations.

Another choice, in place of minimizing the sum of the squares of the errors, is to minimize the maximum deviation. This leads to

$$\frac{x_{\text{max}} + x_{\text{min}}}{2} = x_{\text{mid-range}}$$

which is the usual *mid-range* estimate.

17.4 THE NORMAL LAW OF ERRORS

There appears to be a widespread belief that the principle of least squares implies the *normal law of errors*, which states that the probability of an error ϵ_i in the interval $x, x + \Delta x$ is given by

$$\frac{k}{\sqrt{\pi}} e^{-k^2x^2} \Delta x \qquad (17.4\text{-}1)$$

This belief is false, and it is worth some discussion of the topic at this point to clarify matters. Another belief sometimes met is that the normal law is a law of nature. This situation has been paraphrased by the remark "mathematicians believe it to be a physical law, while physicists believe it to be a mathematical law."

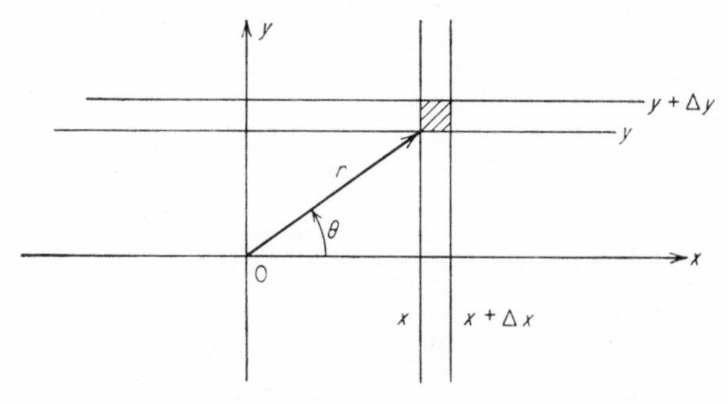

FIG. 17.4-1

One highly seductive derivation (due to Herschel) of the normal law goes as follows:

Consider dropping a dart from a height while aiming at some point O on the (horizontal) floor (Fig. 17.4-1). We now *assume* that the errors do not depend on the coordinate system but only on the distance from O and that larger errors are less likely than smaller ones. These assumptions seem to be quite reasonable, since the coordinate system appears to be perfectly arbitrary.

Let the probability of falling in the strip x, $x + \Delta x$ be given (approximately) by

$$f(x)\,\Delta x$$

while that of falling in the strip y, $y + \Delta y$ is

$$f(y)\,\Delta y$$

Owing to the assumption that the errors in the two coordinates are independent, the probability that it falls in the common rectangle is then

$$f(x)f(y)\,\Delta x\,\Delta y \tag{17.4-2}$$

Let us now choose polar coordinates. Since the probability is assumed to be independent of direction, the probability that the dart falls in the element of area is given by

$$g(r)\,\Delta x\,\Delta y \tag{17.4-3}$$

The probabilities (17.4-2) and (17.4-3) are the same, and we have

$$g(r) = f(x)f(y) \tag{17.4-4}$$

The left-hand side does not depend on θ, and, differentiating with respect to θ, we have

$$\frac{\partial g(r)}{\partial \theta} = 0 = f(x)\,\frac{\partial f(y)}{\partial \theta} + f(y)\,\frac{\partial f(x)}{\partial \theta} \tag{17.4-5}$$

Using the relations

$$x = r\cos\theta$$
$$y = r\sin\theta$$

we have

$$\begin{aligned} \frac{\partial f(x)}{\partial \theta} &= \frac{\partial f(x)}{\partial x}\frac{\partial x}{\partial \theta} = f'(x)(-y) \\ \frac{\partial f(y)}{\partial \theta} &= \frac{\partial f(y)}{\partial y}\frac{\partial y}{\partial \theta} = f'(y)(x) \end{aligned} \tag{17.4-6}$$

Combining (17.4-6) and (17.4-5), we have

$$f(x)f'(y)(x) + f(y)f'(x)(-y) = 0$$

or

$$\frac{f'(x)}{xf(x)} = \frac{f'(y)}{yf(y)}$$

Since x and y are assumed to be independent, each side must be a constant, say K, and

$$\frac{f'(x)}{xf(x)} = K = \frac{f'(y)}{yf(y)}$$

or, what is the same thing,

$$\frac{df(x)}{f(x)} = K \cdot x \cdot dx \qquad \text{and} \qquad \frac{df(y)}{f(y)} = K \cdot y \cdot dy$$

Integrating the first of these, we get

$$\ln f(x) = \frac{Kx^2}{2} + c$$

or

$$f(x) = Ae^{\frac{1}{2}Kx^2}$$

We assumed that large errors were less likely than small ones; hence K must be negative, say

$$K = -2k^2$$

and so we have

$$f(x) = Ae^{-k^2x^2}$$

$$f(y) = Ae^{-k^2y^2}$$

and, from (17.4-4),

$$g(r) = A^2e^{-k^2(x^2+y^2)}$$

Now the dart must fall somewhere; hence, using polar coordinates,

$$\int_0^{2\pi}\int_0^{\infty} g(r)r\,dr\,d\theta = 1$$

$$A^2 \cdot 2\pi \int_0^{\infty} re^{-k^2r^2}\,dr = 1$$

$$A^2\pi \left.\frac{e^{-k^2r^2}}{-k^2}\right|_0^{\infty} = \frac{A^2\pi}{k^2} = 1$$

$$A = \frac{k}{\sqrt{\pi}}$$

and finally we get [Eq. (17.4-1)]

$$f(x) = \frac{k}{\sqrt{\pi}}e^{-k^2x^2}$$

which is the normal (gaussian) distribution function.

It should be noted that this derivation is only one way of obtaining the normal law. The central-limit theorem, which states, loosely, that the distribution of the sum of a large number of small errors becomes normal, provides another approach. In either case, it is not the derivation that matters so much as whether or not the model fits experience.

The normal law has been found in practice to be a useful model in many applications. Deviations from it usually occur from having more values in the "tail" of the distribution when $|x|$ is large than the model indicates there should be. The reason that this is so is that often there is a small effect which has a wide variability. In such cases, a mixture of two normal curves with different k values sometimes is useful. The theory of *quality control* is, in part, based on the observed excesses in the "tails."

EXERCISE

17.4-1. Show that $\sigma^2 = 1/(2k^2)$, where σ is the variance of the distribution $f(x) = (k/\sqrt{\pi})e^{-k^2x^2}$.

17.5 POLYNOMIAL CURVE FITTING

One of the most common situations in which the least-squares approach is used occurs when N observations (x_i,y_i) $(i = 1, \ldots, N)$ are made and we wish to approximate the data by a polynomial of degree $M < N$,

$$y(x) = a_0 + a_1x + a_2x^2 + \cdots + a_Mx^M \tag{17.5-1}$$

The calculated curve $y(x)$, in some sense, gives a smoothed set of values $y(x_i)$ which are different, in general, from the observed y_i. The least-squares criterion states that a good estimate for the polynomial is the one which minimizes

$$\sum_{i=1}^{N} [y_i - y(x_i)]^2 = f(a_0,a_1, \ldots ,a_M) \tag{17.5-2}$$

Up to now we have discussed the principle of least squares only in one dimension; it should be evident that there is no fundamental change in going to $M + 1$ dimensions and regarding the choice of $a_0, a_1, \ldots, a_M$ as choosing the coordinates of a single point in an $(M + 1)$ dimensional euclidean space.

To find the minimum, we proceed as in the calculus and differentiate (17.5-2) with respect to the unknowns a_k,

$$\frac{\partial f}{\partial a_k} = -2 \sum_i [y_i - y(x_i)]x_i^k = 0 \qquad (k = 0, 1, \ldots, M)$$

or

$$\sum_i y_ix_i^k = a_0 \sum x_i^k + a_1 \sum x_i^{k+1} + \cdots + a_M \sum x_i^{k+M}$$

$$= \sum_{j=0}^{M} a_j \sum_{i=1}^{N} x_i^{k+j} \tag{17.5-3}$$

For ease in notation, we introduce

$$\sum_{i=1}^{N} x_i^k = S_k \qquad \sum_{i=1}^{N} y_ix_i^k = T_k$$

Equations (17.5-3) become

$$\sum_{j=0}^{M} a_jS_{k+j} = T_k \qquad (k = 0, 1, \ldots, M) \tag{17.5-4}$$

and are called "the normal equations."

These are $M + 1$ linear equations whose determinant is

$$\Delta = |S_{k+j}|$$

We want to show that $\Delta \neq 0$. If Δ were equal to zero, then the homogeneous equations [corresponding to (17.5-4)]

$$\sum_{j=0}^{M} a_j S_{j+k} = 0$$

would have a nonzero solution. Multiply the kth of these equations by a_k and add for all k,

$$0 = \sum_{k=0}^{M} a_k \sum_{j=0}^{M} a_j \sum_{i=1}^{N} (x_i^j x_i^k) = \sum_{i=1}^{N} \left(\sum_{k=0}^{M} a_k x_i^k\right) \left(\sum_{j=0}^{M} a_j x_i^j\right)$$

$$= \sum_{i=1}^{N} \left(\sum_{k=0}^{M} a_k x_i^k\right)^2 = \sum_{i=1}^{N} y^2(x_i) = 0$$

If this is to be zero, then each $y(x_i) = 0$ ($i = 1, 2, \ldots, N$). Since $N > M$, this is impossible by the fundamental theorem of algebra: A polynomial of degree M cannot vanish for $N > M$ values unless all the $a_k = 0$. Thus there is no nonzero solution, and $\Delta \neq 0$.

In principle, our problem is solved; in practice, it is not easy to solve Eqs. (17.5-4) since the determinant is often quite close to zero. To see why this can happen, suppose that the x_i are more or less uniformly distributed in the interval $0 \leq x_i \leq 1$. Then

$$S_k = \sum_{i=1}^{N} x_i^k \sim N \int_0^1 x^k \, dx = \frac{1}{k+1} N$$

The resulting determinant (suppressing the factor N^{M+1})

$$\left| \frac{1}{i+j+1} \right| \qquad (i,j = 0, 1, \ldots, M)$$

is known as a Hilbert determinant.

The Hilbert determinant of order n has the value

$$H_n = \frac{[1!2!3! \cdots (n-1)!]^3}{n!(n+1)! \cdots (2n-1)!}$$

which approaches zero quite rapidly. Table 17.5-1 gives some values of H_n.

The way out of this difficulty in solving a system of equations whose determinant is very small is not to let the S_k be computed but rather to compute some equivalent, more manageable information. This leads to

TABLE 17.5-1. VALUES OF THE HILBERT DETERMINANT

n	H_n
1	1
2	8.3×10^{-2}
3	4.6×10^{-4}
4	1.7×10^{-7}
5	3.7×10^{-12}
6	5.4×10^{-18}
7	4.8×10^{-25}
8	2.7×10^{-33}
9	9.7×10^{-43}

the topic of orthogonal functions, especially orthogonal polynomials, which has been extensively studied and has a vast literature.

As an aside, we note that Eqs. (17.5-4) can be written in another notation. From (17.5-1), we have

$$a_0 + a_1x_i + a_2x_i^2 + \cdots + a_Mx_i^M = y_i \qquad (i = 1, \ldots, N)$$

or, in matrix notation,

$$Xa = y$$

Let X^T be the transpose of X. Then (17.5-4) is

$$X^TXa = X^Ty$$

This is not a practical way to compute Eqs. (17.5-4).

EXERCISE

17.5-1. Given

x	0	1	2	3	4
y	1	2	1	0	4

find the least-squares line fitting the data.

Ans. $y = \frac{2}{5}(x + 2)$

17.6 ORTHOGONAL FUNCTIONS

In two dimensions, two lines are said to be "orthogonal," or perpendicular, if

$$\tan\theta_1 = -\frac{1}{\tan\theta_2}$$

This may be written

$$\sin\theta_1 \sin\theta_2 + \cos\theta_1 \cos\theta_2 = 0$$

If we now use the complementary angles φ_1 and φ_2 in the sine terms, we get

$$\cos \varphi_1 \cos \varphi_2 + \cos \theta_1 \cos \theta_2 = 0$$

In three dimensions, with angles α, β, γ, we have

$$\cos \alpha_1 \cos \alpha_2 + \cos \beta_1 \cos \beta_2 + \cos \gamma_1 \cos \gamma_2 = 0$$

as the condition for perpendicularity or orthogonality.

It is an easy generalization to n dimensions with angles $\alpha^{(1)}, \alpha^{(2)}, \ldots, \alpha^{(n)}$ to write

$$\cos \alpha_1^{(1)} \cos \alpha_2^{(1)} + \cos \alpha_1^{(2)} \cos \alpha_2^{(2)} + \cdots + \cos \alpha_1^{(n)} \cos \alpha_2^{(n)} = 0$$

as the condition for orthogonality.

In practice, it has been found that the *direction cosines*

$$\cos \alpha_i^{(k)} = \lambda_i(k)$$

are more fundamental than the angles themselves, so that we usually have the form

$$\sum_{k=1}^{n} \lambda_1(k)\lambda_2(k) = 0$$

as the condition for orthogonality. If we now let $k \to \infty$ in a suitable fashion, we can see that it is reasonable to call

$$\int \lambda_1(k)\lambda_2(k)\, dk = 0$$

a condition for orthogonality.

With this motivation, we shall say that two functions $f_1(x)$ and $f_2(x)$ are orthogonal over the interval $a \leq x \leq b$ if

$$\int_a^b f_1(x) f_2(x)\, dx = 0$$

A set of functions $f_1, f_2, \ldots, f_m$ is (mutually) orthogonal if[1]

$$\begin{aligned} \int_a^b f_i(x) f_j(x)\, dx &= 0 \qquad (i \neq j) \\ \int_a^b f_i^2(x)\, dx &= \lambda_i > 0 \qquad (i = j) \end{aligned} \tag{17.6-1}$$

[We are assuming that the $f_i(x)$ are real, continuous, and not identically zero.] Let

$$g_i = \frac{f_i}{\sqrt{\lambda_i}}$$

Then
$$\int_a^b g_i(x) g_j(x)\, dx = \begin{cases} 0 & i \neq j \\ 1 & i = j \end{cases} \tag{17.6-2}$$

[1] It is unfortunately customary to use the notation λ_i here; it should not be confused with the direction cosines.

The functions $g_i(x)$ are called "orthonormal," and the process of going from f_i to the g_i is called "normalization."

The classic set of orthogonal functions is that of the Fourier series

$$\begin{matrix} 1, & \cos x, & \cos 2x, & \ldots, & \cos Mx, & \ldots \\ & \sin x, & \sin 2x, & \ldots, & \sin Mx, & \ldots \end{matrix}$$

in the interval $0 \le x < 2\pi$ (or $-\pi \le x \le \pi$). If the data are given at $2N$ equally spaced points, then the terms end at $\frac{1}{2} \cos Nx$ and $\sin (N - 1)x$, as we saw in Chap. 6.

To see how the orthogonal function approach avoids the bad matrix of the unknown coefficients, let us recall how the expansion goes. We try to represent

$$\begin{aligned} y_i = \frac{a_0}{2} + a_1 \cos x_i + \cdots + a_{N-1} \cos (N-1)x_i + \frac{a_N}{2} \cos Nx_i \\ + b_1 \sin x_i + \cdots + b_{N-1} \sin (N-1)x_i \end{aligned}$$

We multiply by $\cos mx_i$ or $\sin mx_i$ and sum over the $x_i = (\pi/N)i$:

$$a_m = \frac{1}{N} \sum_{i=0}^{2N-1} y_i \cos mx_i$$

$$b_m = \frac{1}{N} \sum_{i=0}^{2N-1} y_i \sin mx_i$$

[using (6.2-3)]. This system of equations for the unknown coefficients a_m and b_m is trivial to solve—they are already solved.

Similarly, if we have a set of orthogonal functions

$$\{f_i(x)\}$$

we write

$$f(x) = a_0 f_0(x) + \cdots + a_N f_N(x)$$

We then multiply by $f_j(x)$ and integrate:

$$\int_a^b f(x) f_j(x)\, dx = \lambda_j a_j \tag{17.6-3}$$

The solution is again trivial. The a_j determined in this way are called *the Fourier coefficients.*

If we know the functions only at isolated points x_m, then the integrals are replaced by sums, and the orthogonality condition becomes

$$\sum_m f_i(x_m) f_j(x_m) = \begin{cases} 0 & i \neq j \\ \lambda_j & i = j \end{cases}$$

Note that $\lambda_j \neq 0$. If the data have varying weights of importance, then a weighting function $\rho(x) \geq 0$ can be included in the integral,

$$\int_a^b \rho(x) f_i(x) f_j(x)\, dx \tag{17.6-4}$$

where the coefficients are given by

$$\int_a^b \rho(x) f(x) f_j(x)\, dx = \lambda_j a_j \tag{17.6-5}$$

If there is a mixture of information in the continuous and discrete forms, then Stieltjes integrals can be used.

It is convenient to develop the theory in the continuous form, even though much of the time we shall actually need the discrete case.

EXERCISE

17.6-1. Show that $P_0 = 1$, $P_1 = x$, $P_2 = \frac{1}{2}(3x^2 - 1)$, $P_3 = \frac{1}{2}(5x^3 - 3x)$ are orthogonal in the interval $-1 \leq x \leq 1$.

17.7 GENERAL PROPERTIES OF ORTHOGONAL FUNCTIONS

Before making use of orthogonal functions, we shall develop some general theorems to give the reader a feeling for the behavior of orthogonal functions.

The idea of linear independence is one of the basic ideas in mathematics. *A set of functions $f_i(x)$ is said to be linearly independent in an interval (a,b) if the equation*

$$a_0 f_0(x) + a_1 f_1(x) + \cdots + a_m f_m(x) \equiv 0$$

in the whole interval implies that all the coefficients a_i are zero;[1] *otherwise they are linearly dependent.*

A simple and important example of a set of linearly independent functions in an arbitrary interval is

$$1, \quad x, \quad x^2, \quad \ldots, \quad x^m$$

since, by the fundamental theorem of algebra,

$$a_0 \cdot 1 + a_1 x + a_2 x^2 + \cdots + a_m x^m \equiv 0$$

implies that all the a_i are zero.

Any set of continuous orthogonal functions in an interval is linearly independent. The proof is quite easy; we merely assume that

$$a_0 f_0 + a_1 f_1 + \cdots + a_m f_m \equiv 0$$

[1] We are excluding the function $f(x) \equiv 0$ from this and subsequent discussions.

and compute the Fourier coefficients (17.6-5)

$$a_j = \frac{1}{\lambda_j} \int_a^b 0 \cdot \rho(x) f_j(x)\, dx = 0$$

The converse, that from a set of linearly independent functions we can construct an orthogonal set, can be shown by the Schmidt process. Let the given set of linearly independent functions be $f_i(x)$. We compute

$$\int_a^b \rho(x) f_0^2(x)\, dx = \lambda_0 > 0 \qquad [\rho(x) \geq 0]$$

Then

$$g_0(x) = \frac{f_0(x)}{\sqrt{\lambda_0}}$$

defines the first orthonormal function $g_0(x)$. Using mathematical induction, we assume that we have constructed the first j orthonormal functions $g_i(x)$ $(i = 0, 1, \ldots, j-1)$. We set

$$F_j(x) = a_0 g_0 + a_1 g_1 + \cdots + a_{j-1} g_{j-1} + f_j(x) \tag{17.7-1}$$

Now $F_j(x) \not\equiv 0$, since the functions $f_i(x)$ are linearly independent and each $g_i(x)$ is a linear combination of the $f_k(x)$ for $k \leq i$. We need

$$\int_a^b \rho(x) F_j(x) g_i(x)\, dx = 0 \qquad 0 \leq i \leq j-1$$

But this is, using the definition of $F_j(x)$,

$$a_i + \int_a^b \rho(x) g_i(x) f_j(x)\, dx = 0$$

which determines the a_i and hence $F_j(x)$. To normalize $F_j(x)$, we need to compute

$$\int_a^b \rho(x) F_j^2(x)\, dx = \lambda_j \qquad [\rho(x) \geq 0]$$

and then set

$$g_j(x) = \frac{F_j(x)}{\sqrt{\lambda_j}}$$

Thus we have taken one more step in the induction.

If we have only a finite number N of samples x_m, then there are at most N linearly independent functions

$$f_j(x_m)$$

That there are N follows from the particular set

$$g_j(x_m) = \begin{cases} 0 & m \neq j \\ 1 & m = j \end{cases} \qquad (j = 1, \ldots, N)$$

since no set of these N functions $g_j(x_m)$ can be linearly dependent.

It is worth noting that in the construction process the set that we obtain is not unique unless we choose the particular $\sqrt{\lambda_i} > 0$ and also choose the set of linearly independent functions in a fixed order. Also, they obviously depend on the interval (a,b) and the weight function $\rho(x)$.

EXERCISE

17.7-1. Given $P_0 = 1$, $P_1 = x$, $\rho(x) = 1$, construct the orthogonal functions $P_2(x)$, $P_3(x)$ over the set of points $-2, -1, 0, 1, 2$, where P_2 and P_3 are polynomials of degree 2 and 3, respectively.

17.8 BESSEL'S INEQUALITY AND COMPLETENESS

The Fourier coefficients

$$a_j = \int_a^b \rho(x)F(x)g_j(x)\,dx$$

of an orthonormal set $g_j(x)$ *satisfy the Bessel inequality*

$$\int_a^b \rho(x)F^2(x)\,dx \geq \sum_{j=0}^{M} a_j^2 \qquad (\rho \geq 0) \tag{17.8-1}$$

The proof is straightforward. We write

$$\int_a^b \rho(x)\left[F(x) - \sum_{i=0}^{M} a_i g_i(x)\right]^2 dx \geq 0$$

and multiply out:

$$0 \leq \int_a^b \rho(x)F^2(x)\,dx - 2\int_a^b \rho(x)F(x)\sum_{i=0}^{M} a_i g_i(x)\,dx + \int_a^b \sum_{i=0}^{M}\sum_{j=0}^{M} a_i a_j \rho(x) g_i g_j\,dx$$

Using the definition of the a_i and the orthogonality of the g_i,

$$\int_a^b \rho(x)F^2(x)\,dx \geq 2\sum_{i=0}^{M} a_i^2 - \sum_{i=0}^{M} a_i^2 = \sum_{i=0}^{M} a_i^2$$

for all M.

In the continuous case, if the equality holds for every continuous function $F(x)$ in the interval a,b, then the infinite set of functions is said to be "complete," and the equality

$$\int_a^b \rho(x)F^2(x)\,dx = \sum_{i=0}^{\infty} a_i^2 \tag{17.8-2}$$

is called *Parseval's equality.*

In the finite case we have, of course, no trouble in the process of interchanging limits to get Parseval's equality, and any set of N functions linearly independent at a set of N points is complete over the same set of points (compare Sec. 6.3).

17.9 LEAST-SQUARES FIT AND THE FOURIER COEFFICIENTS

The Fourier coefficients a_j give the best least-squares fit when a function $F(x)$ is expanded in terms of an orthonormal set of functions $g_j(x)$.

To prove this we want to minimize an arbitrary expansion

$$\begin{aligned} m &= \int_a^b \rho(x)\left[F(x) - \sum_{j=0}^{M} c_j g_j(x)\right]^2 dx \\ &= \int \rho(x)F^2(x)\,dx - 2\sum_{j=0}^{M} c_j \int \rho(x)F(x)g_j(x)\,dx \\ &\qquad\qquad + \sum_{i=0}^{M}\sum_{j=0}^{M} c_j c_i \int \rho(x) g_i g_j\,dx \\ &= \int \rho(x)F^2(x)\,dx - 2\sum_{i=0}^{M} c_i a_i + \sum_{i=0}^{M} c_i^2 \\ &= \int \rho(x)F^2(x)\,dx - \sum_{i=0}^{M} a_i^2 + \sum_{i=0}^{M} (a_i - c_i)^2 \end{aligned}$$

But this is minimum only if each $c_i = a_i$ as required.

This property, that in the best least-squares fit with orthogonal functions each coefficient a_i is determined independently of all the others and that if we decide to change the number of functions $g_i(x)$ that we are using we need not redetermine any of the coefficients that we have already found, is a remarkable and very useful property. We therefore examine the converse problem.

If the coefficients c_i in a least-squares fit of a function $F(x)$ in terms of a set $\mu_i(x)$ are not to change as we change the number of $\mu_i(x)$ that we use, then the $\mu_i(x)$ must be orthogonal. Set

$$g(c_0,c_1, \ldots ,c_n) = \int_a^b \rho(x)\left[F(x) - \sum_0^M c_i\mu_i(x)\right]^2 dx$$

Since g is to be minimized,

$$\frac{\partial g}{\partial c_j} = 0 = -2\int_a^b \rho(x)\left[F(x) - \sum_0^M c_i\mu_i\right]\mu_j(x)\,dx$$

or

$$\int \rho(x)F(x)\mu_j\,dx = \sum_0^M c_i \int_a^b \rho(x)\mu_i\mu_j\,dx \qquad (17.9\text{-}1)$$

This must also be true for $M + 1$, if the property is to be true for all M.

$$\int \rho(x)F(x)\mu_j(x)\,dx = \sum_0^{M+1} c_i \int_a^b \rho(x)\mu_i\mu_j\,dx \qquad (17.9\text{-}2)$$

Subtracting (17.9-1) from (17.9-2), we get

$$c_{M+1}\int_a^b \rho(x)\mu_{M+1}\mu_j\,dx = 0$$

for any j, that is, the μ_j are orthogonal to μ_{M+1} (and M was arbitrary).

Thus we see that orthogonal functions, the Fourier determination of the coefficients in an expansion, and the idea of least-squares approximation are all bound together.

17.10 ORTHOGONAL POLYNOMIALS

A very important subclass of orthogonal functions is that of the orthogonal polynomials, for which the kth polynomial is a polynomial of degree k $(k = 0, 1, \ldots)$.

It is easy to show that the kth orthogonal polynomial $y_k(x)$ has exactly k real distinct roots in the range of integration. For let us suppose that there were $r < k$. We form the product of the distinct, odd-multiplicity zeros

$$\pi(x) = (x - x_1)(x - x_2)\cdots(x - x_r) \qquad (r < k)$$

Then

$$\int_a^b \rho(x)\pi(x)y_k(x)\,dx = 0 \qquad [\rho(x) \geq 0]$$

because $\pi(x)$ can be written as a sum of $y_0, y_1, \ldots, y_r$ and $y_k(x)$ is orthogonal to all of them. But this is impossible since the integrand does not change sign in the interval. Hence there are k real distinct roots in (a,b).

The orthogonal polynomials $y_k(x)$ satisfy a three-term recurrence relation of the form

$$a_ky_{k+1}(x) + (b_k - x)y_k(x) + c_ky_{k-1}(x) = 0 \qquad (k \geq 1) \qquad (17.10\text{-}1)$$

To show this, we set

$$y_i(x) = \alpha_ix^i + \cdots \qquad (\alpha_i > 0)$$

(The choice of α_i as any particular positive number is a matter of convenience only.) Then

$$a_ky_{k+1} - xy_k$$

is a polynomial of degree k, provided that we choose $a_k = \alpha_k/\alpha_{k+1}$. Hence

$$a_ky_{k+1} - xy_k = \gamma_ky_k + \gamma_{k-1}y_{k-1} + \cdots + \gamma_0y_0 \qquad (17.10\text{-}2)$$

We multiply by $\rho(x)y_m(x)$ and integrate. We obtain

$$\int_a^b \rho(x)(a_k y_{k+1} - xy_k)y_m(x)\,dx = \gamma_m\lambda_m \tag{17.10-3}$$

For $m = 0, 1, \ldots, k$, y_m is orthogonal to y_{k+1}, and for $m = 0, 1, \ldots, k-2$, xy_m is a polynomial of degree less than k and hence is orthogonal to y_k. Thus $\gamma_0 = \gamma_1 = \cdots = \gamma_{k-2} = 0$. For $m = k-1$, (17.10-3) becomes

$$\begin{aligned} -\int_a^b \rho y_k(xy_{k-1})\,dx &= -\int_a^b \rho y_k\left(\frac{\alpha_{k-1}}{\alpha_k} y_k + C'_{k-1}y_{k-1} + \cdots\right) dx \\ &= -\frac{\alpha_{k-1}}{\alpha_k}\lambda_k = \gamma_{k-1}\lambda_{k-1} \end{aligned}$$

or, using (17.10-1),

$$-c_k = \gamma_{k-1} = -\frac{\alpha_{k-1}\lambda_k}{\alpha_k\lambda_{k-1}} \neq 0$$

For $m = k$, (17.10-2) becomes [using Eq. (17.10-1)]

$$\int_a^b \rho(a_k y_{k+1} - xy_k)y_k\,dx = -\int_a^b \rho x y_k^2\,dx = \gamma_k\lambda_k = -b_k\lambda_k$$

Thus (17.10-2) can be written

$$\frac{\alpha_k}{\alpha_{k+1}} y_{k+1} - xy_k = \gamma_k y_k + \gamma_{k-1}y_{k-1}$$

or

$$y_{k+1} = \frac{\alpha_{k+1}(\gamma_k + x)}{\alpha_k} y_k - \frac{\alpha_{k-1}\alpha_{k+1}}{\alpha_k^2}\frac{\lambda_k}{\lambda_{k-1}} y_{k-1} \tag{17.10-4}$$

Several important results follow from this formula. We first show that the k zeros of $y_k(x)$ are separated by the $k-1$ zeros of $y_{k-1}(x)$, provided that $\rho(x) \geq 0$. The proof is by induction. As a basis for the induction, we have $y_0(x) = \alpha_0 > 0$ and $y_1(x) = \alpha_1(x + \gamma_0)$. We know that $y_1(x)$ has a real zero in the range of integration, since

$$\int_a^b \rho(x)y_0(x)y_1(x)\,dx = 0$$

We now assume the induction hypothesis that the zeros of $y_k(x)$ are interlaced by those of $y_{k-1}(x)$. At the zeros of $y_k(x_i)$, (17.10-4) becomes

$$y_{k+1}(x_i) = -\frac{\alpha_{k-1}\alpha_{k+1}}{\alpha_k^2}\frac{\lambda_k}{\lambda_{k-1}} y_{k-1}(x_i) \tag{17.10-5}$$

At the top of the range, $x = b$, both $y_{k+1}(b)$ and $y_{k-1}(b)$ are of positive sign, since we chose the leading coefficients $\alpha_i > 0$, and all the zeros of the functions lie inside the interval. At the largest zero of $y_k(x_i) = 0$,

$y_{k-1}(x_i)$ is still positive; hence $y_{k+1}(x_i)$ is negative, by (17.10-5). At the next smaller zero, y_{k-1} has, owing to the induction hypothesis, changed sign; hence $y_{k+1}(x)$ is positive. And so it proceeds; as we pass from zero to zero of $y_k(x_i)$, $y_{k-1}(x_i)$ changes sign; hence $y_{k+1}(x_i)$ changes sign. Thus we have shown that the zeros of $y_k(x)$ interlace those of $y_{k+1}(x)$, except for the smallest zero of y_{k+1}. But since we have just shown that $y_{k+1}(x)$ has $k+1$ real, distinct roots in (a,b), the last zero of $y_{k+1}(x)$ is smaller than all those of $y_k(x)$, and the induction proof is complete.

There is another important consequence of the three-term recurrence relations (17.10-1) or (17.10-2). Once we know the coefficients of the recurrence relation as functions of k and know $y_0(x)$ and $y_1(x)$, we can then compute the $y_k(x)$ one at a time by this relation rather than by the cumbersome Schmidt process. (See Sec. 18.2.)

17.11 THE CLASSICAL ORTHOGONAL POLYNOMIALS

There are three sets of orthogonal polynomials which have been extensively studied: the Legendre, $P_n(x)$,

$$\int_{-1}^{1} P_m(x)P_n(x)\,dx = \begin{cases} 0 & m \neq n \\ 2/(2n+1) & m = n \end{cases}$$

the Laguerre, $L_n(x)$,

$$\int_{0}^{\infty} e^{-x}L_m(x)L_n(x)\,dx = \begin{cases} 0 & m \neq n \\ n! & m = n \end{cases}$$

and the Hermite, $H_n(x)$,

$$\int_{-\infty}^{\infty} e^{-x^2}H_m(x)H_n(x)\,dx = \begin{cases} 0 & m \neq n \\ 2^n n!\sqrt{\pi} & m = n \end{cases}$$

They have the corresponding three-term recurrence relations [see (17.10-4)]

$$\begin{aligned} (n+1)P_{n+1}(x) - (2n+1)xP_n(x) + nP_{n-1}(x) &= 0 \\ (n+1)L_{n+1}(x) - (2n+1-x)L_n(x) + n^2L_{n-1}(x) &= 0 \qquad (n \geq 1) \\ H_{n+1}(x) - 2xH_n(x) + 2nH_{n-1}(x) &= 0 \end{aligned}$$

Many other relations involving these functions can be found in standard references. The zeros of these polynomials, for example, turn out to be the sample points of the Gauss quadrature formulas of Chap. 10.

Corresponding to these orthogonal functions defined for integration, there are orthogonal functions for discrete sets: Of these the most common is the set corresponding to the Legendre polynomials. Sometimes the interval is chosen $-1 \leq x_i \leq 1$ and sometimes $0 \leq x_i \leq 1$ for the equally spaced sample points x_i, depending on convenience, and the reader should be careful to ascertain which set is used in a table.

EXERCISE

17.11-1. If $P_0 = 1$, $P_1 = x$, show that for $-1 \leq x \leq 1$

$$P_2 = \tfrac{1}{2}(3x^2 - 1)$$
$$P_3 = \tfrac{1}{2}(5x^3 - 3x)$$
$$P_4 = \tfrac{1}{8}(35x^4 - 30x^2 + 3)$$

17.12 A COMPARISON OF LEAST SQUARES AND POWER SERIES EXPANSION

The power series expansion of a function $y(x)$,

$$y(x) = \sum_{n=0}^{\infty} \frac{y^{(n)}(0)x^n}{n!}$$

when truncated to

$$y_N(x) = \sum_{n=0}^{N} \frac{y^{(n)}(0)x^n}{n!}$$

gives a very good approximation near $x = 0$, but as x increases the approximation tends to get worse and worse. On the other hand, the least-squares fit tries to find a more or less uniform [depending on the weight function $\rho(x)$] approximation *in an interval.* And this is one of the fundamental differences: The power series fits at a point whereas the least squares fits in an interval.

When it comes to numerical computation, it is not generally possible to estimate accurately the derivatives at a point from the samples scattered in an interval; the exact-matching interpolating polynomial is frequently used in place of the truncated power series.

For the least-squares approach there is often a choice. We can imagine that we have a continuous orthogonal polynomial, and when we need to calculate a Fourier coefficient by an integral, we can estimate the integral by some numerical process. Or we can start with discrete samples and operate using the discrete orthogonal set of functions. It is not easy to say which approach is better in most cases; probably the choice depends on circumstances, including the mental attitude of the person using the results.

It should be noted that in the Fourier series case these two approaches produce *exactly* the same computations. The viewpoint, in this case, is much like that of "alias" and "alibi" in analytic geometry (see Sec. 1.1), and we can shift from one to the other as we see fit, provided that we adopt a consistent view at any one time.

17.13 LEAST SQUARES WITH RESTRAINTS; CONTINUATION OF THE EXAMPLE IN SEC. 1.9

In the example in Sec. 1.9 we fitted an exact-matching polynomial to 11 data points $x = 0(0.1)1$, using Newton's interpolation formula. The success of the computation resulted in new laboratory equipment being built and much better and more extensive data being gathered.

A visual examination of the new data suggested about a sixth-order polynomial, and a plot of the logs of the values near $x = 1$ produced a slope near 6, thus tending to confirm the impression. With no further thought, a least-squares polynomial was computed, and the results were what one would expect—stupid, thoughtless computing produced foolish results! The polynomial began with a small positive value at $x = 0$, when the experiment clearly would give zero, and went slightly negative for a short distance, which physically was impossible.

The obvious remedy was to omit the constant term of the polynomial. Discussion with the physicist revealed that we could *not* expect the curve to be tangent to the x axis at $x = 0$, and so we could not leave out the x term. This more carefully considered computation produced an acceptable polynomial, and the rest of the computation was straightforward as before.

Another obvious way of approaching the problem of fitting the data at $x = 0$ would have been to give the value there a large weight, say 1,000, where the other data points had weight 1.

Many variations occur on the theme of placing extra conditions on a least-squares fit. For example, suppose that we require the solution to take on two specified values, one at a and one at b, and are using orthogonal polynomials. We may use the classic method of Lagrange multipliers. If the values at a and b were zero, we could use a weight function $(x - a)^2(x - b)^2$ and construct the polynomials $P_n(x)$. Then the

$$Q_n(x) = (x - a)(x - b)P_n(x)$$

are orthogonal polynomials passing through zero at $x = a$ and $x = b$.

EXERCISE

17.13-1. Show that in the orthogonal-polynomial approach a straight line may be subtracted from the data first, to reduce two conditions $f(a) = A$, $f(b) = B$ to $\bar{f}(a) = \bar{f}(b) = 0$.

17.14 CONCLUDING REMARKS ON LEAST SQUARES

Having been shown how to produce a least-squares polynomial, the reader probably expects the text to go on to the problems of integrals and differential equations. While this could be done, the facts are that seldom is a least-squares approximation used as a basis for integration. Integration itself is a "smoothing operation." When there is a "noisy" term in a differential equation, it is customary to smooth it and then use an exact-matching polynomial for integrating the equation.

Another idea which may occur to the reader, but which has not been adequately explored, is the following: In the exact-matching process, we have used $E_0, E_1, \ldots, E_{m-1} = 0$, and let the rest, $E_m, E_{m+1}, \ldots$, fall where they may. Suppose that we were to try to minimize

$$m(a_0,a_1, \ldots) = \sum_{k=0}^{\infty} a_k E_k^2$$

The exact matching used $a_m = a_{m+1} = \cdots = 0$, but we could taper the a_k and thus allow for an improved fit for some of the higher powers at the cost of some exactness for the lower powers.

This can be viewed as a proposal to fit in error space rather than in the function space. An adequate theory for the error term is not known at present.

An example of this proposal to sacrifice a small amount of exactness in fitting at some value for a large gain at some higher value of k is given by the idea of throwback, discussed briefly in Sec. 9.5 as applied to Everett's interpolation formula. In the particular case, we gave up a small amount of accuracy in x^2 (modified Δ^2) for a large gain in fitting x^3 and x^4.

CHAPTER 18

Least Squares: Practice

18.1 GENERAL REMARKS ON THE POLYNOMIAL SITUATION

The major use of the least-squares method is in situations in which the coefficients to be determined occur in a linear fashion, especially as coefficients of a polynomial. We shall take up the polynomial situation first and then pass on to the other cases.

As was indicated in the previous chapter, the determinant of the normal equations (17.5-4) tends to be very small; hence the solution for coefficients is likely to be uncertain. It is necessary to distinguish, however, between two things: the accuracy of the coefficients and the smallness of the sum of the squares of the errors. When the determinant is small, the coefficients may be poorly determined, but still the sum of the squares of the errors can be close to the minimum. Generally speaking, when there are up to five or six coefficients to be determined, the direct solution of the normal equations is usually satisfactory, but above this there is likely to be trouble.

The theory indicates that the expansion in orthogonal polynomials is a substitute for the direct solution. However, experience shows that if the orthogonalization is attempted by means of the Schmidt process, then the same difficulty arises in a different disguise. In the Schmidt process, the construction of the mth polynomial is done by first subtracting all the components of the vector x^m, that is, $(x_1{}^m, x_2{}^m, \ldots, x_N{}^m)$, which lie in the direction of the previously determined polynomials. The result is apt to be small when m is fairly large, and it is this remainder that is normalized as the last step. Thus the normalization increases the errors, because the normalization factor is usually much greater than 1 in size.

Loosely speaking, the trouble may be expressed by the remark that for large n the vector x^n points in about the same direction as does x^{n-1}, x^{n-2}, etc. Thus the equations leading to the Hilbert determinant are almost linearly dependent; hence the determinant is small.

18.2 THE THREE-TERM RECURRENCE RELATION

If we insist on using the orthogonal polynomials, then we can apparently avoid the troubles mentioned in the previous section by using the three-term recurrence relation (17.10-2) to generate the orthogonal polynomials.[1]

We write (17.10-1) in the form

$$\begin{aligned} p_0(x) &= 1 \\ p_1(x) &= xp_0(x) - \alpha_1 p_0(x) \\ p_{k+1}(x) &= xp_k(x) - \alpha_{k+1}p_k(x) - \beta_{k+1}p_{k-1}(x) \qquad (k > 1) \end{aligned} \tag{18.2-1}$$

where α_{k+1} and β_{k+1} are to be determined.

We first determine α_1. We know that

$$\int_a^b \rho(x)p_0(x)p_1(x)\,dx = 0$$

Hence, by (18.2-1),

$$\int_a^b \rho(x)x\,dx = \alpha_1 \int_a^b \rho(x)\,dx$$

We now suppose that we have $p_0(x)$, $p_1(x)$, . . . , $p_k(x)$ and that they are orthogonal to one another. We need to compute $p_{k+1}(x)$, the next polynomial of the set. We require first

$$\begin{aligned} \int \rho(x)p_{k+1}(x)p_k(x)\,dx &= 0 \\ \int \rho(x)p_{k+1}(x)p_{k-1}(x)\,dx &= 0 \end{aligned}$$

and these suffice to determine α_{n+1} and β_{n+1}. Using the definition of $p_{k+1}(x)$ [Eqs. (18.2-1)], we get

$$\begin{aligned} \int \rho(x)xp_k{}^2(x)\,dx &= \alpha_{k+1}\int \rho(x)p_k{}^2(x)\,dx + \beta_{k+1}\int \rho(x)p_k(x)p_{k-1}(x)\,dx \\ \int \rho(x)xp_k(x)p_{k-1}(x)\,dx &= \alpha_{k+1}\int \rho(x)p_k(x)p_{k-1}(x)\,dx + \beta_{k+1}\int \rho(x)p_{k-1}^2(x)\,dx \end{aligned}$$

Since $p_k(x)$ and $p_{k-1}(x)$ are orthogonal, we have

$$\begin{aligned} \alpha_{k+1} &= \frac{\int \rho(x)xp_k{}^2(x)\,dx}{\int \rho(x)p_k{}^2(x)\,dx} \\ \beta_{k+1} &= \frac{\int \rho(x)xp_k(x)p_{k-1}(x)\,dx}{\int \rho(x)p_{k-1}^2(x)\,dx} \end{aligned}$$

The denominator of the β_{k+1} equation was computed during the previous step when $\alpha_k(x)$ was determined; thus we have three integrals to compute on each step.

That the $p_{k+1}(x)$ is orthogonal to all $p_i(x)$ $(i < k - 1)$ follows from the defining equations (18.2-1)

$$p_{k+1}(x) = xp_k(x) - \alpha_{k+1}p_k(x) - \beta_{k+1}p_{k-1}(x)$$

[1] G. E. Forsythe, Generation and Use of Orthogonal Polynomials for Data-fitting with a Digital Computer; *J. Soc. Ind. Appl. Math.*, vol. 5, no. 2, June, 1957; M. Ascher and G. E. Forsythe, SWAC Experiments on the Use of Orthogonal Polynomials for Data Fitting; *J. Assoc. Computing Machinery*, vol. 5, no. 1, January, 1958.

since, on multiplying by $\rho(x)p_i(x)$ and integrating, we have

$$\int \rho p_k(x)[xp_i(x)]\,dx - \alpha_{k+1}\int \rho(x)p_k(x)p_i(x)\,dx - \beta_{k+1}\int \rho(x)p_{k-1}(x)p_i(x)\,dx = 0$$

The last two integrals are zero because of orthogonality, and in the first $xp_i(x)$ is a polynomial of degree less than k; hence the integral is also zero.

When we try to construct polynomials over a discrete set of points, x_j ($j = 1, \ldots, N$), the integrals are replaced by sums. If we attempt to construct more than N polynomials [that is, above $p_{N-1}(x)$], then the equation determining α_N fails, as it should.

EXERCISE

18.2-1. Orthogonalize $1, x, x^2, x^3$ over the interval $0 \leq x \leq 1$, using the three-term recurrence method.

18.3 THE CONSTRUCTION OF QUASI-ORTHOGONAL POLYNOMIALS

The direct approach leads to normal equations which have nonzero terms in places off the main diagonal of the system. The use of orthogonal polynomials leads to equations for which all the terms off the main diagonal are zero; hence the solution of the equations is trivial.

If we attempt to construct some orthogonal polynomials by the Schmidt process (or by the three-term recurrence relation) and do not determine them accurately, then the resulting equations to be solved have large terms on the main diagonal, while those off the diagonal are proportional to the deviation from orthogonality. If this is not large, then the system of equations is fairly easy to solve.

This observation suggests that we try our hand at constructing some quasi-orthogonal polynomials without doing much computation. In so far as we make them orthogonal, that far will the terms off the main diagonal of the system of normal equations be zero, and in so far as we miss orthogonality, that far away will they be from zero.

Some meditating on this question suggests that the fundamental property that we should try to keep is that of the interlacing zeros. Hence we choose the first polynomial as $p_0(x) = 1$. The second we take as a straight line of convenient slope but with a zero near the median of the set of points. We next pick a quadratic whose zeros are on each side of that of the straight line. And so we proceed, at each step choosing the zeros of our next polynomial so that they are separated by the preceding polynomial and also leave room for the next polynomials. Unless we are trying to determine a polynomial whose degree is near that of the exact-matching polynomial through all the points, and this is rarely if ever the case, then there is enough freedom to select the zeros at convenient places to make the polynomials have convenient coefficients and thus lighten the burden of subsequent calculation.

Experience with this method shows that it is fairly effective; usually the off-diagonal terms of the system of equations are fairly small, and the equations are easy to solve.

EXERCISE

18.3-1. Examine the quasi-orthogonal polynomials $(-1 \leq x \leq 1)$, $\rho = 1$, $p_0 = 1$, $p_1 = x$, $p_2 = x^2 - \frac{1}{4}$, $p_3 = x(x^2 - \frac{9}{16})$.

18.4 THE NONPOLYNOMIAL CASE

Sometimes we are given a set of functions $f_i(x)$ and asked to find the least-squares approximation to some data $f(x)$ of the form

$$f(x) = a_1 f_1(x) + a_2 f_2(x) + \cdots + a_n f_n(x)$$

We proceed as before. We set up the error ϵ_i at x_i and form the sum of the squares. Next we differentiate with respect to the coefficients (which are to be found and hence are the variables of the problem). The resulting normal equations are formally like those of the polynomial case, except that we now define

$$S_{k,j} = S_{j,k} = \sum_i f_j(x_i) f_k(x_i) \qquad T_k = \sum_i f(x_i) f_k(x_i)$$

If the $f_i(x)$ are strongly linearly independent, by which we mean that no one of them with unit coefficient can be expressed by a linear combination of the others plus some small corrections, then we can expect to be able to solve the normal equations without much trouble. If they are close to linear dependence, then some similar scheme to orthogonalization can be used, although there will be, generally speaking, no three-term recurrence relation and the quasi-orthogonal method may be difficult to picture if the $f_i(x)$ have many zeros.

18.5 NONLINEAR PARAMETERS

It frequently happens that theory supplies the form that some data are to fit, and the data are to be used to determine the coefficients of the form, using the least-squares criterion. For example, suppose that the given form is

$$y(x) = a + be^{cx} \tag{18.5-1}$$

and we are to determine a, b, c such that for the data (x_i, y_i) $(i = 1, \ldots, N)$

$$\sum_{i=1}^{N} [y_i - (a + be^{cx_1})]^2 = \min = m \tag{18.5-2}$$

When we differentiate with respect to a, b, and c, respectively, and set the result equal to zero, the equations are difficult to solve.

Suppose that we plot the data and make some guess for the value $c = c_1$. We now determine the a and b in the usual way. We can then compute $m(c_1)$ from (18.5-2) for these values.

Next we try some other likely value of $c = c_2$ and again compute $m = m(c_2)$. An examination of these two values

c	$m(c)$
c_1	$m(c_1)$
c_2	$m(c_2)$

will suggest (by linear approximation) a new value of c, say c_3.

In this way we can approach the minimum value of $m(c)$. The details of the strategy for searching will be discussed in a later chapter. It should be evident, however, that we shall use a great deal of machine computation as compared with the situation where *all* the parameters to be determined occur linearly. The more parameters that occur nonlinearly, the very much more the computation required to find the least-squares fit. Experience shows that when the number of nonlinear parameters reaches four or five the process can be exceedingly painful and slow.

CHAPTER 19

Chebyshev Polynomials

19.1 INTRODUCTION

The Fourier series has a number of remarkable properties. Among them are the following:

1. The individual functions are equal-ripple functions, that is, their alternating maxima and minima are all of the same size.
2. The same analytic expressions (sine and cosine) are orthogonal over *both* the continuous and *any* equally spaced set of discrete sample points (Chap. 6).

To see how remarkable the second property is, consider the Legendre polynomials $P_n(x)$. If the first m of these, $m = 0, 1, \ldots, n-1$, are to be orthogonal over a discrete set of n points, there will be $n(n-1)/2$ equations of the form

$$\sum_{i=1}^{n} P_j(x_i)P_k(x_i) = 0 \qquad (j \neq k)$$

to determine the locations of the n sample points x_i (allowing any spacing whatever). For n moderately large, the unlikeliness of this is evident. This is not, of course, to say that there is not some set of polynomials which are orthogonal over the samples, but the members of the set depend, in general, on n, the number of points being used. As n approaches infinity, the members of the set which are orthogonal over the discrete set of points approach the corresponding members of the set orthogonal over the continuous interval.

In the class of orthogonal functions, the subclass of orthogonal polynomials has a number of special properties such as the following:

1. They satisfy a three-term recurrence relation.
2. They are easy to compute and convert to a power-series form.
3. Their zeros interlace each other.

The Chebyshev polynomials have all the properties of both the Fourier series and the orthogonal polynomials; they are, in fact, the

Fourier functions $\cos n\theta$ in the disguise of a simple transformation of the variable

$$\theta = \arccos x \tag{19.1-1}$$

Thus the Chebyshev polynomials naturally play a unique role in the field of orthogonal functions. The Fourier expressions for orthogonality become

$$\int_0^\pi \cos m\theta \cos n\theta \, d\theta = \begin{Bmatrix} 0 & (m \neq n) \\ \frac{\pi}{2} & (m = n) \\ \pi & (m = n = 0) \end{Bmatrix}$$
$$= \int_{-1}^{1} T_m(x) T_n(x) \frac{dx}{\sqrt{1 - x^2}} \tag{19.1-2}$$
$$\sum_{j=0}^{N-1} \cos m\theta_j \cos n\theta_j = \begin{Bmatrix} 0 & (m \neq n) \\ \frac{N}{2} & (m = n) \\ N & (m = n = 0) \end{Bmatrix} = \sum_{j=0}^{N-1} T_m(x_j) T_n(x_j)$$

where we have written[1]

$$T_n(x) = \cos (n \arccos x) \tag{19.1-3}$$

and used the orthogonality of the $\cos nx$ over the interval $(0 \leq x \leq \pi)$ (see Sec. 6.7).

We need to show that the $T_n(x)$ are polynomials. By de Moivre's theorem,

$$\cos n\theta + i \sin n\theta = (\cos \theta + i \sin \theta)^n$$

Expanding the binomial, taking the real parts of both sides, and replacing the even powers of $\sin \theta$ by

$$(\sin^2 \theta)^k = (1 - \cos^2 \theta)^k$$

we see that

$$\cos n\theta = \text{polynomial of degree } n \text{ in } \cos \theta$$

But

$$\cos (\arccos x) = x$$

Hence

$$T_n(x) = \cos (n \arccos x) = \text{polynomial of degree } n \text{ in } x$$

The transformation (19.1-1) can be viewed as the projection of the intersections of a semicircle with a set of lines having equal angles between them (Fig. 19.1-1). Thus the set of points x_i over which the set of Chebyshev polynomials $T_n(x)$ is orthogonal is given by

$$x_j = \cos \frac{\pi}{N} j \qquad (j = 0, 1, \ldots, N - 1) \tag{19.1-4}$$

[1] The notation $T_n(x)$ is due to the old French spelling of Chebyshev as Tschebycheff.

This unequal spacing, which tends to crowd up the x_j at each end of the interval $(-1 \leq x \leq 1)$, compensates for the weight function $(1 - x^2)^{-1/2}$ in the continuous case (19.1-2). Thus we almost have property 2 of the Fourier series; we only give up the equal spacing of the sample points. Since the $T_n(x)$ are $\cos n\theta$ in disguise, they are still equal-ripple functions,

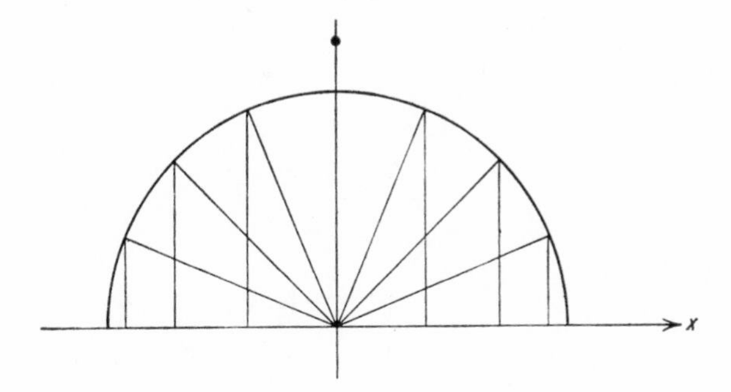

FIG. 19.1-1. The Chebyshev sample points.

and since they are polynomials, they have all the properties of the orthogonal polynomials.

19.2 SOME IDENTITIES

Many of the properties of the Chebyshev polynomials follow from the corresponding identities among the trigonometric functions. For example, the identity

$$\cos (n + 1)\theta + \cos (n - 1)\theta = 2 \cos \theta \cos n\theta$$

becomes, using (19.1-3),

$$T_{n+1}(x) + T_{n-1}(x) = 2xT_n(x) \qquad (n \geq 1) \tag{19.2-1}$$

which is the appropriate three-term recurrence relation (17.10-1).

From the identity

$$\cos (m + n)\theta + \cos (m - n)\theta = 2 \cos m\theta \cos n\theta$$

we get

$$T_{m+n}(x) + T_{m-n}(x) = 2T_m(x)T_n(x)$$

Putting $m = n$, we have

$$T_{2n}(x) = 2T_n^2(x) - 1 \tag{19.2-2}$$

which is occasionally useful for producing a single high-order polynomial.

It is easy to deduce from (19.2-1) that the leading term of $T_n(x)$ is

$$T_n(x) = 2^{n-1}x^n + \cdots \qquad (n \geq 1) \tag{19.2-3}$$
$$T_0(x) = 1 \qquad (n = 0)$$

and that the $T_n(x)$ are of odd or even degree according as n is odd or even.

Since $$T_{n+1} = \cos [(n+1) \arccos x]$$

we have $$\frac{1}{n+1} \frac{dT_{n+1}}{dx} = \frac{-\sin [(n+1) \arccos x]}{-\sqrt{1-x^2}}$$

and $$\frac{1}{n+1} \frac{dT_{n+1}}{dx} - \frac{1}{n-1} \frac{dT_{n-1}}{dx} = \frac{T'_{n+1}}{n+1} - \frac{T'_{n-1}}{n-1}$$

$$= \frac{\sin (n+1)\theta - \sin (n-1)\theta}{\sin \theta}$$

$$= \frac{2 \cos n\theta \sin \theta}{\sin \theta} = 2T_n \qquad (19.2\text{-}4)$$

If we write this in the form

$$\frac{T'_n}{n} = 2T_{n-1} + \frac{T'_{n-2}}{n-2} \qquad (n > 2)$$

we are led to

$$\frac{T'_{2n+1}}{2n+1} = 2(T_{2n} + T_{2n-2} + T_{2n-4} + \cdots + T_2) + 1$$
$$\frac{T'_{2n}}{2n} = 2(T_{2n-1} + T_{2n-3} + \cdots + T_1) \qquad (19.2\text{-}5)$$

for the odd- and even-order cases, respectively.

EXERCISES

19.2-1. Derive Eq. (19.2-5).
19.2-2. From (19.2-1), prove

$$x^k T_n(x) = \frac{T_{n+k} + C(n > k)T_{n+k-2} + C(k,2)T_{n+k-4} + \cdots + T_{n-k}}{2^k}$$

19.2-3. Prove

$$\frac{T_{2k+1}(x)}{x} = 2T_{2k} - 2T_{2k-2} + 2T_{2k-4} + \cdots \pm T_0$$

19.3 THE CHEBYSHEV CRITERION

Chebyshev showed that of all polynomials $p_n(x)$ of degree n, having a leading coefficient equal to 1, the polynomial

$$\frac{T_n(x)}{2^{n-1}}$$

has the smallest least upper bound for its absolute value in the interval $(-1 \le x \le 1)$. Since the upper bound of $|T_n(x)|$ is 1, then this upper bound is $1/2^{n-1}$.

The proof of this remarkable property follows from an examination of the difference

$$\varphi_{n-1}(x) = \frac{T_n(x)}{2^{n-1}} - p_n(x)$$

which is a polynomial of degree $n - 1$ {since x^n drops out [see Eq. (19.2-3)]}. From the fact that $T_n(x)$ is $\cos n\theta$ in disguise, we see that in the interval $T_n(x)$ takes its extreme value $n + 1$ times, alternately positive and negative. If the extreme value of $p_n(x)$ is less than that of $T_n(x)/2^{n-1}$, then at these $n + 1$ extreme points $\varphi_{n-1}(x)$ is alternately positive and negative; hence it must have n real zeros between these extremes. Since $\varphi_{n-1}(x)$ is of degree $n - 1$, we conclude that $\varphi_n(x) \equiv 0$ and

$$p_n = \frac{T_n(x)}{2^{n-1}}$$

This property is of great interest in numerical computation. If some error can be expressed as a Chebyshev polynomial of degree n, then any other form of the error which is a polynomial of degree n and having the same leading coefficient will have, in the range $(-1 \leq x \leq 1)$, a greater extreme error than the Chebyshev error representation. Thus the expression "Chebyshev approximation" is associated with those approximations[1] which try to keep the maximum error down to a minimum. This is sometimes called the *minimax principle.* Least-squares approximation keeps the *average* square error down, but in doing so isolated extreme errors are permitted; Chebyshev keeps the extreme errors down but allows a larger average square error.

As a simple illustration, consider the problem of interpolating in the interval $(-1 \leq x \leq 1)$ with an exact-matching polynomial of degree n. The error term (8.6-1) has the form

$$\frac{(x - x_1)(x - x_2) \cdots (x - x_{n+1})y^{(n+1)}(\bar{x})}{(n + 1)!}$$

If we want to minimize the maximum error due to the factor

$$(x - x_1)(x - x_2) \cdots (x - x_{n+1})$$

we can choose the sample points as the zeros of $T_{n+1}(x) = 0$. Thus we have for the part of the error that we can control easily the equal-ripple polynomial which has the least maximum error.

19.4 ECONOMIZATION

Another simple but very important illustration of the use of the Chebyshev approach is that of "economization," due mainly to Lanczos.

[1] Chebyshev approximation should not be confused with Chebyshev integration; he had many good ideas, some of which have been named after him.

Suppose that we have a truncated power series of a function

$$f(x) = a_0 + a_1x + a_2x^2 + \cdots + a_Nx^N \tag{19.4-1}$$

in the interval $(-1 \le x \le 1)$. For a power series the error tends to be large at the ends of the interval and small in the middle. As the first step in the process of economization, we use the table

$$\begin{aligned} 1 &= T_0 \\ x &= T_1 \\ x^2 &= \tfrac{1}{2}(T_0 + T_2) \\ x^3 &= \tfrac{1}{4}(3T_1 + T_3) \\ x^4 &= \tfrac{1}{8}(3T_0 + 4T_2 + T_4) \\ x^5 &= \tfrac{1}{16}(10T_1 + 5T_3 + T_5) \\ &\cdots\cdots\cdots\cdots\cdots \end{aligned} \tag{19.4-2}$$

and convert the power series to an expansion in Chebyshev polynomials:

$$f(x) = b_0 + b_1T_1(x) + b_2T_2(x) + \cdots + b_NT_N(x) \tag{19.4-3}$$

This is an expansion in orthogonal polynomials. For a broad class of functions the expansion in Chebyshev polynomials converges more rapidly than the expansion in any other set of orthogonal polynomials (see Sec. 19.5 for the reason). Thus we expect to find the b_k of (19.4-3) becoming small very rapidly as compared with the a_k of (19.4-1).

To illustrate this, consider the trivial example (trivial so that it can be followed easily)

$$y = \ln(1 + x) \simeq x - \frac{x^2}{2} + \frac{x^3}{3} - \frac{x^4}{4} + \frac{x^5}{5}$$

Using (19.4-2), we get

$$\begin{aligned} y &= T_1 - \tfrac{1}{4}(T_0 + T_2) + \tfrac{1}{12}(3T_1 + T_3) - \tfrac{1}{32}(3T_0 + 4T_2 + T_4) \\ &\qquad + \tfrac{1}{80}(10T_1 + 5T_3 + T_5) \\ &= -\tfrac{11}{32}T_0 + \tfrac{11}{8}T_1(x) - \tfrac{3}{8}T_2(x) + \tfrac{7}{48}T_3(x) - \tfrac{1}{32}T_4(x) \\ &\qquad + \tfrac{1}{80}T_5(x) \end{aligned} \tag{19.4-4}$$

In the interval $(0 \le x \le 1)$, dropping the last term of the power series (19.4-1) would produce a change of $\tfrac{1}{5}$ (for $x = 1$), while dropping the last *three* terms of (19.4-4) would produce a change of less than

$$\frac{7}{48} + \frac{1}{32} + \frac{1}{80} = \frac{91}{15 \cdot 32} < \frac{1}{5.2}$$

since we know that $|T_k(x)| \le 1$ for $(-1 \le x \le 1)$. Thus we can represent

$$y = \ln(1 + x) \simeq -\tfrac{11}{32}T_0 + \tfrac{11}{8}T_1(x) - \tfrac{3}{8}T_2(x)$$

as a quadratic almost as accurately in the Chebyshev form as we can as a quartic truncated power series. The Chebyshev expansion may be converted back to a polynomial, using the table

$$\begin{aligned} T_0 &= 1 \\ T_1 &= x \\ T_2 &= 2x^2 - 1 \\ T_3 &= 4x^3 - 3x \\ T_4 &= 8x^4 - 8x^2 + 1 \\ T_5 &= 16x^5 - 20x^3 + 5x \\ &\cdots\cdots\cdots \end{aligned} \tag{19.4-5}$$

Thus

$$\begin{aligned} y = \ln(1+x) &\simeq -\tfrac{11}{32} + \tfrac{11}{8}(x) - \tfrac{3}{8}(2x^2 - 1) \\ &\simeq \frac{1}{32} + \frac{11x}{8} - \frac{3x^2}{4} \qquad (0 \le x \le 1) \end{aligned}$$

In general, we may expect that, if we take a power series consisting of many terms and convert it to a Chebyshev expansion, then we can obtain a much lower-order polynomial approximation by dropping many of the later Chebyshev terms without greatly increasing the error over that of the error due to originally taking a finite number of terms in the power series.

19.5 MECHANIZATION OF ECONOMIZATION

The process of economization as described used two tables, (19.4-2) and (19.4-5). The first table is really the well-known trigonometric identity in disguise:

$$\begin{aligned} (\cos\theta)^k &= \left(\frac{e^{i\theta} + e^{-i\theta}}{2}\right)^k \\ &= \frac{1}{2^{k-1}}\left[\frac{e^{ik\theta} + e^{-ik\theta}}{2} + C(k,1)\,\frac{e^{i(k-2)\theta} + e^{-i(k-2)\theta}}{2}\right. \\ &\qquad\left. + C(k,2)\,\frac{e^{i(k-4)\theta} + e^{-i(k-4)\theta}}{2} + \cdots\right] \end{aligned}$$

ending with

$$\begin{aligned} &C(k,m)\cos\theta && \text{if } k = 2m + 1 \\ &\tfrac{1}{2}C(k,m) && \text{if } k = 2m \end{aligned}$$

We can also construct the first table using (19.2-1) for each line. The second table can be constructed from the three-term recurrence relation (19.2-1) used in the opposite fashion.

However, instead of using tables, it is probably better to program the process in a simple way. One such approach is based on the substitution

$x = \cos\theta$ in the original power series

$$\begin{aligned} y &= \sum_{k=0}^{N} a_k x^k = \sum_{k=0}^{N} a_k \cos^k \theta \\ &= a_0 + \cos\theta\{a_1 + \cos\theta[a_2 + \cdots + (a_{N-1} + \cos\theta \cdot a_N) \cdots]\} \end{aligned}$$

Starting with the trivial Fourier series in the parentheses

$$a_{N-1} + a_N \cos\theta$$

we multiply by $\cos\theta$ to get

$$\tfrac{1}{2}a_N + a_{N-1}\cos\theta + \tfrac{1}{2}a_N \cos 2\theta$$

which is again a Fourier series.

In general, if we have a Fourier series at the kth stage

$$\alpha_0^{(k)} + \alpha_1^{(k)} \cos\theta + \alpha_2^{(k)} \cos 2\theta + \cdots + \alpha_k^{(k)} \cos k\theta$$

and multiply by $\cos\theta$, we get

$$\frac{\alpha_1^{(k)}}{2} + \left(\alpha_0^{(k)} + \frac{\alpha_2^{(k)}}{2}\right)\cos\theta + \left(\frac{\alpha_1^{(k)}}{2} + \frac{\alpha_3^{(k)}}{2}\right)\cos 2\theta + \cdots + \frac{\alpha_k^{(k)}}{2}\cos(k+1)\theta$$

which is again a Fourier series but contains one higher frequency. In words, the constant coefficient of the kth stage appears in the $(k + 1)$st stage as part of the coefficient of $\cos\theta$. Each coefficient except the first is divided by 2 and appears both in the coefficient of the frequency just below and in the frequency just above where it was. This simple routine suffices to compute the coefficients of the Chebyshev expansion from the power series expansion. This process shows that

$$b_N = \frac{a_N}{2^{N-1}}$$

Speaking loosely, if we view the coefficients as masses, then the process has the property of the conservation of mass. A closer examination of the process shows why the b_k's for large k tend to zero more rapidly than the a_k's. This is the justification for the assertion that the Chebyshev expansion tends to converge more rapidly than does the power series, but it is not a proof of it.

The whole process can be deduced directly from the recurrence relation (19.2-1) used in the form

$$\begin{aligned} xT_n &= \tfrac{1}{2}T_{n+1} + \tfrac{1}{2}T_{n-1} \qquad & n \geq 1 \\ xT_0 &= T_1 & n = 0 \end{aligned}$$

but our derivation has also shown how to go from an expansion in powers of cos θ to the corresponding Fourier expansion.

To proceed with the economization process, we now examine the coefficients b_k of the Chebyshev expansion and drop all those whose sum is less than the allowable error (in addition to the error of the truncated power series).

To convert back to the polynomial form, we reverse the above process. The ability to reverse the process depends on the observation that the highest frequency term comes from only one place in the previous line of computation. Starting at the top frequency of the $(k + 1)$st line, we can compute the top frequency of the kth line. Note that this modifies the frequency two below the top of the $(k + 1)$st line. We then take the next lower frequency and progress down the line. (See Fig. 19.5-1.) The last step determines the corresponding $a_0^{(k)}$ of the power series expansion.

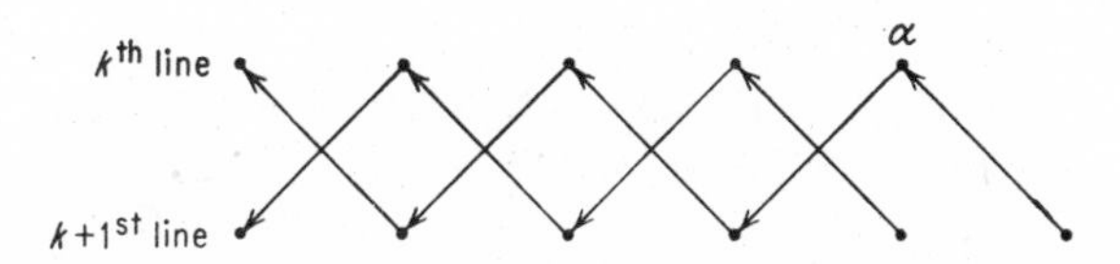

FIG. 19.5-1. Process for converting to the polynomial form.

In both the forward and backward steps the sum of the coefficients is preserved during each cycle, and this can be used as a check on roundoff, since

$$\sum_{k=0}^{N} a_k = \sum_{k=0}^{N} b_k$$

as can be seen by putting $x = 1$ $[T_k(1) = 1]$.

Thus we see that the economization process is easily mechanized without the use of extensive tables.

EXERCISE

19.5-1. Draw a flow diagram for the process of economization.

19.6 THE SHIFTED CHEBYSHEV POLYNOMIALS

It is often convenient to use the interval $(0 \leq x \leq 1)$ instead of the interval $(-1 \leq x \leq 1)$. For this purpose the *shifted Chebyshev polynomials*

$$T_n^*(x) = T_n(2x - 1) \tag{19.6-1}$$

are used. Thus we have

$$\begin{aligned}
T_0^*(x) &= 1 \\
T_1^*(x) &= 2x - 1 \\
T_2^*(x) &= 8x^2 - 8x + 1 \\
T_3^*(x) &= 32x^3 - 48x^2 + 18x - 1 \\
T_4^*(x) &= 128x^4 - 256x^3 + 160x^2 - 32x + 1 \\
&\cdots
\end{aligned}$$

and

$$\begin{aligned}
1 &= T_0^* \\
x &= \tfrac{1}{2}(T_0^* + T_1^*) \\
x^2 &= \tfrac{1}{8}(3T_0^* + 4T_1^* + T_2^*) \\
x^3 &= \tfrac{1}{32}(10T_0^* + 15T_1^* + 6T_2^* + T_3^*) \\
x^4 &= \tfrac{1}{128}(35T_0^* + 56T_1^* + 28T_2^* + 8T_3^* + T_4^*) \\
&\cdots
\end{aligned}$$

More extensive tables are given in the National Bureau of Standards.[1]

The recurrence relation for the shifted polynomials is

$$T_{n+1}^*(x) = (4x - 2)T_n^*(x) - T_{n-1}^*(x) \qquad T_0^* = 1$$

or

$$xT_n^*(x) = \tfrac{1}{4}T_{n+1}^*(x) + \tfrac{1}{2}T_n^*(x) + \tfrac{1}{4}T_{n-1}^*(x) \qquad (19.6\text{-}2)$$

where

$$T_n^*(x) = \cos\,[n \arccos\,(2x - 1)] = T_n(2x - 1) \qquad (19.6\text{-}3)$$

EXERCISES

19.6-1. Apply the shifted polynomials T_n^* to the example in Sec. 19.4. (You will have to extend the tables of T_n^* and x^n one line in the process.)

19.6-2. Discuss the mechanization for the shifted polynomials $T_n^*(x)$ along the lines of Sec. 19.4. [Use Eq. (19.6-2).]

19.6-3. Find the equivalent of Eqs. (19.2-5) for the shifted polynomials.

19.7 LANCZOS' τ PROCESS

As still another example of the use of Chebyshev polynomials, we briefly examine the τ process devised by Lanczos.

The idea behind the τ process for solving a linear differential equation with polynomial coefficients is simple but important. Usually it is difficult to find the error of a computed solution of a problem, but the reverse question "What slightly changed problem would have the answer that I just computed as its correct solution?" is frequently easy to answer. This is one answer to the fourth basic question of Sec. 7.1, "What accuracy?", and many times it is the correct one.

To illustrate the τ method, suppose that we have the simple differential equation

$$y' + y = 0 \qquad y(0) = 1$$

[1] Tables of Chebyshev Polynomials $S_n(x)$ and $C_n(x)$, *Natl. Bur. Standards (U.S.), Appl. Math. Series* 9, 1952.

and want a polynomial solution; that is, we expect to terminate the power series solution. Evidently no polynomial will exactly satisfy the equation. Let us consider changing the right-hand side slightly by the addition of a Chebyshev polynomial, using τ as the size of the maximum change,

$$y' + y = \tau T_n^*(x)$$

To keep the computation easy to follow, we set $n = 4$ and try

$$y = a + bx + cx^2 + dx^3 + ex^4$$

We get, equating coefficients of like powers of x,

$$\begin{aligned} b + a &= \tau \\ 2c + b &= -32\tau \\ 3d + c &= 160\tau \\ 4e + d &= -256\tau \\ e &= 128\tau \end{aligned}$$

The initial condition fixes $a = 1$. Hence, in turn, we obtain

$$\begin{aligned} b &= \tau - 1 \\ c &= \frac{1 - 33\tau}{2} \\ d &= \frac{1}{3}\left(160\tau + \frac{33\tau - 1}{2}\right) = \frac{353\tau - 1}{6} \\ e &= \frac{1}{4}\left(-256\tau + \frac{1 - 353\tau}{6}\right) = \frac{1}{24}(1 - 1{,}889\tau) \\ e &= 128\tau = 1/24(1 - 1{,}889\tau) \\ \tau &= \frac{1}{4{,}961} \end{aligned}$$

Thus the error that we have put in the original equation is

$$\left|\frac{T_4^*(x)}{4{,}961}\right| \leq \frac{1}{4{,}961}$$

and for this equation we have an exact solution

$$y = 1 - \frac{4{,}960}{4{,}961}x + \frac{2{,}464}{4{,}961}x^2 - \frac{768}{4{,}961}x^3 + \frac{128}{4{,}961}x^4$$

as compared with the series solution

$$y = 1 - x + \frac{x^2}{2} - \frac{x^3}{6} + \frac{x^4}{24}$$

whose maximum error is about

$$\frac{1}{5!} = \frac{1}{120}$$

In this case it is easy to compute a bound on the error due to the addition of the Chebyshev term to the differential equation

$$y' + y = \tau T_4^*(x)$$

$$y(x) = e^{-x} \cdot 1 + \tau e^{-x} \int_0^x T_4^*(\theta) e^{\theta} \, d\theta$$

For $(0 \leq x \leq 1)$

$$|y - e^{-x}| \leq \left| \tau e^{-x} \int_0^x T_4^*(\theta) e^{\theta} \, d\theta \right|$$

$$\leq \tau e^{-x} \int_0^x e^{\theta} \, d\theta = \tau(1 - e^{-x}) \leq \frac{0.665}{4{,}961} \simeq 1.34 \times 10^{-4}$$

which is much less than $1/5!$.

We have illustrated the τ method with a simple example. Lanczos' book [23] gives a much more extensive discussion of the method. However, the main idea is the same; we slightly alter the problem and obtain an exact solution of the new problem. The alteration being in the domain of the physics of the original problem, we can usually understand the change more readily than we can some approximation made in the middle of some numerical computation. The τ method illustrates the importance of carefully answering the fourth basic question, "What accuracy?"

EXERCISE

19.7-1. Apply the τ method to

$$xy'' + y = 0 \qquad \begin{cases} y(0) = 0 \\ y'(0) = 1 \end{cases}$$

for $(0 \leq x \leq 1)$, using $n = 4$.

19.8 A VARIATION OF THE τ METHOD

Instead of proceeding as Lanczos did, suppose that we try the direct approach and represent the solution of the differential equation (having polynomial coefficients) as a series in Chebyshev polynomials.

$$y(x) = \sum_{k=0}^{\infty} a_k T_k(x) \tag{19.8-1}$$

Then

$$y'(x) = \sum_{k=0}^{\infty} a_k T_k'(x)$$

and we now use (19.2-5)

$$y'(x) = \sum_{k=0}^{\infty} ka_k \left[2T_{k-1} + 2T_{k-3} + \cdots + \binom{2T_1}{1} \right] \tag{19.8-2}$$

and similar expressions for higher derivatives, to eliminate all the derivatives.

We next use (19.2-1)

$$xT_n(x) = \tfrac{1}{2}[T_{n+1}(x) + T_{n-1}(x)] \tag{19.8-3}$$

and similar expressions for $x^k T_n(x)$ (see Exercise 19.2-2) to eliminate all powers of x. Thus we are reduced to a series in Chebyshev polynomials $T_n(x)$.

The method of equating coefficients of the same powers of x on both sides of an identity in x rests on the fact that the various powers of x are linearly independent. Once we have eliminated the derivatives and terms in x so that we have expansions in $T_n(x)$ *only*, then we may similarly equate coefficients of the same Chebyshev polynomials. We therefore proceed exactly as we would were we dealing with a power series representation of the solution. Whenever we decide to truncate the series, then we have an error that is approximately the first Chebyshev polynomial which we neglect.

We can make the truncated series an exact solution by adding the proper Chebyshev polynomials to the right-hand side to produce exact cancellation. The result is the same as with the original τ method, but we have much more flexibility in deciding when to truncate the solution.

We have discussed the process in terms of the standard Chebyshev polynomials, but we could just as well have discussed it in terms of the shifted polynomials $T_n^*(x)$.

We have included what amounts to two treatments of the same τ method to illustrate the point that, if we want the error in the form of a Chebyshev polynomial, then it is probably best to start by representing the solution as a series in Chebyshev polynomials, rather than trying a mixture of standard polynomials and Chebyshev polynomials. Identities among the polynomials can usually be found to do what is needed.

EXERCISES

19.8-1. Discuss the variation of the τ method using the shifted Chebyshev polynomials $T_n^*(x)$.

19.8-2. Show that

$$\int T_n(x)\,dx = \frac{1}{2}\left[\frac{T_{n+1}(x)}{n+1} - \frac{T_{n-1}(x)}{n-1}\right] \qquad (n > 1)$$

$$\int T_1(x)\,dx = \tfrac{1}{4}T_2(x)$$

$$\int T_0(x)\,dx = T_1(x)$$

so that integration can be done directly in terms of Chebyshev polynomials.

19.8-3. Carry out the details of the example $y' = -y''$ in Sec. 19.7 by the direct τ method (Sec. 19.8).

19.8-4. Apply the direct method to Exercise 19.7-1.

19.9 SOME REMARKS ON CHEBYSHEV APPROXIMATION

Chebyshev approximation has caused some stir in table-making circles where there is great emphasis on having exact knowledge of the maximum error. Thus, instead of using exact-matching polynomials for interpolating, it has been proposed to make tables using Chebyshev polynomials for interpolation between entries.

Chebyshev approximation and least-squares approximation both can be brought into a common framework of ideas. Suppose that we minimize

$$\min_k = \left(\sum_{i=1}^{N} |\epsilon_i|^k\right)^{1/k}$$

For $k = 2$ this is least squares; for $k = \infty$ it is Chebyshev. This provides an alternative approach to Chebyshev approximation. We first find a least-squares fit. Then, using the square of the error as a weight function, we again find a least-squares fit. The fourth power of the new error we use as a weight factor and repeat. In this way we gradually approximate m_∞.

The expansion in Chebyshev polynomials gives a weighted least-squares fit. If the convergence of the expansion is rapid enough, then the first term neglected can be viewed as the error of a Chebyshev fit. In any case, the expansion provides a first approximation to a Chebyshev fit.

19.10 THE CRITERION OF MATCHING MOMENTS

We have introduced three criteria so far for selecting from the class of polynomials a particular function to be used in analytic substitution:

1. Exact matching at the sample points
2. Least squares
3. Chebyshev, or minimax

Another criterion, which we used in Sec. 2.9 and which is widely used in statistics, is that of fitting the approximating function by matching some of the moments of the distribution. In statistics the moments (above the first) are usually measured about the mean (the first moment), but if all the moments match up to the kth, then the moments about the mean also match, and so we get the same result whether we match "moments" or "moments about the mean," although the roundoff errors may be quite different.

This concept provides an excellent example of the way to choose a criterion. In statistics moments are often used; hence it is easy to see why matching moments in an analytic substitution of one function for another is likely to be a good choice in statistical problems.

As an example of the method of matching moments, consider approximating

$$y(x) = \frac{1}{\sigma\sqrt{2\pi}} e^{-\frac{1}{2}(x^2/\sigma^2)}$$

by a quadratic polynomial. Generally $y(x)$ represents a probability distribution with, of course,

$$\int_{-\infty}^{\infty} y(x)\,dx = 1$$

If we were to take the first two terms of a power series expansion in an exact-matching polynomial (only in the region where it is above the x axis), then we would not have a total probability of 1 for the approximation. Thus we are led to abandon a power series approximation and to select the parabola

$$\bar{y}_0 = A(1 - b^2x^2) \qquad \left(-\frac{1}{b} \le x \le \frac{1}{b}\right)$$

by matching moments.

The moments of the original distribution

$$\frac{1}{\sigma\sqrt{2\pi}} \int_{-\infty}^{\infty} x^k e^{-\frac{1}{2}(x^2/\sigma^2)}\,dx = m_k$$

give

$$m_0 = 1$$
$$m_1 = 0$$
$$m_2 = \sigma^2$$

For the approximation we get

$$\bar{m}_k = A\int_{-1/b}^{1/b} x^k(1 - b^2x^2)\,dx = A\left(\frac{x^{k+1}}{k+1} - \frac{b^2x^{k+3}}{k+3}\right)\Bigg|_{-1/b}^{1/b}$$
$$= \frac{2A[1 + (-1)^k]}{b^{k+1}(k+1)(k+3)}$$

Hence

$$\bar{m}_0 = \frac{4A}{3b}$$

$$\bar{m}_1 = 0$$

$$\bar{m}_2 = \frac{4A}{15b^3}$$

We therefore have

$$1 = \frac{4A}{3b}$$

$$\sigma^2 = \frac{4A}{15b^3}$$

Hence

$$b = \frac{1}{(\sqrt{5})\sigma}$$

$$A = \frac{3}{(4\sqrt{5})\sigma}$$

Thus the parabola

$$\bar{y} = \frac{3}{4\sqrt{5}\,\sigma}\left[1 - \left(\frac{x^2}{5\sigma^2}\right)\right]$$

is the approximation to the normal error distribution which has the same zero, first, and second moments.

CHAPTER 20

Rational Functions

20.1 INTRODUCTION

Up to this point in Part II we have assumed that between the sample points the function behaved as a polynomial. In this chapter we shall assume that the function behaves as the quotient of two polynomials, which is usually called a *rational function*. Rational functions have the two fundamental properties of still being a rational function under translation and scale change of the independent variable.

Rational functions have a number of properties that are frequently needed for an approximation leading to the analytic substitution of a new function in place of the given one. Most noticeable is the fact that rational functions can approximate functions having infinite values y_i for finite values x_i. Another feature is that they can approximate straight lines, especially the x axis, for large values of x_i, which is something that nontrivial polynomials cannot do.

Inside a computer, rational functions are easy and rapid to evaluate; hence they are often used to provide a computable approximation to some more difficult function. For such purposes a Chebyshev type of minimax fit is likely to be needed.

The theory of approximation by rational functions is in a confused, rapidly evolving state, and we shall give only the simplest approach. One reason for not going farther is that, in computing, the average person is not likely to do many rational approximations in a year, and more efficient methods, which save a few seconds of a fast computer, do not seem to be worth the time in an elementary course.

20.2 THE DIRECT APPROACH

Suppose that we have either a function from which we can compute some samples (x_i,y_i) or the samples themselves, and we wish to fit a rational function to the data. Thus we want

$$y = f(x) \simeq \frac{N(x)}{D(x)}$$

where $N(x)$ and $D(x)$ are polynomials.

The first step is to decide on the polynomial form to try. In one particular case, plotting

$$\log y_i \text{ versus } \log x_i$$

for small $x_i > 0$ produced a slope of approximately 3, whence we deduced that we should try

$$y = f(x) \simeq \frac{ax^3 + \cdots}{1 + \cdots}$$

For large x the original data fell on what appeared to be a straight line of positive slope. This determined that the degree of $N(x)$ should be 1 greater than the degree of $D(x)$. Let the degree of $D(x)$ be k. Then

$$y = f(x) = \frac{a_3x^3 + a_4x^4 + \cdots + a_{k+1}x^{k+1}}{1 + b_1x + b_2x^2 + \cdots + b_kx^k} \tag{20.2-1}$$

It remains to choose k. It is hard to say just how to choose k, except by noting the number of parameters that will be available and comparing that (subjectively) with the "complexity" of the data.

When k is chosen, we then know the number of parameters M. Note that we must fix one coefficient because only the ratio of $N(x)$ and $D(x)$ occurs; hence one coefficient (known to be not zero) must be chosen as a fixed number.[1] We either select or compute the appropriate number M of samples, located where they seem to be most important. We then have

$$D(x_i)y_i = N(x_i) \qquad (i = 1, \ldots, M)$$

as a set of M equations to be solved. In the example [Eq. (20.2-1)] we have

$$y_i + b_1x_iy_i + b_2x_i^2y_i + \cdots + b_kx_i^ky_i = a_3x_i^3 + a_4x_i^4 + \cdots + a_{k+1}x_i^{k+1} \tag{20.2-2}$$

for $(i = 1, 2, \ldots, 2k - 1)$.

These are linear equations in the unknowns a_i and b_i and can be solved by a library routine.

The point is often raised that such a system is of high degree. As far as can be determined, seldom does the number of unknowns exceed 10 or 12, and thus the solution involves no more than 1,000 to 2,000 operations at most; this is hardly serious on a fast machine, if it is to be done once in a problem.

[1] In practice, it is usually best to select either the highest power of the numerator or of the denominator as 1, since this saves one multiplication when evaluating the rational function.

It is well to compute many values of the approximating function before accepting it. Once in the author's experience the denominator $D(x)$ had a real zero between two sample points, thus producing an infinity in the approximation. Between the same two sample points the numerator $N(x)$ also had a real zero, thus masking the behavior. If the approximation is not satisfactory, then the following are all possible changes to try:

1. Moving the sample points
2. Changing the form
3. Changing k

Which one to try depends on why the approximation is not satisfactory.

EXERCISES

20.2-1. What form would you assume for a rational function if $y(x) \geq 0$ and $y(x) \to 0$ as $x \to \infty$?

20.2-2. What form would you assume for a rational function if $y(x)$ had a zero at $x = a$? A zero of $D(x)$ at $x = b$? Both?

20.3 CHEBYSHEV APPROXIMATION BY RATIONAL FUNCTIONS

As indicated in Sec. 20.1, frequently a rational function is used as an easily computable approximation to a transcendental function. When such an approximation is to be used many times, either in some frequently recurring problem or in a special-purpose computer, then a Chebyshev, equal-ripple (or minimax) approximation is likely to be desired.

We begin by choosing a set of samples x_i and compute the corresponding y_i. From these we find the corresponding rational function of the appropriate form (as in Sec. 20.2).

We then plot the error curve by computing the error at many points. For intervals where the error is large we want to move the sample points closer together, and where the local extreme error is small we want to move the samples farther apart. With this new choice of sample points, we repeat the process. Seldom are as many as 10 trials necessary to come quite close to an equal-ripple error curve. The speed of the approach to the equal-ripple error curve depends, among other things, on the method used for respacing the samples.

We offer here no proof that this heuristic method will work, but experience indicates that a number of simple schemes work reasonably well. A little "low cunning" in arranging the computation is usually necessary to avoid the loss of too many figures.

*20.4 RECIPROCAL DIFFERENCES

In the previous section we used an iterative scheme which required a number of trials, each of which involved the solution of the system of linear equations (20.2-2). This suggests that we examine the possibilities of systematizing the solution.

Following Milne,[1] we consider the special case

$$y = \frac{a_0 + a_1x + a_2x^2 + a_3x^3}{b_0 + b_1x + b_2x^2}$$

which uses six data points (x_i,y_i) $(i = 1, \ldots, 6)$. Multiplying, we get

$$a_0 - b_0y + a_1x - b_1xy + a_2x^2 - b_2x^2y + a_3x^3 = 0$$

Using the method that we used to find (8.2-3), it is easy to see that the determinant

$$\begin{vmatrix} 1 & y & x & xy & x^2 & x^2y & x^3 \\ 1 & y_1 & x_1 & x_1y_1 & x_1^2 & x_1^2y_1 & x_1^3 \\ 1 & y_2 & x_2 & x_2y_2 & x_2^2 & x_2^2y_2 & x_2^3 \\ \cdot & \cdot & \cdot & \cdot & \cdot & \cdot & \cdot \\ 1 & y_6 & x_6 & x_6y_6 & x_6^2 & x_6^2y_6 & x_6^3 \end{vmatrix} = 0 \qquad (20.4\text{-}1)$$

is the required solution.

We first reduce the second row to the form 1, 0, 0, 0, 0, 0, 0. We multiply:

1. Column 1 by y_1 and subtract from column 2
2. Column 3 by y_1 and subtract from column 4
3. Column 5 by y_1 and subtract from column 6
4. Column 5 by x_1 and subtract from column 7
5. Column 3 by x_1 and subtract from column 5
6. Column 1 by x_1 and subtract from column 3

After expanding by the second row, we divide:

7. Row 1 by $y - y_1$
8. Row 2 by $y_2 - y_1$
   Row 3 by $y_3 - y_1$
   . . . .
9. Row 7 by $y_6 - y_1$

This division can lead to trouble if any $y_i = y_1$ $(i \neq 1)$ or even if they are close. In practice, the line to reduce to 1, 0, 0, 0, 0, 0, 0 should be chosen with this division process in mind and with the intent to avoid large cancellations in $y_i - y_1$ (and hence division by small, inaccurate numbers).

[1] See Ref. 25. We change the notation somewhat.

We now write

$$\frac{x - x_1}{y - y_1} = \rho_1(x,x_1) \qquad \frac{x_i - x_1}{y_i - y_1} = \rho_1(x_i,x_1)$$

and the determinant (20.4-1) becomes

$$\begin{vmatrix} 1 & \rho_1(x,x_1) & x & x\rho_1(x,x_1) & x^2 & x^2\rho_1(x,x_1) \\ 1 & \rho_1(x_2,x_1) & x_2 & x_2\rho_1(x_2,x_1) & x_2^2 & x_2^2\rho_1(x_2,x_1) \\ \cdot & \cdot & \cdot & \cdot & \cdot & \cdot \\ 1 & \rho_1(x_6,x_1) & x_6 & x_6\rho_1(x_6,x_1) & x_6^2 & x_6^2\rho_1(x_6,x_1) \end{vmatrix} = 0 \qquad (20.4\text{-}2)$$

which is of the same form as before except that it is of sixth order.

We repeat the process. Multiply:

1. Column 1 by $\rho_1(x_2,x_1)$ and subtract from column 2
2. Column 3 by $\rho_1(x_2,x_1)$ and subtract from column 4
3. Column 5 by $\rho_1(x_2,x_1)$ and subtract from column 6
4. Column 3 by x_2 and subtract from column 5
5. Column 1 by x_2 and subtract from column 3

Reduce the determinant and divide:

Row 1 by $\rho_1(x,x_1) - \rho_1(x_2,x_1)$
Row 2 by $\rho_1(x_3,x_1) - \rho_1(x_2,x_1)$
. . . .

Again a wise choice of the row in which we produce 1, 0, 0, 0, 0, 0 can help to keep accuracy. If we write

$$u_x = \frac{x - x_2}{\rho_1(x,x_1) - \rho_1(x_2,x_1)}$$

we get for (20.4-2)

$$\begin{vmatrix} 1 & u_x & x & xu_x & x^2 \\ 1 & u_3 & x_3 & x_3u_3 & x_3^2 \\ 1 & u_4 & x_4 & x_4u_4 & x_4^2 \\ 1 & u_5 & x_5 & x_5u_5 & x_5^2 \\ 1 & u_6 & x_6 & x_6u_6 & x_6^2 \end{vmatrix} = 0 \qquad (20.4\text{-}3)$$

The quantity $\rho_1(x_i,x_j)$ was clearly symmetric in its variables. But this is not true of the quantities that we have labeled u_i. It is customary to use the symmetric quantities

$$\frac{x - x_2}{(x - x_1)/(y - y_1) - (x_2 - x_1)/(y_2 - y_1)} + y_1$$
$$\equiv \frac{x - x_1}{(x - x_2)/(y - y_2) - (x_2 - x_1)/(y_2 - y_1)} + y_2$$
$$\equiv \frac{x_2 - x_1}{(x_2 - x)/(y_2 - y) - (x - x_1)/(y - y_1)} + y$$

and call them $\rho_2(x,x_1,x_2)$

To convert the determinant to this form, multiply:

Column 1 by y_1 and add to column 2
Column 3 by y_1 and add to column 4

We have

$$\begin{vmatrix} 1 & \rho_2(x,x_2,x_1) & x & x\rho_2(x,x_2,x_1) & x^2 \\ 1 & \rho_2(x_3,x_2,x_1) & x_3 & x_3\rho_2(x_3,x_2,x_1) & x_3^2 \\ \cdots & \cdots & \cdots & \cdots & \cdots \\ 1 & \rho_2(x_6,x_2,x_1) & x_6 & x_6\rho_2(x_6,x_2,x_1) & x_6^2 \end{vmatrix} = 0 \tag{20.4-4}$$

The value of this symmetry for practical computing is open to debate, but it makes the theory much easier to follow, and we shall keep it.

If we repeat this process, we finally have

$$\begin{vmatrix} 1 & \rho_5(x,x_5,x_4,x_3,x_2,x_1) \\ 1 & \rho_5(x_6,x_5,x_4,x_3,x_2,x_1) \end{vmatrix} = 0 \tag{20.4-5}$$

Just as Newton's formula was seen to be an identity (Sec. 8.4), so too we can successively write

$$\begin{aligned} y &= y_1 + \frac{x - x_1}{\rho_1(x,x_1)} \\ &= y_1 + \frac{x - x_1}{\rho_1(x_2,x_1) + (x - x_2)/[\rho_2(x,x_2,x_1) - y_1]} \\ &= y_1 + \frac{x - x_1}{\rho_1(x_2,x_1) + (x - x_2)/\rho_2(x_3,x_2,x_1) - y_1 + (x - x_3)/[\rho_3(x,x_3,x_2,x_1) - \rho_1(x_2,x_1)]} \\ &\cdots\cdots\cdots \end{aligned} \tag{20.4-6}$$

In the particular case that we are examining, we end with

$$\frac{x - x_5}{\rho_5(x_6,x_5,x_4,x_3,x_2,x_1) - \rho_3(x_4,x_3,x_2,x_1)}$$

since $\rho_5(x,x_5,x_4,x_3,x_2,x_1) = \rho_5(x_6,x_5,x_4,x_3,x_2,x_1)$

The quantities ρ_i are called reciprocal differences, and it is convenient to think in terms of a reciprocal difference table:

x_1	y_1			
		$\rho_1(x_2,x_1)$		
x_2	y_2		$\rho_2(x_3,x_2,x_1)$	
		$\rho_1(x_3,x_1)$		$\rho_3(x_4,x_3,x_2,x_1)$
x_3	y_3		$\rho_2(x_4,x_2,x_1)$	
		$\rho_1(x_4,x_1)$		
x_4	y_4			
. . . .				

*20.5 AN EXAMPLE

Let us pick the favorite function

$$y = \frac{1}{1 + x^2}$$

and construct a table of values. As with divided differences, we may compute the reciprocal differences of adjacent lines rather than use the first and the kth; we use only the top diagonal values. Thus we get Table 20.5-1. The column of constants (or almost constants) plays the same

TABLE 20.5-1

x	y				
0	1				
		-2			
1	$1/2$		-1		
		$-10/3$		0	
2	$1/5$		$-1/10$		0
		$-50/5$		40	
3	$1/10$		$-1/25$		0
		$-170/7$		140	
4	$1/17$		$-1/46$		0
		$-442/9$		324	
5	$1/26$		$-1/73$		
		$-962/11$			
6	$1/37$				

role as a column of zeros in the normal difference table; it indicates how far to go. We now get from (20.4-6) by direct substitution

$$y = 1 + \cfrac{x - 0}{-2 + \cfrac{x - 1}{-2 + \cfrac{x - 2}{2 + (x - 3)/1}}}$$

which, by reduction from the *continued fraction* form to a rational function, gives

$$y = \frac{1}{1 + x^2}$$

We shall not go farther in this highly developed field but merely refer the reader to standard books such as Milne [25], Milne-Thomson [27], and Hildebrand [14].

PART III

Nonpolynomial Approximation

CHAPTER 21

Periodic Functions—Fourier Approximation

21.1 PURPOSE OF THIS THEORY

The art of computing has three phases in practice: planning, execution, and interpretation. These three phases may be compared with the classical three phases in chess: the opening, the mid-game, and the end game. In chess each of the three phases has a distinct character, although each blends into the next and is determined by the preceding, and on occasions the game may end during mid-game or even in the opening phase. Computing is very much the same.

In the planning phase of a computation the calculations actually made are usually limited to the "back-of-the-envelope type" used to estimate the amount of computing to be done, the time needed on the machine, etc. Typically, during the planning phase, decisions are made such as the following: "We first compute the forcing function at the following set of points; we then replace the integral in the integral equation by a Gauss integration formula using seven points, solve the resulting simultaneous linear equations by the elimination process, combine the resulting solutions according to such and such a formula, and print out all the following results." The planning phase includes estimates of coding effort and time, of checking and debugging plans, of machine time to be used, of when the results will be available, and a description of how the results will be used.

In the normal evolution of a problem there is usually a good deal of feedback between the problem and the computation so that an exact formulation of the entire problem cannot be made before the computing begins; nevertheless, the computer should be approached each time only after careful planning.

In the execution phase the plan is often modified. For example, when the forcing function mentioned is actually computed, the results may suggest that a 10-point Gauss formula should be used rather than the seven-point one, and this may double the amount of labor to solve the resulting system of equations. The solution of the equations may reveal large residuals and an iteration procedure may be added to improve the solutions, thus increasing the coding and running expenses.

In the interpretation phase, not only are the answers to be discussed and explained, but some parts of the answers must be correctly identified as arising from the computation and not from the physical model. Further, the changes in the plan must be explained. The forcing function, when finally "seen" by the analyst, is different from that expected; what does this say about the model? The equations were hard to solve; what does this imply? Does it mean that the results are very sensitive to the particular forcing function, to certain details of the model? Some investigation of this question is necessary.

It is customary in mathematical circles to neglect the first and third phases as not being the proper domain of mathematics. Thus these two phases, especially the third, have been greatly neglected, although they are of major importance to the whole project.

The polynomial methods of computation which were discussed in Part II are not well suited to the planning and interpreting phases. The errors are expressed in terms of high-order derivatives and hence rarely can be estimated accurately in the planning phase, nor do later estimates of them shed much light on the original problem during the interpretation phase.

It is the purpose of this and the next four chapters to develop the elements of a theory of approximation by band-limited functions which, for a broad class of problems, provides models that permit the analyst to act at least in a semi-intelligent manner during the planning and interpreting phases. As a result, realistic estimates of costs and time can be made during the planning phase, and what happens during the computation can shed new light on the physical situation being studied. The theory is not completely developed; at times it involves somewhat more computation than the polynomial model; and it does not answer all questions. However, experience with it over the years suggests that often the model of a band-limited function provides badly needed help both during the planning of the computation and during the interpretation. To repeat the motto of this book:

THE PURPOSE OF COMPUTING IS INSIGHT, NOT NUMBERS

21.2 ALIASING; THE EFFECT OF SAMPLING

In most computation the sampling is done at a set of equally spaced points. Let this spacing be chosen as the unit (of time); that is,

$h = \Delta t = 1$. We are first going to examine the effect of this sampling on sinusoidal functions.

Corresponding to the sinusoid

$$\cos [\pi(n + \epsilon)t + \varphi]$$

there is another sinusoid

$$\cos [\pi(n - \epsilon)t - \varphi]$$

which has exactly the same sample values at the sample points

$$t = 0, 1, 2, \ldots$$

This can be seen from the trigonometric identity

$$\cos [\pi(n + \epsilon)t + \varphi] - \cos [\pi(n - \epsilon)t - \varphi] = -2 \sin (\pi nt) \sin (\pi\epsilon t + \varphi) = 0$$

(for t and n integers). Thus the sampling process confuses the two frequencies $\pi(n + \epsilon)$ and $\pi(n - \epsilon)$—both appear as if they were the same frequency—provided that the phase angles are properly related to each other.

This is a familiar effect to movie and television viewers. In both cases the moving picture is sampled at around 20 frames per second. As the stagecoach wheel in a Western picture starts rolling, the viewer first sees it going faster and faster, but then it begins to slow down, stop, and finally go backward. Further increases in the actual speed cause the wheel to appear as if it again were going forward, slowing down, stopping, reversing, etc.

This well-known *stroboscopic effect* is frequently used to analyze periodic phenomena. Flashes of light which sample the phenomenon are arranged to occur at a frequency slightly less than that of the period of the phenomenon. Thus each sample shows the periodic phenomenon at a slightly later phase of its cycle. There are many frequencies which can be used for the stroboscopic light which show the same results to the viewer. Thus if one frequency shows a certain apparent rate of rotation, approximately one-half that frequency will show the same rate to the human eye, although the quality of the picture suffers somewhat because of a low-frequency flicker.

In both the above examples the human eye sees, as it were, both the x and the y components of motion and can detect some forward and backward rotations as being different. If only one component can be seen, then some forward rotations cannot be distinguished from some backward rotations.

It is customary to refer to this confusion of frequencies as "aliasing"[1]—various different frequencies adopting the name of one particular frequency, which is usually chosen to be the lowest positive frequency of all

[1] A term apparently introduced by J. W. Tukey.

those which are aliased. It is obvious that this aliasing is an inevitable consequence of sampling at equal intervals, is a fundamental fact of computing, *and should not be ignored.*

EXERCISE

21.2-1. Graph the functions $\cos(1.1\pi t)$ and $\cos(0.9\pi t)$ for $0 \le t \le 5$ and note the behavior at the sample points $t = 0, 1, 2, 3, 4, 5$.

21.3 FOURIER SERIES; PERIODIC PHENOMENA

If $y = y(t)$ is a periodic function of t of period $2N$, then the theory of Fourier series states that (see Chap. 6)

$$y(t) \sim \frac{a_0}{2} + \sum_{k=1}^{\infty} a_k \cos\frac{\pi}{N}kt + \sum_{k=1}^{\infty} b_k \sin\frac{\pi}{N}kt \tag{21.3-1}$$

provided that $y(t)$ satisfies some rather weak restraints (weak when viewed from the point of view of computing). The a_k and b_k are defined by

$$\begin{aligned} a_k &= \frac{1}{N}\int_0^{2N} y(t)\cos\frac{\pi}{N}kt\,dt \\ b_k &= \frac{1}{N}\int_0^{2N} y(t)\sin\frac{\pi}{N}kt\,dt \end{aligned} \tag{21.3-2}$$

In Chap. 6 the finite Fourier series based on $2N$ points was developed. When we compute the finite Fourier series based on the $2N$ sample points $0, 1, 2, \ldots, 2N-1$, we get

$$y(t) \sim \frac{A_0}{2} + \sum_{k=1}^{N-1} A_k \cos\frac{\pi}{N}kt + \sum_{k=1}^{N-1} B_k \sin\frac{\pi}{N}kt + \frac{A_N}{2}\cos\pi t \tag{21.3-3}$$

where $$A_k = \frac{1}{N}\sum_{t=0}^{2N-1} y(t)\cos\frac{\pi}{N}kt \qquad B_k = \frac{1}{N}\sum_{t=0}^{2N-1} y(t)\sin\frac{\pi}{N}kt \tag{21.3-4}$$

We shall now show that the following relationships hold between the discrete and the continuous coefficients:

$$\begin{aligned} A_0 &= a_0 + 2\sum_{j=1}^{\infty} a_{2Nj} \\ A_k &= a_k + \sum_{j=1}^{\infty} (a_{2Nj-k} + a_{2Nj+k}) \\ B_k &= b_k + \sum_{j=1}^{\infty} (-b_{2Nj-k} + b_{2Nj+k}) \end{aligned} \tag{21.3-5}$$

To prove these relationships, we first sum both sides of (21.3-1) for $t = 0, 1, 2, \ldots, 2N - 1$. We get

$$\sum_{t=0}^{2N-1} y(t) = \frac{a_0}{2} \sum_{t=0}^{2N-1} 1 + \sum_{k=1}^{\infty} a_k \sum_{t=0}^{2N-1} \cos \frac{\pi}{N} kt + \sum_{k=1}^{\infty} b_k \sum_{t=0}^{2N-1} \sin \frac{\pi}{N} kt$$

If $k \neq 0, 2N, 4N, \ldots$, then

$$\sum_{t=0}^{2N-1} \cos \frac{\pi}{N} kt = 0 \qquad \sum_{t=0}^{2N-1} \sin \frac{\pi}{N} kt = 0$$

If $k = 0, 2N, 4N, \ldots$, then, setting $k = 2Nj$,

$$\sum_{t=0}^{2N-1} \cos \frac{\pi}{N} (2Nj)t = \sum_{t=0}^{2N-1} 1 = 2N$$

$$\sum_{t=0}^{2N-1} \sin \frac{\pi}{N} (2Nj)t = \sum_{t=0}^{2N-1} 0 = 0$$

Hence we have, using (21.3-2) and (21.3-4),

$$2N \frac{A_0}{2} = 2N \frac{a_0}{2} + 2N \sum_{j=1}^{\infty} a_{2Nj}$$

which is the first of Eqs. (21.3-5).

If we now multiply (21.3-1) by $\cos (\pi m/N)t$ $(0 < m < N)$ and sum, we get, in the same way,

$$NA_m = \sum_{k=1}^{\infty} a_k \sum_{t=0}^{2N-1} \cos \frac{\pi k}{N} t \cos \frac{\pi m}{N} t$$

$$= \sum_{k=1}^{\infty} a_k \sum_{t=0}^{2N-1} \frac{1}{2} \left[\cos \frac{\pi(k+m)t}{N} + \cos \frac{\pi(k-m)t}{N} \right]$$

from which the second of Eqs. (21.3-5) follows. The third can be found by multiplying by $\sin (\pi m/N)t$ and summing.

These relationships (21.3-5) between the coefficients of the two expansions, the continuous and the discrete sampled, show clearly the "aliasing" of the frequencies. If we represent the frequencies as points along a line running from zero to infinity, we can represent the aliasing by folding the line back and forth on itself. The first frequency at which a fold occurs is called *the folding frequency*, or the *Nyquist*[1] *frequency*, and is $k = N$. Subsequent folds occur at this same spacing. All points on the curve (Fig. 21.3-1) above the same location on the frequency axis appear as the same frequency owing to the effect of sampling, and *once we have sampled, we cannot again separate the different frequencies that have been aliased together.* At the folding frequency itself, we are sampling

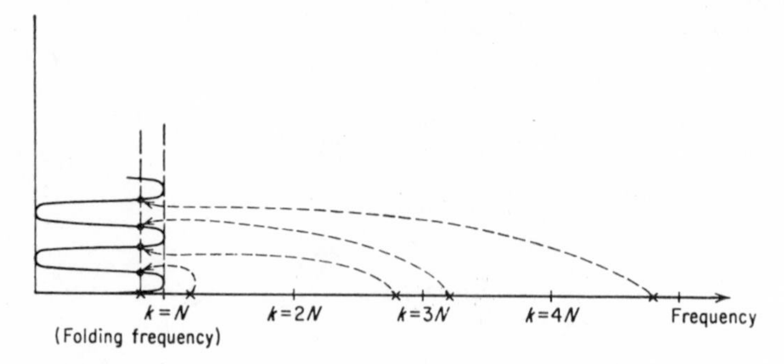

FIG. 21.3-1. Aliasing of frequencies.

at the rate of two samples per cycle, and $\sin \pi t \equiv 0$, $\cos \pi t = (-1)^t$ (t = an integer).

EXERCISE

21.3-1. Carry out the details for the derivation of the third equation of (21.3-5).
21.3-2. Find the equation for A_N in (21.3-5).

21.4 INTERPOLATION OF PERIODIC FUNCTIONS

Part II developed the theory of polynomial approximation, and we now face the problem of developing a similar theory for the Fourier series approximation of a periodic function.

While the problem may be discussed for unequally spaced points, and some results are known, both theory and practice are usually based on equally spaced points. For $2N$ equally spaced sample points the interpolation problem is solved by using the finite Fourier series (Chap. 6) to estimate the function between the sample points.

As in Part II, once we have the formula we next want to know the error of the approximation. The error can be divided into two parts: that due to taking only a finite number of terms in the series and that due to the sampling effect. We have already studied the effects of sampling, and we now examine the effect of taking only a finite number of terms of a Fourier series.

[1] Named after Harry Nyquist, a famous scientist in electrical engineering.

If the continuous Fourier series is of the form (21.3-1), then, substituting the formulas for the coefficients (21.3-2), we get for the first $2N$ terms, where we take one-half the last cosine term as we would for a discrete series,

$$y_{2N}(t) = \frac{1}{N}\int_0^{2N} y(s)\left\{\frac{1}{2} + \sum_{k=1}^{N-1}\left[\cos\frac{\pi}{N}kt\cos\frac{\pi}{N}ks + \sin\frac{\pi}{N}kt\sin\frac{\pi}{N}ks\right] + \frac{1}{2}\cos\pi t\cos\pi s\right\}ds$$

$$= \frac{1}{N}\int_0^{2N} y(s)\left\{\frac{1}{2} + \sum_{k=1}^{N-1}\left[\cos\frac{\pi}{N}k(t-s)\right] + \frac{1}{2}\cos\pi(t-s)\right\}ds$$

In order to sum[1] the quantity in the braces, we multiply by

$$\sin\frac{\pi}{2N}(t-s)$$

and get

$$\begin{aligned}\frac{1}{2}\Bigg\{&\sin\frac{\pi}{2N}(t-s)\\ &+\left[\sin\frac{\pi}{N}\left(1+\frac{1}{2}\right)(t-s) - \sin\frac{\pi}{N}\left(1-\frac{1}{2}\right)(t-s)\right]\\ &+\left[\sin\frac{\pi}{N}\left(2+\frac{1}{2}\right)(t-s) - \sin\frac{\pi}{N}\left(2-\frac{1}{2}\right)(t-s)\right]\\ &\cdots\cdots\cdots\cdots\cdots\cdots\cdots\cdots\\ &+\left[\sin\frac{\pi}{N}\left(N-1+\frac{1}{2}\right)(t-s) - \sin\frac{\pi}{N}\left(N-1-\frac{1}{2}\right)(t-s)\right]\\ &+\sin\frac{\pi}{2N}(t-s)\cos\pi(t-s)\Bigg\}\end{aligned} \tag{21.4-1}$$

Because of cancellation, all that we have left are two terms

$$\frac{1}{2}\left\{\sin\left[\frac{\pi}{N}\left(N-\frac{1}{2}\right)(t-s)\right] + \sin\frac{\pi}{2N}(t-s)\cos\pi(t-s)\right\}$$

We now expand the first of these terms in the form

$$\sin\pi(t-s)\cos\frac{\pi}{2N}(t-s) - \cos\pi(t-s)\sin\frac{\pi}{2N}(t-s)$$

and, because of cancellation of the second of these terms with the second

[1] We could clearly use the methods of Chap. 3 to sum this series more elegantly, but it is likely that the reader has forgotten such details by now, and so we shall use the more clumsy but familiar trigonometric methods.

term that we had left, we have finally for the braces in the integral

$$\frac{\sin \pi(t-s) \cos (\pi/2N)(t-s)}{2 \sin (\pi/2N)(t-s)}$$

Thus $$y_{2N}(t) = \frac{1}{N} \int_0^{2N} y(s) \frac{\sin \pi(t-s) \cos (\pi/2N)(t-s)}{2 \sin (\pi/2N)(t-s)} ds \qquad (21.4\text{-}2)$$

Since $y(s)$ is assumed to be periodic, we can shift the limits to any interval of length $2N$ and set $t - s = \theta$:

$$y_{2N}(t) = \frac{1}{2N} \int_{-N}^{N} y(t-\theta) \frac{\sin \pi\theta \cos [(\pi/2N)\theta]}{\sin [N(\pi/2)\theta]} d\theta \qquad (21.4\text{-}3)$$

If $y(t) \equiv 1$, then we would have

$$1 = \frac{1}{2N} \int_{-N}^{N} \frac{\sin \pi\theta \cos (\pi\theta/2N)}{\sin (\pi\theta/2N)} d\theta$$

Hence, since the variable of integration is θ,

$$y(t) = \frac{1}{2N} \int_{-N}^{N} y(t) \frac{\sin \pi\theta \cos (\pi\theta/2N)}{\sin (\pi\theta/2N)} d\theta \qquad (21.4\text{-}4)$$

and, subtracting (21.4-3) from (21.4-4), we get the error when we use $2N$ terms of the series:

$$\epsilon_{2N} = y(t) - y_{2N}(t) = \frac{1}{2N} \int_{-N}^{N} [y(t) - y(t-\theta)] \frac{\sin \pi\theta \cos (\pi\theta/2N)}{\sin (\pi\theta/2N)} d\theta \qquad (21.4\text{-}5)$$

It is worth noting that most books on Fourier series sum an odd number of terms, and in place of (21.4-5) they get a closely related, less useful form.[1] The form (21.4-5), as it stands, is a bit deceptive; the more samples we take, the longer the range. In practice, of course, the range is fixed, and we change the sampling interval. Hence we make the change of variables and notation (which should cause only a little confusion):

$$\theta = 2N\varphi \qquad t = 2N\tau \qquad y(2Nx) = y(x)$$

$$\epsilon_{2N} = \int_{-1/2}^{1/2} [y(\tau) - y(\tau - \varphi)] \left\{ \frac{\sin 2N\pi\varphi \cos \pi\varphi}{\sin \pi\varphi} \right\} d\varphi \qquad (21.4\text{-}6)$$

The quantity in braces is often called the kernel, $K(\varphi)$. Clearly $K(0) = 2N$. For large N the $\sin (2N\pi\varphi)$ oscillates rapidly while the $\cot \pi\varphi$ term dampens (modulates) the magnitude of the oscillation so that

[1] But see Ref. 44, p. 50, where Zygmund refers to the result we obtain as a "modified kernel."

$K(\pm\frac{1}{2}) = 0$. The envelope of the modulation is $\cot \pi\varphi$ and

$$K(\varphi) = K(-\varphi)$$

If the function $y(x)$ is smooth, then the major contribution to the integral, as far as the kernel is concerned, comes from the immediate neighborhood of $\varphi = 0$, say from $-1/2N$ to $1/2N$ between the first zeros on each side. Thus the error in the approximation tends to be a local property (see Fig. 21.4-1). As N increases, the peak becomes higher and narrower.

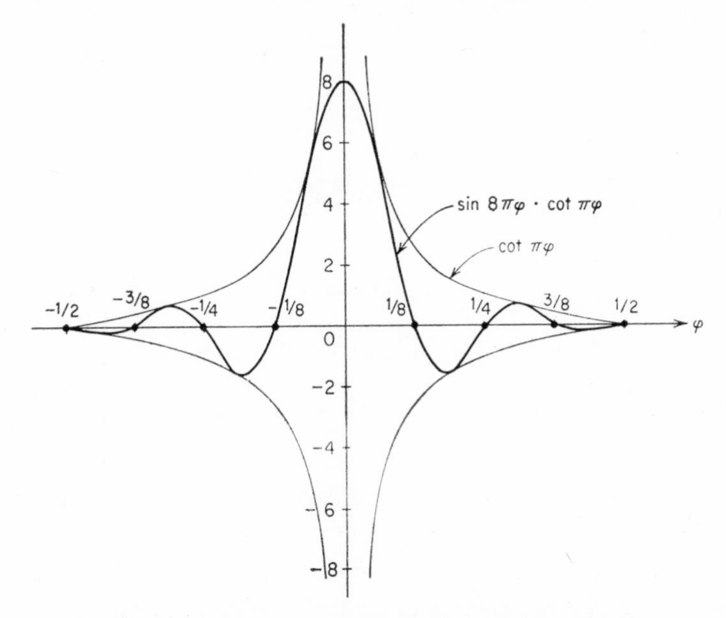

FIG. 21.4-1. The kernel of the Fourier series for $N = 4$.

If, on the other hand, there are rapid changes in $y(x)$, then the tendency of the damped oscillations approximately to cancel out, and thus contribute only small amounts, is no longer true, and sudden changes in $y(x)$ remote from the point of interpolation can produce errors.

Note that the integrand involves the function rather than a high-order derivative as in polynomial approximation.

EXERCISES

21.4-1. We have approximated trigonometric functions by polynomials. Approximate $y(t) = (1 - t^2)^2 \ (-1 \le t \le 1)$ by a Fourier series using $N = 3$, that is, samples at $t = -\frac{2}{3}, -\frac{1}{3}, 0, \frac{1}{3}, \frac{2}{3}, 1$, and interpolate for the value at $t = \frac{1}{2}$.

21.4-2. In this section we have developed the equivalent of Lagrange interpolation; examine the equivalent of Hermite interpolation which matches the function and derivative at each sample point. *Ans.* Not possible

21.5 INTEGRATION

As in Part II, having examined the question of interpolation and its errors, we next turn to the problem of integration. If the coefficients of the continuous Fourier series of $y(t)$ fall off fairly rapidly as $n \to \infty$, then the A_0 coefficient of the finite Fourier series gives a good estimate of the integral

$$\int_0^{2N} y(t)\, dt = N a_0$$

since by (21.3-5)

$$A_0 = \frac{1}{N} \sum_{t=0}^{2N-1} y(t) = a_0 + 2 \sum_{j=1}^{\infty} a_{2Nj}$$

Thus the error involved in estimating the integral from $2N$ equally spaced, equally weighted samples is

$$2 \sum_{j=1}^{\infty} a_{2Nj}$$

and represents the inevitable aliasing due to the discrete sampling.

For periodic functions the formula for integration which uses $2N$ equally spaced, equally weighted samples corresponds to the following:

Newton-Cotes, because it is equally spaced
Chebyshev, because it is equally weighted
Gauss, because $2N$ samples give exactness for $4N$ functions

Let us illustrate the accuracy of this formula for integrating periodic functions by a pair of examples. Consider the two elliptic integrals

$$I_1 = \int_0^{\pi/2} \sqrt{1 - k^2 \sin^2 \theta}\, d\theta$$

$$I_2 = \int_0^{\pi/2} \frac{1}{\sqrt{1 - k^2 \sin^2 \theta}}\, d\theta$$

where $|k| \leq 1$ is a parameter. We can imagine expanding the square root as a binomial. If k^2 is small, then the series will converge rapidly. Converting the powers $\sin^{2n} \theta$ to multiple angles

$$\left(\frac{e^{i\theta} - e^{-i\theta}}{2i}\right)^{2n} = \frac{1}{2^{2n} i^{-2n}} [e^{2ni\theta} - C(2n,1)e^{2(n-1)i\theta} + \cdots]$$

$$= \frac{1}{2^{2n-1} i^{-2n}} [\cos 2n\theta - C(2n,1) \cos (2n - 2)\theta + \cdots]$$

we get a rapidly converging Fourier series. Thus for small k^2 we expect accurate answers; for k^2 near 1 we do not expect accurate answers.

As an experiment, we chose sample points at 0°, 30°, 60°, 90°, although to get periodicity we had to assume six points in the interval of periodicity $0 \leq t \leq \pi$. Thus we actually computed only four sample points, one of which ($\theta = 0$) was trivial. The results (comparing them with those from a table) are given in Tables 21.5-1 and 21.5-2.

TABLE 21.5-1. ELLIPTIC INTEGRAL $I_1(k^2)$

k^2	I_1 (calc.)	I_1 (table)	Error
$k^2 = \frac{1}{4}$	1.46746	1.4675	
$k^2 = \frac{1}{2}$	1.35064	1.3506	
$k^2 = \frac{3}{4}$	1.21099	1.2111	0.0001

TABLE 21.5-2. ELLIPTIC INTEGRAL $I_2(k^2)$

k^2	I_2 (calc.)	I_2 (table)	Error
$k^2 = \frac{1}{4}$	1.68575	1.6858	
$k^2 = \frac{1}{2}$	1.85410	1.8541	
$k^2 = \frac{3}{4}$	2.15789	2.1565	0.0014

This example shows two things. First, it is often possible to make reasonable estimates of the size of the coefficients of a Fourier series; hence we can estimate the error of the computation *before* starting it. Second, the method of integration which uses $2N$ equally spaced, equally weighted samples gives very accurate results for periodic functions. This approach has been used to compute the Bessel function $J_0(x)$ from the formula

$$J_0(x) = \frac{1}{\pi} \int_0^{\pi} \cos (x \sin \varphi)\, d\varphi$$

EXERCISE

21.5-1. Evaluate $J_0(\frac{1}{2})$ from the integral form, using six samples.

21.6 THE GENERAL-OPERATOR APPROACH

In Chap. 10 of Part II, we described a general method (third method) for finding formulas which were exact for $1, x, \ldots, x^n$. In the same way we can find formulas which are exact for

$$\begin{aligned} &1, \quad \cos \frac{\pi}{N} t, \quad \cos \frac{2\pi}{N} t, \quad \cos \frac{3\pi}{N} t, \quad \ldots, \quad \cos \frac{(N-1)\pi}{N} t, \quad \cos \pi t \\ &\sin \frac{\pi}{N} t, \quad \sin \frac{2\pi}{N} t, \quad \sin \frac{3\pi}{N} t, \quad \ldots, \quad \sin \frac{(N-1)\pi}{N} t \end{aligned} \tag{21.6-1}$$

Such conditions are the natural ones to apply to a function that is periodic and for which we use $2N$ equally spaced samples.

Let the result of the linear operator L operating on the function $f(t)$, $L(f)$, be expressed in terms of the set of $2N$ equally spaced samples of the function $t = 0, 1, 2, \ldots, 2N - 1$; that is,

$$L[f(t)] = w_0 f(0) + w_1 f(1) + \cdots + w_{2N-1} f(2N - 1) \quad (21.6\text{-}2)$$

The defining equations are, for $1, \cos \frac{\pi}{N} m, \cos \frac{2\pi}{N} m, \ldots, \cos \frac{\pi(N-1)}{N} m,$ $\cos \pi m, \sin \frac{\pi}{N} m, \ldots, \sin \frac{\pi(N-1)}{N} m$:

$$M_0 = w_0 + w_1 + w_2 + \cdots + w_{2N-1}$$

$$M_1 = w_0 + w_1 \cos \frac{\pi}{N} + w_2 \cos \frac{2\pi}{N} + \cdots + w_{2N-1} \cos \frac{2N-1}{N} \pi$$

$$M_2 = w_0 + w_1 \cos \frac{2\pi}{N} + w_2 \cos \frac{4\pi}{N} + \cdots + w_{2N-1} \cos \frac{2N-1}{N} 2\pi$$

. . . .

$$M_{N-1} = w_0 + w_1 \cos \frac{N-1}{N} \pi + w_2 \cos \frac{(N-1)2\pi}{N} + \cdots + w_{2N-1} \cos \frac{(2N-1)(N-1)\pi}{N}$$

$$M_N = w_0 - w_1 + w_2 - \cdots - w_{2N-1}$$

$$M_{N+1} = 0 + w_1 \sin \frac{\pi}{N} + w_2 \sin \frac{2\pi}{N} + \cdots + w_{2N-1} \sin \frac{(2N-1)\pi}{N}$$

. . . .

$$M_{2N-1} = 0 + w_1 \sin \frac{(N-1)\pi}{N} + \sin \frac{2(N-1)\pi}{N} + \cdots + w_{2N-1} \sin \frac{(2N-1)(N-1)\pi}{N}$$

Thus the matrix of the unknown weights w_i is

$$\begin{pmatrix} 1 & 1 & 1 & \cdots & 1 \\ 1 & \cos \frac{\pi}{N} & \cos \frac{2\pi}{N} & \cdots & \cos \frac{2N-1}{N} \pi \\ \cdots & \cdots & \cdots & \cdots & \cdots \\ 1 & \cos \frac{N-1}{N} \pi & \cos \frac{(N-1)2\pi}{N} & \cdots & \cos \frac{(N-1)(2N-1)\pi}{N} \\ 1 & -1 & 1 & \cdots & -1 \\ 0 & \sin \frac{\pi}{N} & \sin \frac{2\pi}{N} & \cdots & \sin \frac{2N-1}{N} \pi \\ \cdots & \cdots & \cdots & \cdots & \cdots \\ 0 & \sin \frac{N-1}{N} \pi & \sin \frac{N-1}{N} 2\pi & \cdots & \sin \frac{(N-1)(2N-1)\pi}{N} \end{pmatrix} \quad (21.6\text{-}3)$$

Because of the orthogonality relations (6.2-3)

$$\sum_{k=0}^{2N-1} \cos \frac{m\pi}{N} k \cos \frac{n\pi}{N} k = N\delta_{m,n} \qquad (m, n \neq 0, N)$$

$$\sum_{0}^{2N-1} \sin \frac{m\pi}{N} k \cos \frac{n\pi}{N} k = 0$$

$$\sum_{0}^{2N-1} \sin \frac{m\pi}{N} k \sin \frac{n\pi}{N} k = N\delta_{m,n}$$

$$\sum_{0}^{2N-1} 1 = \sum_{0}^{2N-1} \cos^2 \pi k = 2N$$

the inverse matrix is the transpose of (21.6-3) except that two of the columns, the first and the $(N + 1)$st, corresponding to the functions 1 and $\cos \pi m$, have been divided by 2.

$$\left(\frac{1}{N}\right) \left\{ \begin{matrix} \frac{1}{2} & 1 & \cdots & 1 \\ \frac{1}{2} & \cos \frac{\pi}{N} & \cdots & \cos \frac{N-1}{N} \pi \\ \frac{1}{2} & \cos \frac{2\pi}{N} & \cdots & \cos \frac{N-1}{N} 2\pi \\ \cdots & \cdots & \cdots & \cdots \\ \frac{1}{2} & \cos \frac{2N-1}{N} \pi & \cdots & \cos \frac{(N-1)(2N-1)}{N} \pi \end{matrix} \right.$$

$$\left. \begin{matrix} \frac{1}{2} & 0 & \cdots & 0 \\ -\frac{1}{2} & \sin \frac{\pi}{N} & \cdots & \sin \frac{N-1}{N} \pi \\ \frac{1}{2} & \sin \frac{2\pi}{N} & \cdots & \sin \frac{N-1}{N} 2\pi \\ \cdots & \cdots & \cdots & \cdots \\ -\frac{1}{2} & \sin \frac{2N-1}{N} \pi & \cdots & \sin \frac{(N-1)(2N-1)}{N} \pi \end{matrix} \right\} \qquad (21.6\text{-}4)$$

Here we have produced the postmultiplier inverse, while in Chap. 10 we produced the premultiplier inverse; they are, of course, the same, but in each case it is harder to think about the other inverse. Interchanging the order of the two matrices gives a different set of orthogonal relations, which, at times, can be useful.

This general approach of using the third method may be illustrated by an example. Suppose that we wish to estimate the first derivative of the periodic function at the point $t = 0$. To be specific, suppose that we examine the case $N = 3$. The components of the "moments" M are, using 1, $\cos (\pi/3)t$, $\cos (2\pi/3)t$, $\cos \pi t$, $\sin (\pi/3)t$, $\sin (2\pi/3)t$,

$$0, \quad 0, \quad 0, \quad 0, \quad \frac{\pi}{3}, \quad \frac{2\pi}{3}$$

The inverse matrix, times the moment vector, gives

$$(\tfrac{1}{2})(\tfrac{1}{3}) \begin{pmatrix} 1 & 2 & 2 & 1 & 0 & 0 \\ 1 & 1 & -1 & -1 & \sqrt{3} & \sqrt{3} \\ 1 & -1 & -1 & 1 & \sqrt{3} & -\sqrt{3} \\ 1 & -2 & 2 & -1 & 0 & 0 \\ 1 & -1 & -1 & 1 & -\sqrt{3} & \sqrt{3} \\ 1 & 1 & -1 & -1 & -\sqrt{3} & -\sqrt{3} \end{pmatrix} \begin{pmatrix} 0 \\ 0 \\ 0 \\ 0 \\ \frac{\pi}{3} \\ \frac{2\pi}{3} \end{pmatrix} = \tfrac{1}{6} \begin{pmatrix} 0 \\ \sqrt{3}\,\pi \\ -\frac{\sqrt{3}}{3}\pi \\ 0 \\ \frac{\sqrt{3}}{3}\pi \\ -\sqrt{3}\,\pi \end{pmatrix} \tag{21.6-5}$$

The last vector is the weight vector; thus the formula is

$$f'(0) = \frac{\sqrt{3}\,\pi}{18} [3f(1) - f(2) + f(4) - 3f(5)]$$

EXERCISES

21.6-1. Using the matrix (21.6-5) of the illustrative example, find a formula for $f''(0)$.

21.6-2. Using (21.6-5), find a formula for estimating $\int_0^{2N} \int_0^t f(t)\, dt\, dt$.

21.7 SOME REMARKS ON THE GENERAL METHOD

In the case of polynomial approximation, it was efficient to use the general matrix approach. In the case of approximation with a Fourier series, it is not. The difference is due to the fact that we can find the Fourier expansion of a function with much less labor than performing the matrix multiplication to get the weights (see Secs. 6.4 to 6.6). Once we have the Fourier coefficients of the function, it is easy to multiply them by the "moments" to get the final answer. Thus if we have the Fourier series

$$f(t) = \frac{a_0}{2} + \sum_{k=1}^{N-1} a_k \cos \frac{\pi}{N} kt + \frac{a_N}{2} \cos \pi t + \sum_{k=1}^{N-1} b_k \sin \frac{\pi}{N} kt$$

applying the operator $L(\ )$, we have

$$L[f(t)] = \frac{a_0}{2} L(1) + \sum_{k=1}^{N-1} a_k L\left(\cos\frac{\pi}{N}kt\right) + \frac{a_N}{2} L(\cos \pi t) + \sum_{k=1}^{N-1} b_k L\left(\sin\frac{\pi}{N}kt\right)$$

$$= \frac{a_0}{2} M_0 + \sum_{k=1}^{N-1} a_k M_k + \frac{a_N}{2} M_N + \sum_{k=1}^{N-1} b_k M_{N+k}$$

The reason that we can find the coefficients easily is mainly the fact that the sines and cosines are orthogonal and can be found independently of each other. Thus the inverse matrix approach, while of theoretical interest for Fourier series approximation, is not of much practical importance.

It is necessary to point out again a distinction which first arose in Sec. 13.3, which covers indefinite integration. Whether we make a formula such as

$$y = \int_a^b f(x)\, dx$$

exact for $f(x) = 1, x, x^2, \ldots$ or for $y = 1, x, x^2, \ldots$ makes a difference. As was pointed out in Sec. 13.3, exactness for $f(x) = 1, x, x^2, \ldots$ is the same, for this formula, as exactness for $y = x, x^2, x^3, \ldots$ since the integrals of powers of x are powers of x one higher. We found it convenient at that time to *add* the condition of exactness for $y = 1$ and thus restore the equivalence of the two choices.

For the Fourier series approach, if we first approximate the integrand and then integrate, the terms cos kx and sin kx go into each other, but the constant term 1 goes into x. Thus the equivalence of approximating the integrand or the solution by a Fourier series is lost. We must, therefore, before we start decide on what part of the problem we are going to approximate by the set of functions we are using—approximations at various stages of the problem are not necessarily equivalent. In practice it is usual to know properties of the input function or the output function, and hence these are the ones usually chosen for the approximation.

This effect of nonequivalence will be seen again in Chap. 26 on exponential functions where upon integration the exponential $e^{\alpha_i x}$ goes into itself (within a multiplicative constant) *except* when $\alpha_i = 0$.

CHAPTER 22

The Convergence of Fourier Series

22.1 CONVERGENCE OF POWER SERIES AND FOURIER SERIES

The rapidity of convergence of a power series and the rapidity with which an exact-matching polynomial approximation approaches the function as the number of sample points increases are both determined by the location of the singularities of the function in the complex plane. This is not to say that, in principle, the behavior along the real axis does not determine the rapidity but rather that it is difficult to move from the values on the real axis to the convergence rate except via the singularities. The singularities in the complex plane are often unobtainable from a practical point of view; even crude estimates are rarely available. As an example, consider the problem of computing a trajectory to the moon. How are we to estimate the location of nearby singularities in the complex plane?

The convergence of a Fourier series, on the other hand, can be found readily from the values of the function along the real axis; indeed, there are simple relations between the two. The purpose of this chapter is to bring out these relations and to show how to take advantage of them. We also propose to show how to estimate the coefficients of the Fourier series *before* starting a computation, so that reasonable estimates can be made of how many terms will be needed for a given accuracy (cf. Sec. 21.5). Thus, in the planning phase of a computation, the Fourier approach offers significant advantages over the polynomial approach.

22.2 FUNCTIONS WITH A SIMPLE DISCONTINUITY

We begin with the example (see Fig. 22.2-1)

$$y = \frac{t}{2\pi} \qquad (-\pi < t < \pi) \tag{22.2-1}$$

which has a single discontinuity of unit height at $t = \pi$ [we are assuming that $y(t)$ is periodic]. Since the function is odd [that is, $f(-t) = -f(t)$] all the cosine terms vanish, including the constant term. The sine coefficients are given by

$$\begin{aligned} b_k &= \frac{1}{\pi} \int_{-\pi}^{\pi} y(t) \sin kt \, dt = \frac{2}{\pi} \int_0^{\pi} \frac{t}{2\pi} \sin kt \, dt \\ &= \frac{1}{\pi^2} t \frac{-\cos kt}{k} \Big|_0^{\pi} + \frac{1}{k\pi^2} \int_0^{\pi} \cos kt \, dt \\ &= \frac{\pi}{\pi^2 k} (-1) \cos \pi k = \frac{(-1)^{k-1}}{\pi k} \end{aligned}$$

Hence
$$y(t) = \frac{1}{\pi} \left(\sin t - \frac{\sin 2t}{2} + \frac{\sin 3t}{3} - \frac{\sin 4t}{4} + \cdots \right) \tag{22.2-2}$$

This series converges to $y(t)$ for all values of t *except* at the discontinuity

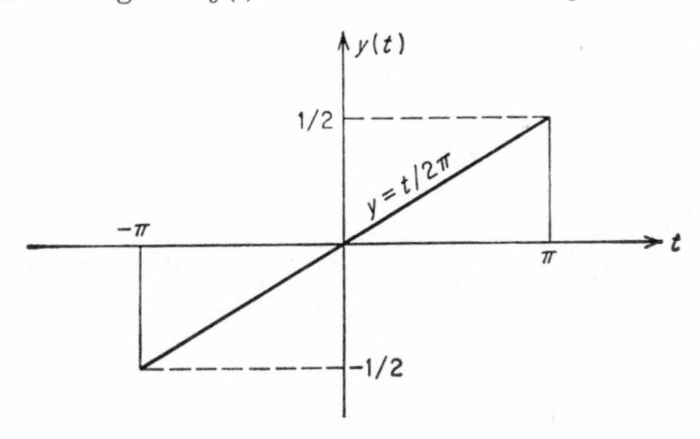

FIG. 22.2-1

$(t = \pi)$ where clearly it converges to zero (the average of the two values ½ and $-$½ which are the limit values as $t \to \pi$ or $-\pi$ from the origin).

If we set

$$t = t' - a + \pi$$

we translate the discontinuity to $t' = a$. In place of (22.2-2) we have (dropping the prime on t), for a function $y(t)$ having a jump at $t = a$,

$$\begin{aligned} y(t) &= \frac{1}{\pi} \sum_{k=1}^{\infty} \frac{(-1)^{k-1}}{k} \sin [(kt - ka) + k\pi] \\ &= \frac{1}{\pi} \sum_{k=1}^{\infty} \frac{(-1)^{k-1}}{k} \sin (kt - ka) \cos k\pi = \frac{1}{\pi} \sum_{k=1}^{\infty} \frac{\sin (ka - kt)}{k} \\ &= \frac{1}{\pi} \sum_{k=1}^{\infty} \frac{\sin ka}{k} \cos kt - \frac{1}{\pi} \sum_{k=1}^{\infty} \frac{\cos ka}{k} \sin kt \end{aligned} \tag{22.2-3}$$

In this Fourier series the coefficients again fall off as $1/k$.

If we now consider a function with a finite number, M, of simple discontinuities and with straight lines between the discontinuities, then if the discontinuity at $t = a_i$ is of size y_i, we take the appropriate linear combination of (22.2-3) (including possibly an a_0 term)

$$y(t) = \frac{1}{\pi} \sum_{i=1}^{M} y_i \left(\sum_{k=1}^{\infty} \frac{\sin ka_i}{k} \cos kt - \sum_{k=1}^{\infty} \frac{\cos ka_i}{k} \sin kt \right)$$

$$= \frac{1}{\pi} \sum_{k=1}^{\infty} \left(\sum_{i=1}^{M} y_i \sin ka_i \right) \frac{\cos kt}{k} - \frac{1}{\pi} \sum_{k=1}^{\infty} \left(\sum_{i=1}^{M} y_i \cos ka_i \right) \frac{\sin kt}{k}$$

where we have formally interchanged the summation processes.

Thus we see that functions with simple discontinuities and with straight-line segments between them produce Fourier series whose coefficients fall off as $1/k$; we also have a simple method for constructing such series.

For practical purposes the converse is true; if the coefficients fall off as $1/k$, then there are simple discontinuities in the function.[1]

The *continuous*[2] broken-line situation is of frequent occurrence and is worth examining. We calculate the coefficients of the cosine terms first, using the values t_i to mark the ends of the line segments. The cosine coefficients are

$$\pi a_k = \int_{-\pi}^{\pi} f(t) \cos kt \, dt = \sum_i \int_{t_i}^{t_{i+1}} f(t) \cos kt \, dt$$

$$= \sum_i \frac{1}{k} [f(t_{i+1}) \sin k(t_{i+1}) - f(t_i) \sin kt_i] - \sum_i \int_{t_i}^{t_{i+1}} \frac{f'(t) \sin kt \, dt}{k}$$

When the summation is performed on the integrated piece, all the terms cancel, because $f(t)$ was assumed to be continuous. In the integral term $f'(t)$ is constant in each interval; hence the result of integrating the sin kt leads to another factor $1/k$. The result is that the coefficients of the cosine terms fall off as $1/k^2$. The same occurs for the sine terms. Thus the continuous broken-line function has coefficients that fall off as $1/k^2$.

EXERCISE

22.2-1. Find the continuous Fourier series for the saw-tooth function

$$f(t) = \begin{cases} 1 + t & \text{for } (-1 < t < 0) \\ t & \text{for } (0 < t < 1) \end{cases}$$

[1] G. Raisbeck, Order of Magnitude of Fourier Coefficients, *Am. Math. Monthly*, vol. 62, pp. 149–154, March, 1955.

[2] That is, we allow no discontinuities in the function which is composed of pieces of straight lines. Remember that $f(\pi) = f(-\pi)$ is necessary for continuity.

22.3 FUNCTION HAVING CONTINUOUS HIGHER DERIVATIVES

If the function has a continuous first derivative and, except for a finite number of places, a continuous second derivative, then we can proceed as in Sec. 22.2, except that we integrate twice, and show that the coefficients fall off as $1/k^2$:

$$a_k = \frac{1}{\pi}\int_{-\pi}^{\pi} f(t)\cos kt\,dt = -\frac{1}{\pi k}\int_{-\pi}^{\pi} f'(t)\sin kt\,dt$$

since, because of continuity, the integrated term vanishes. We now break up the integral into pieces and integrate again:

$$a_k = -\frac{1}{\pi k}\sum_{i=0}^{M}\int_{t_i}^{t_{i+1}} f'(t)\sin kt\,dt$$

$$= -\frac{1}{\pi k^2}\sum_{i=0}^{M}\int_{-\pi}^{\pi} f''(t)\cos kt\,dt$$

$$|a_k| \le \frac{M_2}{\pi k^2}2\pi$$

where M_2 is the maximum of the second derivative. We treat b_k similarly.

If the coefficients fall off as $1/k^2$, then we can conclude that the Fourier series converges everywhere, since

$$|a_k \cos kt + b_k \sin kt| \le \frac{4M_2}{k^2}$$

The same pattern clearly applies to functions having a continuous mth derivative; the coefficients fall off as $1/k^{m+1}$ (m = integer).

EXERCISES

22.3-1. If

$$\begin{aligned} f(t) &= t(\pi - t) && \text{for } 0 \le t \le \pi \\ &= -f(t) && \text{for } 0 \ge t \ge -\pi \end{aligned}$$

find the Fourier series of $f(t)$.

22.3-2. Let $f(t) = 1 - t^2/\pi^2$ $(-\pi \le t \le \pi)$. How do the coefficients behave?

22.4 IMPROVING THE CONVERGENCE OF A FOURIER SERIES

Fourier series expansions of functions with discontinuities or other singularities on the real axis, especially at the ends of the interval, occur frequently in practice. Since we now know that the singularities on the real axis control the rapidity of convergence, it is natural to proceed much as we did in Sec. 4.2 and from a given series subtract a suitably chosen one having the same slow convergence. Thus we shall have left a more rapidly converging series to compute. The idea being obvious, we need not develop it further here.

The coefficients a_k and b_k are given by integrals which often cannot be evaluated, and we are reduced to numerical evaluation from some samples of the integrand. Thus we face the problems of aliasing. Before doing this, however, we subtract "canonical" expansions of singularities such as (22.2-3), as well as integrals of it, to compensate for singularities in the function and in the higher derivatives. Thus we are reduced to numerically evaluating the coefficients of a rapidly converging series of a function which is quite smooth, and the aliasing is no longer a serious question. Thus "subtracting the singularities" not only decreases the numerical work but also improves the accuracy.

EXERCISE

22.4-1. Compute the first three integrals of Eq. (22.2-3) to get canonical forms for the corresponding singularities.

22.5 THE POWER SPECTRUM

In both pure and applied mathematics it is common to look for the invariants of a representation—invariants with respect to a class of transformations. In the case of periodic functions, a translation of the axis

$$t = t' + b$$

should not change those things in the representation of a function that do not depend on the coordinate system. It is immediately evident that the Fourier coefficients a_k and b_k do change when the axis is shifted, that is, when the origin of time is changed. On setting $t = t' + b$ and making use of the periodicity of $f(t)$ to shift the limits in the integral, we have

$$\begin{aligned} a_k &= \frac{1}{\pi}\int_{-\pi}^{\pi} f(t)\cos kt\,dt = \frac{1}{\pi}\int_{-\pi}^{\pi} f(t'+b)\cos k(t'+b)\,dt' \\ &= \left[\frac{1}{\pi}\int_{-\pi}^{\pi} f(t'+b)\cos kt'\,dt'\right]\cos kb - \left[\frac{1}{\pi}\int_{-\pi}^{\pi} f(t'+b)\sin kt'\,dt'\right]\sin kb \\ &= a_k'\cos kb - b_k'\sin kb \end{aligned}$$

Similarly,
$$b_k = a_k'\sin kb + b_k'\cos kb$$

Although the a_k and b_k are not invariant, the quantity

$$\begin{aligned} a_k^2 + b_k^2 &= (a_k'\cos kb - b_k'\sin kb)^2 + (a_k'\sin kb + b_k'\cos kb)^2 \\ &= (a_k')^2 + (b_k')^2 \end{aligned}$$

is clearly invariant. It is customary to refer to the quantity $a_k^2 + b_k^2$ as the power at frequency k and to plot it in the form of a discrete *power spectrum*. There is confusion as to whether

$$a_k^2 + b_k^2 \qquad \text{or} \qquad \sqrt{a_k^2 + b_k^2}$$

gives the *spectrum*, but we shall use the words "power spectrum" and "spectrum" as meaning the same thing, $a_k^2 + b_k^2$.

EXERCISE

22.5-1. Compute the spectrum of the function in Exercise 22.2-1.

22.6 THE GIBBS PHENOMENON[1]

We begin with the special case of a rectangular wave $H(t)$ of period 2π (see Fig. 22.6-1). If we compute the partial sum of the first $2n$ terms

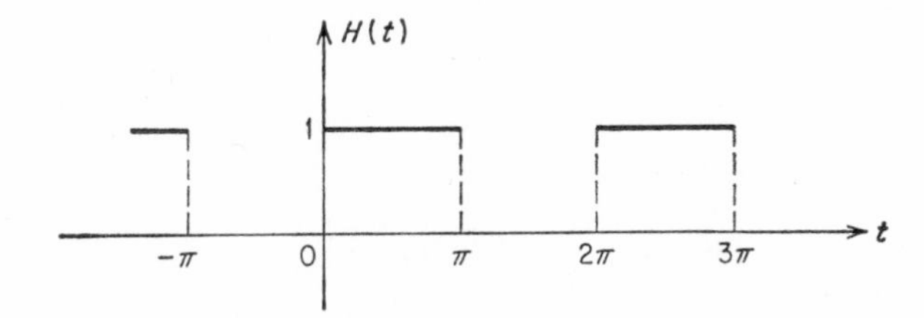

FIG. 22.6-1. Rectangular wave.

the cosine terms are all zero, and we have (in the continuous case)

$$H_{2n}(t) = \frac{1}{2} + \frac{2}{\pi} \sum_{k=1}^{n} \frac{1}{2k-1} \sin (2k-1)t \tag{22.6-1}$$

Gibbs[2] pointed out that the partial sum H_{2n} overshoots the function by a certain amount (see Fig. 22.6-2). More precisely,

$$H_{2n}\left(\frac{\pi}{2n}\right) \to 1.08949 \cdots \qquad \text{as } n \to \infty \tag{22.6-2}$$

Indeed, not only does $H_{2n}(t)$ overshoot the function $H(t)$ but it tends to oscillate about $H(t)$, and the oscillations decrease slowly as we move away from the discontinuity.

[1] Much of this is taken from a Bell Telephone Laboratories memorandum by R. G. Segers, although it is common knowledge in computing circles. See Ref. 23.

[2] J. Willard Gibbs, *Nature*, April 27, 1899.

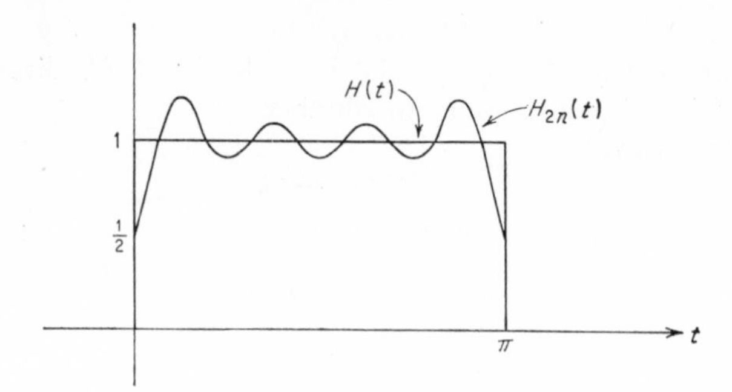

FIG. 22.6-2. The Gibbs phenomenon.

In order to explain the phenomenon, we write (22.6-1) as

$$\begin{aligned} H_{2n}(t) &= \frac{1}{2} + \frac{2}{\pi} \sum_{k=1}^{n} \int_0^t \cos (2k - 1)x \, dx \\ &= \frac{1}{2} + \frac{2}{\pi} \int_0^t \sum_{k=1}^{n} \cos (2k - 1)x \, dx \\ &= \frac{1}{2} + \frac{1}{\pi} \int_0^t \frac{\sin 2nx}{\sin x} \, dx \end{aligned} \tag{22.6-3}$$

where we have used

$$\sum_{k=1}^{n} \cos (2k - 1)x = \frac{\sin 2nx}{2 \sin x}$$

From (22.6-3) it is clear that the maxima and minima occur (for $0 \le t \le \pi$) at

$$\frac{dH_{2n}(t)}{dt} = \frac{1}{\pi} \frac{\sin 2nt}{\sin t} = 0$$

or at

$$t = \frac{m\pi}{2n} \qquad m = 1, 2, \ldots, 2n - 1 \tag{22.6-4}$$

and that they alternate. The size of these has been computed by Carslaw [7, p. 275].

What is true for this special function is clearly true for more general functions since the discontinuity can be regarded as coming from a rectangular wave *added* to a smooth function.

EXERCISE

22.6-1. For the function in Exercise 22.2-1 plot the sum of the first four terms.

22.7 LANCZOS' σ FACTORS

We propose to replace the rapidly oscillating function $H_{2n}(t)$ by a smoothed function

$$\overline{H_{2n}}(t) = \frac{n}{\pi} \int_{t-\pi/2n}^{t+\pi/2n} H_{2n}(\tau)\, d\tau \tag{22.7-1}$$

where we have averaged over one complete oscillation of $H_{2n}(t)$ centered about t.

Using (22.6-1) in (22.7-1), we get

$$\begin{aligned}
\overline{H_{2n}}(t) &= \frac{n}{\pi} \int_{t-\pi/2n}^{t+\pi/2n} \left[\frac{1}{2} + \frac{2}{\pi} \sum_{k=1}^{n} \frac{1}{2k-1} \sin (2k-1)\tau \right] d\tau \\
&= \frac{n}{\pi} \left[\frac{\pi}{2n} + \frac{2}{\pi} \sum_{k=1}^{n} \frac{-1}{(2k-1)^2} \cos (2k-1)\tau \Bigg|_{t-\pi/2n}^{t+\pi/2n} \right] \\
&= \frac{1}{2} + \frac{2}{\pi} \sum_{k=1}^{n} \frac{1}{2k-1} \frac{\sin [(2k-1)(\pi/2n)]}{(2k-1)(\pi/2n)} \sin (2k-1)t
\end{aligned} \tag{22.7-2}$$

If we compare this with $H_{2n}(t)$, we find that we now have an extra factor

$$\sigma_{2k-1} = \frac{\sin (2k-1)(\pi/2n)}{(2k-1)(\pi/2n)} \tag{22.7-3}$$

for each term in the summation.

The effect of this factor σ_k is to reduce the maximum overshoot

from 0.08949 to 0.01187

and the first minimum

from 0.04859 to 0.00473

etc. Thus the Gibbs phenomenon has been greatly reduced by the presence of the σ factors which arose from the smoothing of $H_{2n}(t)$ over a short interval of length π/n.

We have studied the special case of a rectangular wave; we now show that the weight factors σ_k are the same for any Fourier series.

Let $f(t)$ $(0 \le t \le 2\pi)$ be integrable, and let

$$a_k = \frac{1}{\pi} \int_0^{2\pi} f(t) \cos kt\, dt \qquad k = 0, 1, 2, \ldots$$

$$b_k = \frac{1}{\pi} \int_0^{2\pi} f(t) \sin kt\, dt \qquad k = 1, 2, \ldots$$

be the Fourier coefficients. Then[1]

$$f_n(t) = \frac{a_0}{2} + \sum_{k=1}^{n} (a_k \cos kt + b_k \sin kt)$$

We now compute

$$\begin{aligned}
\bar{f}_n(t) &= \frac{1}{2\pi/n} \int_{t-\pi/n}^{t+\pi/n} f_n(\tau)\, d\tau \\
&= \frac{n}{2\pi} \left[\frac{a_0}{2} \frac{\pi}{n} + \sum_{k=1}^{n} \left(a_k \frac{\sin k\tau}{k} - b_k \frac{\cos k\tau}{k} \right) \Bigg|_{t-\pi/n}^{t+\pi/n} \right] \\
&= \frac{n}{2\pi} \left(\frac{\pi}{2n} a_0 + \sum_{k=1}^{n} \left\{ \frac{a_k}{k} \left[\sin k\left(t + \frac{\pi}{n}\right) - \sin k\left(t - \frac{\pi}{n}\right) \right] \right.\right. \\
&\qquad \left.\left. - \frac{b_k}{k} \left[\cos k\left(t + \frac{\pi}{n}\right) - \cos k\left(t - \frac{\pi}{n}\right) \right] \right\} \right) \\
&= \frac{a_0}{2} + \frac{n}{2\pi} \sum_{k=1}^{n} \left(\frac{2a_k}{k} \sin \frac{\pi k}{n} \cos kt + \frac{2b_k}{k} \sin \frac{\pi k}{n} \sin kt \right) \\
&= \frac{a_0}{2} + \sum_{k=1}^{n} \frac{\sin (\pi k/n)}{\pi k/n} (a_k \cos kt + b_k \sin kt)
\end{aligned}$$

and again we obtain the σ factors [Eq. (22.7-3)]

$$\sigma_k = \frac{\sin (\pi k/n)}{\pi k/n} \tag{22.7-4}$$

to be inserted in the various coefficients of the Fourier series. Note that the terms $\sin kt$ and $\cos kt$ both get the factor σ_k.

EXERCISE

22.7-1. Apply the σ factors to Exercise 22.6-1, and plot the result.

22.8 A COMPARISON OF CONVERGENCE METHODS

Besides the σ-factor method of Lanczos there is the well-known Fejer method of using the arithmetic means of the partial sums. Fejer's method completely eliminates the oscillation, while Lanczos' method greatly dampens it. As a result, Fejer's method pays a large price in

[1] For this argument it does not matter how we terminate the series, and so we use the form with the easier notation.

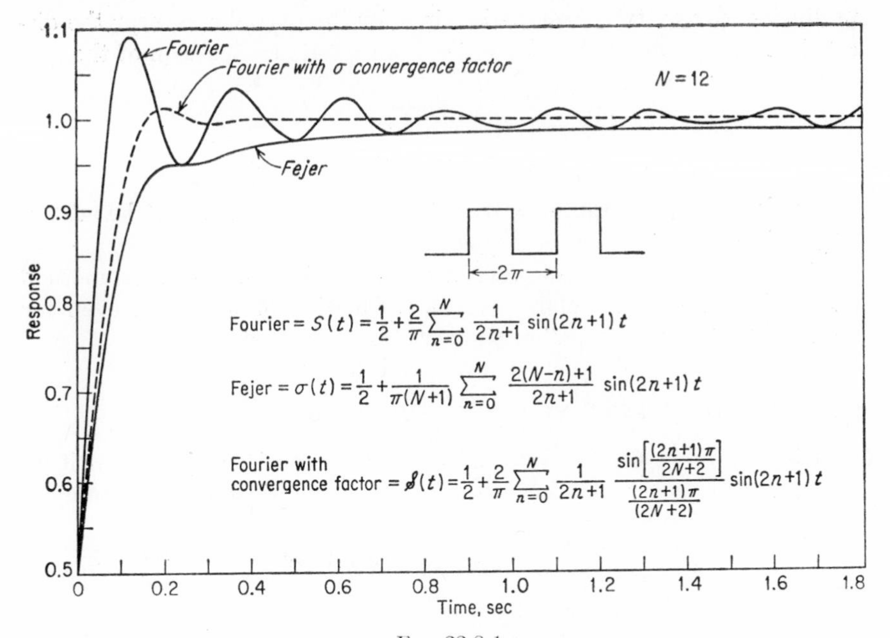

FIG. 22.8-1

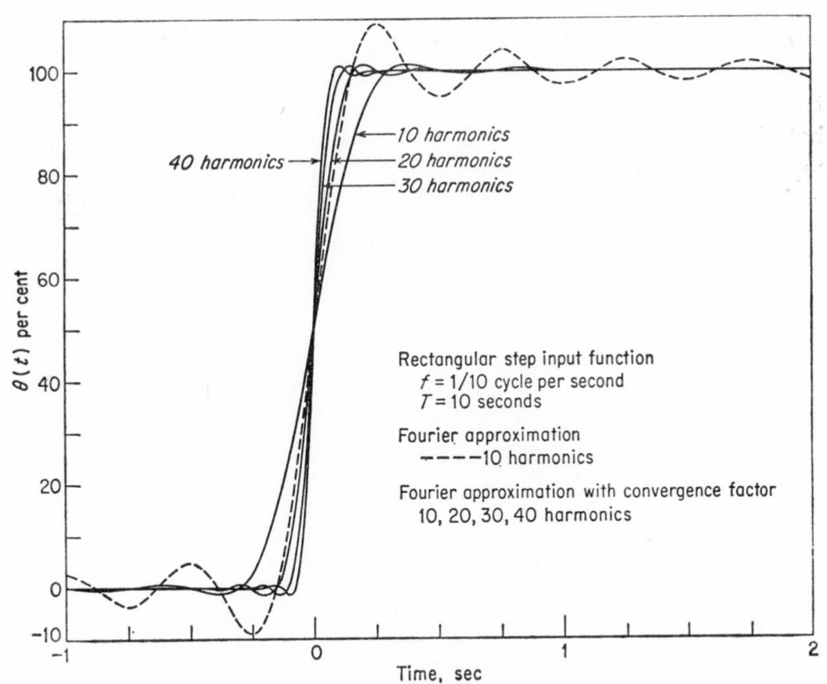

FIG. 22.8-2

producing, for finite n, a slow-rise time for the partial sum near the discontinuity. Figure 22.8-1 shows the appropriate comparison of the

Fourier series
Fejer sum
Lanczos σ-factor

curves for a 12-term approximation to a rectangular wave. The advantages of the σ-factor method are clearly apparent.

Figure 22.8-2 shows the rapidity with which the σ-factor method approaches the rectangular wave as a function of n. Thus even for moderate n the "rise time" of the curve is very short.

22.9 LANCZOS' DIFFERENTIATION TECHNIQUE

It is sometimes necessary to differentiate a Fourier series. But if we do so, we usually get what is not wanted in the physical problem, namely, a large high-frequency oscillation.

Lanczos [23, p. 221] introduced a practical approach, namely, that we use

$$\frac{f(t + \pi/n) - f(t - \pi/n)}{2\pi/n} \tag{22.9-1}$$

as an estimate of the derivative of $f(t)$ at the point t (instead of the limit process which has dubious physical significance) when we differentiate a truncated Fourier-series approximation. Note that the same n appears both in (22.9-1) and in the order of the truncated series. We shall not reproduce the argument here but shall merely observe that *the same σ factors* (22.7-4) put in the formally differentiated series give the appropriate result.

CHAPTER 23

Nonperiodic Functions—The Fourier Integral

23.1 PURPOSE OF THE CHAPTER

In Chaps. 21 and 22 we approximated periodic functions by linear combinations of periodic functions—the sines and cosines. We now turn to the study of nonperiodic functions and their approximation by sines and cosines. An example of a nonperiodic function which is made up of periodic parts is

$$y(t) = \cos t + \cos (\sqrt{2}\, t)$$

[since 1 and $\sqrt{2}$ are not commensurate, $y(t)$ is not periodic]. The approximation of a nonperiodic function by periodic terms is no more unreasonable than the Taylor-series representation of a periodic function by means of nonperiodic terms, namely, 1, x, x^2, ...

In the Fourier series we used an infinite set of discrete frequencies to represent a periodic function; now we shall use all frequencies, or perhaps all frequencies in some band,[1] to represent a nonperiodic function. The basic tool is the Fourier integral, which we shall develop only briefly; the reader is referred to standard texts for a more rigorous treatment.[2]

The purpose of this chapter is to discuss in a nonrigorous fashion a number of questions that arise in computing, such as the following:

1. What is the effect of sampling at equally spaced points?
2. How do we reconstruct a band-limited function from its samples (the so-called sampling theorem of information theory) so that we may interpolate for the missing values or use it in analytic substitution?
3. What is the effect of having only a finite number of samples from an infinitely long function?

[1] A function with all its frequencies in a band is called "a band-limited function."

[2] One of the more recent approaches to the Fourier integral is given in M. J. Lighthill [24].

These are not the kinds of questions that the polynomial approach of Part II can answer in a practical way, and they serve to point out the advantages of the present approach during the planning, execution, and interpretation phases.

23.2 NOTATION AND SUMMARY OF RESULTS

This section is devoted to introducing notation and summarizing some of the results to be obtained and does not contain proofs of most of the statements made.

The continuous Fourier series in an interval $-N \leq t \leq N$ can be written [(21.3-1) and (21.3-2), slightly changed]

$$f(t) = \frac{a_0}{2} + \sum_{k=1}^{\infty} a_k \cos \frac{\pi}{N} kt + \sum_{k=1}^{\infty} b_k \sin \frac{\pi}{N} kt \tag{23.2-1}$$

where

$$\begin{aligned} a_k &= \frac{1}{N} \int_{-N}^{N} f(t) \cos \frac{\pi}{N} kt\, dt \qquad (k = 0, 1, 2, \ldots) \\ b_k &= \frac{1}{N} \int_{-N}^{N} f(t) \sin \frac{\pi}{N} kt\, dt \qquad (k = 1, 2, \ldots) \end{aligned} \tag{23.2-2}$$

If we now use the complex-number representation of the trigonometric function

$$\cos \theta = \frac{e^{i\theta} + e^{-i\theta}}{2}$$

$$\sin \theta = \frac{e^{i\theta} - e^{-i\theta}}{2i}$$

we get

$$f(t) = \sum_{k=-\infty}^{\infty} c_k e^{(i\pi/N)kt} \tag{23.2-3}$$

where

$$c_k = \frac{1}{2N} \int_{-N}^{N} f(t) e^{-(i\pi/N)kt}\, dt \tag{23.2-4}$$

It is easy to see that

$$c_k = \begin{cases} \dfrac{a_k - ib_k}{2} & k > 0 \\ \dfrac{a_0}{2} & k = 0 \\ \dfrac{a_k + ib_k}{2} & k < 0 \end{cases} \tag{23.2-5}$$

This complex form of the Fourier series is much easier to handle in theoretical discussions than the usual one, but, of course, we generally compute with the real forms (23.2-1) and (23.2-2). The ease of formal manipulation with the complex form is apt to conceal considerable computing, and we have been careful up to this point to avoid it. But we are now interested in *understanding* what effects various computing procedures have on our answers rather than doing any actual computing, and for this purpose the complex form is easier to understand and use. Notice that both positive and negative frequencies have meaning in the complex form; we have replaced two functions, sine and cosine, at each positive frequency with a single exponential function at each positive and each negative frequency.

We shall first show (Sec. 23.3) that, corresponding to the Fourier-series [(23.2-3) and (23.2-4)] representation of a periodic function, there is Fourier-integral representation of a general function

$$f(t) = \int_{-\infty}^{\infty} F(\sigma)e^{2\pi i\sigma t}\, d\sigma \tag{23.2-6}$$

where

$$F(\sigma) = \int_{-\infty}^{\infty} f(t)e^{-2\pi i\sigma t}\, dt \tag{23.2-7}$$

The $F(\sigma)\, d\sigma$ corresponds loosely to c_k in the Fourier series (23.2-3) and (23.2-4).

$F(\sigma)$ is the *density function* which describes the amount of the frequency σ that is present in the function $f(t)$ and is called the *transform* of $f(t)$. It is a custom to use capital and small letters to represent the transform and the corresponding function. The change in the sign of the exponent is the only difference between the two functions. We also use Latin and Greek letters to distinguish the transform variables, Latin in the time domain and Greek in the frequency domain.

As we discussed in Sec. 21.2, the effect of sampling is to cause "aliasing" of different frequencies. The discussion in no way depends on the periodicity of the function being sampled but refers to the confusion of the frequencies used to represent the function. This aliasing is the reason that the concept of a band-limited function plays such a leading role in the theory; if all the frequencies are limited to a band

$$-\Omega < \sigma < \Omega$$

of width 2Ω and if we sample at an interval of Δt between samples, then we must have

$$2\Omega\, \Delta t < 1 \tag{23.2-8}$$

to avoid aliasing. The maximum sampling interval $1/2\Omega$ is related to the *Nyquist*, or *folding*, frequency from the role that it plays in Fig. 21.3-1. Another common way of expressing (23.2-8) is to say that, to avoid aliasing, we must have at least two samples in the highest frequency present.

The question naturally arises "Can we reconstruct a band-limited function from its samples if we do obey (23.2-8)?" One intuitive argument that we can do so goes along the lines used in the Lagrange interpolation process (Sec. 8.3). The function

$$\frac{\sin \pi t}{\pi t}$$

has the property that it has the value 1 at $x = 0$ and is 0 for $t = \pm 1, \pm 2, \pm 3, \ldots$. Thus the function

$$\frac{\sin \pi(t-k)}{\pi(t-k)} \tag{23.2-9}$$

plays the role of L_k in the Lagrange interpolation process, since it is 1 at $t = k$ and 0 at all other sample points. Still following the Lagrange approach,

$$f(k) \frac{\sin \pi(t-k)}{\pi(t-k)}$$

takes the value of the function $f(k)$ at $t = k$ and 0 at all other sample points; hence the formal expansion

$$\sum_{k=-\infty}^{\infty} f(k) \frac{\sin \pi(t-k)}{\pi(t-k)}$$

will pass through all the sample points. It is an easy matter to see that

$$\frac{\sin \pi t}{\pi t} = \int_{-1}^{1} \frac{1}{2} e^{i\pi t\sigma}\, d\sigma \tag{23.2-10}$$

Hence, by (23.2-6), the frequencies present in $\sin \pi t/(\pi t)$ fall in a band. This band has σ going from -1 to 1, but to put the exponent in the form of (23.2-6), we must use $\sigma/2 = \sigma'$,

$$\frac{\sin \pi t}{\pi t} = \int_{-1/2}^{1/2} e^{2\pi i\sigma' t}\, d\sigma'$$

and so the band is $-\frac{1}{2} < \sigma' < \frac{1}{2}$. This agrees with what we expect; having used $\Delta t = 1 =$ sampling interval, and expecting two samples for the highest frequency present, we find that the band of frequencies runs from $-\frac{1}{2}$ to $\frac{1}{2}$ cycles per unit time. This is, in essence, the sampling theorem of information theory; given the proper equally spaced samples of a band-limited function, the function can be reconstructed from the samples.

In analogy with Sec. 22.5, the quantity

$$|F(\sigma)|^2 \tag{23.2-11}$$

is often called the *power spectrum;* the word "power" comes from its use in many engineering applications, and its use here is merely following tradition. The power spectrum, or, more simply, *the spectrum,* may be viewed somewhat as the optical case of a prism decomposing the incoming light into various frequencies (colors)—the Fourier integral decomposes the time function $f(t)$ into frequencies of intensity $F(\sigma)$. For a single frequency, we often speak of the *spectral line.*

In Sec. 23.7, we shall show that the effect of taking a sample of finite length from an infinitely long function $(-\infty < t < \infty)$ is to broaden the spectral lines—the shorter the length of data, the greater the broadening.

The above gives the gist of the chapter and indicates its relevance to computing practice. Clearly, both the effect of sampling and the effect of selecting a finite length of the potentially infinite sample can be partially understood in terms of the theory of band-limited functions and cannot be readily understood in terms of the classical polynomial approach.

The purpose of the rest of the chapter is to develop a little of the theory of the Fourier integral in order to give a slightly more adequate development of the above remarks. We do not intend to become involved in questions of mathematical rigor since the class of functions that one seriously tries to approximate on a computing machine is restricted to "well-behaved functions" and excludes the pathological cases with which mathematicians are so often concerned.

For those whose skill in manipulating complex numbers is a bit rusty, we provide the following simple exercises for practice. If they cause trouble, it would be well to study up on the topic before going on.

EXERCISES

23.2-1. Prove $|e^{ix}| = 1$ (for real x).

23.2-2. Prove that the addition formulas of trigonometry follow from $e^{i\alpha}e^{i\beta} = e^{i(\alpha+\beta)}$.

23.2-3. If $1/(a + ib) = u + iv$, find u, v in terms of a and b.

23.2-4. If $u + iv = (a + ib)^{1/2}$, find u, v. (Watch the sign of the u term.)

23.2-5. If $c_k = \bar{c}_{-k}$, prove that $\sum_{k=-\infty}^{\infty} c_k e^{ikt}$ is real.

23.2-6. If m and k are integers, prove that

$$\int_{-\pi}^{\pi} e^{imx}e^{-ikx}\,dx = \begin{cases} 0 & m \neq k \\ 2\pi & m = k \end{cases}$$

23.2-7. If $f(t) = t(-\pi < t < \pi)$, find the Fourier coefficients in $f(t) = \sum_{k=-\infty}^{\infty} c_k e^{ikt}$.

23.2-8. If $f(t) = \int_{-\infty}^{\infty} F(\sigma)e^{2\pi i\sigma t}\,d\sigma$, find the transform of $f'(t)$, $f''(t)$, $f^{(k)}(t)$.

23.2-9. If

$$f(t) = \begin{cases} t & (-\pi < t < \pi) \\ 0 & \text{elsewhere} \end{cases}$$

find $F(\sigma)$.

23.2-10. If

$$F(\sigma) = \begin{cases} \sigma & (-\pi < \sigma < \pi) \\ 0 & \text{elsewhere} \end{cases}$$

find $f(t)$.

23.3 THE FOURIER INTEGRAL

In order to "derive" the Fourier integral from the Fourier series we first eliminate the coefficients c_k by substituting (23.2-4) into (23.2-3):

$$f(t) = \sum_{k=-\infty}^{\infty} \left[\int_{-N}^{N} f(t) e^{-(i\pi/N)kt} \, dt \right] e^{(i\pi/N)kt} \frac{1}{2N}$$

We are assuming that the function $f(t)$ is periodic in the interval $-N \leq t \leq N$. In order to approach a nonperiodic function, we let $N \to \infty$. For this purpose we set

$$\frac{1}{2N} = \Delta\sigma$$

and note that in the summation the spacing of the adjacent exponentials behaves as $\Delta\sigma$, which, as $N \to \infty$, will approach zero. Thus the exponentials in the sum are being packed closer and closer together as $N \to \infty$. We have

$$f(t) = \sum_{k=-\infty}^{\infty} \left[\int_{-N}^{N} f(t) e^{-2\pi ikt\,\Delta\sigma} \, dt \right] e^{2\pi ikt\,\Delta\sigma} \, \Delta\sigma$$

But $k\,\Delta\sigma \to \sigma$ and we have, as $N \to \infty$, the limit of the sum which is the integral

$$f(t) = \int_{-\infty}^{\infty} \left[\int_{-8}^{\infty} f(t) e^{-2\pi it\sigma} \, dt \right] e^{2\pi i\sigma t} \, d\sigma$$

We now set

$$F(\sigma) = \int_{-\infty}^{\infty} f(t) e^{-2\pi it\sigma} \, dt \tag{23.3-1}$$

and we have

$$f(t) = \int_{-\infty}^{\infty} F(\sigma) e^{2\pi i\sigma t} \, d\sigma \tag{23.3-2}$$

The function $F(\sigma)$ is said to be the *Fourier transform* of $f(t)$. The two functions $f(t)$ and $F(\sigma)$ have almost exactly reciprocal relationships to each other; the exception is in the sign of the exponent, $\sigma \to -\sigma$. Both functions contain the same information since each can be found from the other; they merely present the same information in different forms: $f(t)$ in the time domain and $F(\sigma)$ in the frequency domain. These two alternative views of the same information account for much of the value of the Fourier-integral representation of a function.

The above is not a rigorous proof but rather a suggestive argument. The assumption of periodicity in the interval $-N \le t \le N$ becomes, when $N \to \infty$, no assumption of periodicity at all.

23.4 SOME TRANSFORM PAIRS

The almost reciprocal relationship (23.3-1) and (23.3-2) indicates that to each function $f(t)$ there corresponds a transform $F(\sigma)$ and conversely.

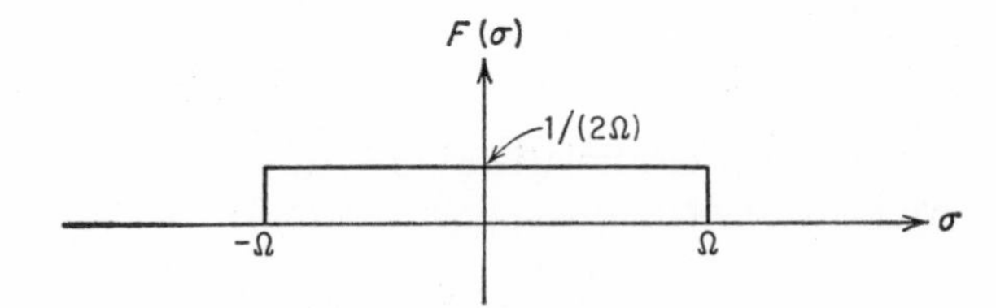

FIG. 23.4-1. Rectangular pulse.

Extensive tables [6, 8] of $f(t)$ and $F(\sigma)$ have been made. We propose to develop only a few which we need for further work.

As a first example, consider the band-limited function

$$f(t) = \int_{-\infty}^{\infty} F(\sigma)e^{2\pi i\sigma t}\,d\sigma \tag{23.4-1}$$

where (Fig. 23.4-1)

$$F(\sigma) = \begin{cases} \dfrac{1}{2\Omega} & \text{for } |\sigma| < \Omega \\ 0 & \text{for } |\sigma| > \Omega \end{cases} \tag{23.4-2}$$

Thus $F(\sigma)$ has unit area, and $f(t)$ has all frequencies in the band $-\Omega < \sigma < \Omega$ and none outside. Using (23.4-2), we can write (23.4-1) in the form

$$\begin{aligned} f(t) &= \frac{1}{2\Omega}\int_{-\Omega}^{\Omega} e^{2\pi i\sigma t}\,d\sigma \\ &= \frac{1}{2\Omega}\,\frac{e^{2\pi i\sigma t}}{2\pi i t}\bigg|_{-\Omega}^{\Omega} = \frac{e^{2\pi i\Omega t} - e^{-2\pi i\Omega t}}{2i}\,\frac{1}{2\Omega\pi t} \\ &= \frac{\sin 2\pi\Omega t}{2\pi\Omega t} \end{aligned} \tag{23.4-3}$$

which corresponds to (23.2-10) (Fig. 23.4-2). Using (23.3-1), we know that the transform is

$$F(\sigma) = \int_{-\infty}^{\infty} \frac{\sin 2\pi\Omega t}{2\pi\Omega t} e^{-2\pi i\sigma t}\, dt \tag{23.4-4}$$

where $F(\sigma)$ is given in (23.4-2).

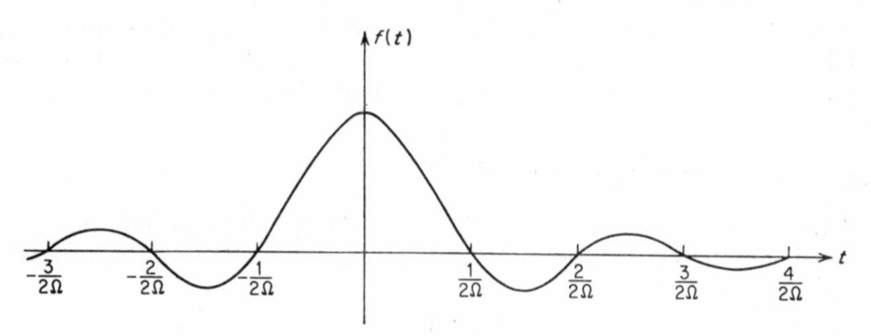

FIG. 23.4-2. The band-limited function ($\sin 2\pi\Omega t/2\pi\Omega t$).

As a second illustration of a relationship between Fourier transforms, suppose that we know the transform pair $f(t)$, $F(\sigma)$; that is,

$$f(t) = \int_{-\infty}^{\infty} F(\sigma) e^{2\pi i\sigma t}\, d\sigma \tag{23.4-5}$$

What is the transform $f_1(t)$ that corresponds to $F(\sigma)e^{2\pi i\sigma y}$? We have

$$\begin{aligned} f_1(t) &= \int_{-\infty}^{\infty} F(\sigma) e^{2\pi i\sigma y} e^{2\pi i\sigma t}\, d\sigma \\ &= \int_{-\infty}^{\infty} F(\sigma) e^{2\pi i\sigma(y+t)}\, d\sigma \end{aligned}$$

Set $y + t = z$; then, from (23.4-5),

$$f_1(t) = f(z) = f(y + t) = \int_{-\infty}^{\infty} [F(\sigma)e^{2\pi i\sigma y}] e^{2\pi i\sigma t}\, d\sigma \tag{23.4-6}$$

Hence the effect of the exponential term is to shift the argument of the transform. Equation (23.4-6) is sometimes known as the "shifting theorem."

23.5 BAND-LIMITED FUNCTIONS AND THE SAMPLING THEOREM

In Sec. 23.2 we gave one rather intuitive derivation of the sampling theorem. Because of its importance in computing, we shall give another derivation of it.

The central idea of the sampling theorem is that of sampling a band-limited function at twice the rate of the highest frequency in the band. The concept of a band-limited function is widely used in many fields of science. In practice it is frequently true that, outside a band, $F(\sigma)$ is very small and the model of a band-limited function provides a useful model. However, it is necessary to give a few words of caution. If $f(t)$ is band-limited, then the mathematical model says that it cannot be "time-limited," that is, vanish for all $|t| > t_0$ for some t_0. In particular, if $f(t)$ represents an electric current, then it must have been flowing for all past time and will flow for all future time. Correspondingly, if $f(t)$ is "time-limited," then it cannot be band-limited. Evidently, the mathematical model should not be pushed too far in applying it to the real world; the mathematical model of a band-limited function is a useful mathematical approximation of, but does not necessarily correspond exactly to, the physical world.

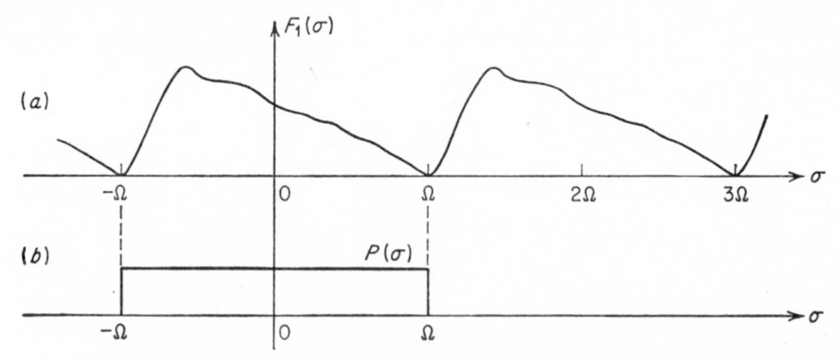

FIG. 23.5-1. (a) $F_1(\sigma)$ is periodic; (b) $P(\sigma)$, the rectangular pulse.

The folding, or Nyquist, frequency is clearly related to the aliasing of frequencies due to discrete, equal-spaced sampling.

To derive the sampling theorem, we start with

$$\begin{aligned} F(\sigma) &= \int_{-\infty}^{\infty} f(t)e^{-2\pi i\sigma t}\,dt \\ f(t) &= \int_{-\infty}^{\infty} F(\sigma)e^{2\pi i\sigma t}\,d\sigma \end{aligned} \tag{23.5-1}$$

If $F_1(\sigma)$ is periodic (Fig. 23.5-1a) with base period $-\Omega < \sigma < \Omega$, then $F_1(\sigma)$ has a Fourier series expansion

$$F_1(\sigma) = \sum_{k=-\infty}^{\infty} c_k e^{(i\pi/\Omega)k\sigma} \tag{23.5-2}$$

where $$c_k = \frac{1}{2\Omega}\int_{-\Omega}^{\Omega} F_1(\sigma)e^{-(i\pi/\Omega)k\sigma}\,d\sigma = \frac{1}{2\Omega} f_1\left(\frac{-k}{2\Omega}\right) = \frac{1}{2\Omega} f\left(\frac{-k}{2\Omega}\right) \tag{23.5-3}$$

Now consider $P(\sigma)$, a rectangular pulse (Fig. 23.5-1b):

$$P(\sigma) = \begin{cases} \dfrac{1}{2\Omega} & \text{for } |\sigma| < \Omega \\ 0 & \text{for } |\sigma| > \Omega \end{cases}$$

By (23.4-3), the transform is

$$p(t) = \frac{\sin 2\pi\Omega t}{2\pi\Omega t} \tag{23.5-4}$$

Finally, since $F(\sigma)$ is assumed to be band-limited to the interval $-\Omega < \sigma < \Omega$, we regard

$$F(\sigma) = F_1(\sigma) \cdot P(\sigma) \cdot 2\Omega$$

and use (23.5-2) and (23.5-3):

$$F(\sigma) = \sum_{k=-\infty}^{\infty} c_k e^{(\pi i/\Omega)k\sigma} P(\sigma) \cdot 2\Omega = \sum_{k=-\infty}^{\infty} f\left(\frac{-k}{2\Omega}\right) P(\sigma) e^{(\pi i/\Omega)k\sigma}$$

We now take the transform back and apply the shifting theorem [Eq. (23.4-6)]:

$$\begin{aligned} f(t) &= \sum_{k=-\infty}^{\infty} f\left(\frac{-k}{2\Omega}\right) \frac{\sin 2\pi\Omega(t + k/2\Omega)}{2\pi\Omega(t + k/2\Omega)} \\ &= \sum_{k=-\infty}^{\infty} f\left(\frac{k}{2\Omega}\right) \frac{\sin \pi(2\Omega t - k)}{\pi(2\Omega t - k)} \end{aligned} \tag{23.5-5}$$

Thus we have the sampling theorem.

At the folding frequency itself we cannot reconstruct the function by using the sampling theorem, since if $\Delta t = 1$ and $f(t) = \sin \pi t$ then we would have all the samples equal to zero and have, from the sampling theorem, $f(t) \equiv 0$.

23.6 THE CONVOLUTION THEOREM

Another useful relation involving Fourier transforms is the convolution theorem. Suppose that we have two functions $f(t)$ and $g(t)$. The *convolution* of $f(t)$ with $g(t)$ is defined as

$$h(t) = \int_{-\infty}^{\infty} f(s)g(t - s)\, ds \tag{23.6-1}$$

Note that, if for a fixed t we write $t - s = s'$,

$$h(t) = \int_{-\infty}^{\infty} f(t - s')g(s')\, ds' \tag{23.6-2}$$

and the convolution of f with g is the same as the convolution of g with f.

We now ask "What is the Fourier transform of the convolution $h(t)$?" By definition, we have

$$H(\sigma) = \int_{-\infty}^{\infty} h(t)e^{-2\pi it\sigma}\,dt$$

Using (23.6-1),

$$\begin{aligned} H(\sigma) &= \int_{-\infty}^{\infty}\int_{-\infty}^{\infty} f(s)g(t-s)\,ds\,e^{-2\pi it\sigma}\,dt \\ &= \int_{-\infty}^{\infty} f(s)e^{-2\pi is\sigma}\left[\int_{-\infty}^{\infty} g(t-s)e^{-2\pi i(t-s)\sigma}\,dt\right]ds \\ &= \int_{-\infty}^{\infty} f(s)e^{-2\pi is\sigma}G(\sigma)\,ds \\ &= F(\sigma)G(\sigma) \end{aligned} \tag{23.6-3}$$

Thus, *the transform of the convolution of two functions is the product of their transforms.* The above formal derivation proves this statement for two time functions, and Exercise 23.6-1 covers the case of the convolution of two frequency functions of σ.

An interesting result that follows from the convolution theorem considers the convolution of $f(t)$ by $f(-t)$:

$$h(t) = \int f(s)f(-t+s)\,ds$$

This is equal to the transform of the product of the transforms

$$h(t) = \int_{-\infty}^{\infty} F(\sigma)F(-\sigma)e^{-2\pi i\sigma t}\,d\sigma$$

Setting $t = 0$,

$$h(0) = \int_{-\infty}^{\infty} f^2(s)\,ds = \int_{-\infty}^{\infty} F(\sigma)F(-\sigma)\,d\sigma \tag{23.6-4}$$

EXERCISES

23.6-1. Prove that if

$$H(\sigma) = \int_{-\infty}^{\infty} F(\tau)G(\sigma-\tau)\,d\tau$$

then

$$h(t) = f(t)g(t)$$

23.6-2. Prove the formula corresponding to (23.6-4), using Exercise 23.6-1.

23.7 THE EFFECT OF A FINITE SAMPLE SIZE

We can regard a finite number of sample points selected from an infinite sequence of sample points in time as the product of the true time function $f(t)$ and a rectangular pulse

$$p(t) = \begin{cases} 1/2T & |t| < T \\ 0 & |t| > T \end{cases}$$

Thus what we have is

$$f_1(t) = p(t)f(t)2T$$

and, by the convolution theorem (Sec. 23.6),

$$F_1(\sigma) = 2T \int_{-\infty}^{\infty} F(\tau) \frac{\sin 2\pi T(\tau - \sigma)}{2\pi T(\tau - \sigma)} d\tau \tag{23.7-1}$$

To see what this means, suppose that we have a function $f(t) = e^{2\pi i \sigma_1 t}$. Then the spectrum of $f(t)$ consists of a single line at $\sigma = \sigma_1$, and the convolution represents a "smearing out" of this line into

$$(2T) \frac{\sin 2\pi T(\sigma_1 - \sigma)}{2\pi T(\sigma_1 - \sigma)} \tag{23.7-2}$$

The larger T is, the narrower is the central spike (Fig. 23.4-2). In a sense, this smearing out represents the "end effects" when we try to determine the frequency from a finite length of a pure sinusoid. If the spectrum is now imagined to consist of many lines, each is smeared out by the same factor (23.7-2), and in the limit of a continuous distribution we get (23.7-1). In the optical analogy, this smearing out corresponds to the ideas of resolving power—the longer the time signal, the better we can resolve the difference between two nearby frequencies.

CHAPTER 24

Linear Filters—Smoothing and Differentiating

24.1 INTRODUCTION

The word "filter," like the expression "power spectrum," arose from the physical situations studied in electrical network theory, but both have achieved much wider application. The filters with which we are concerned remove selected frequencies from a function much as an oil filter removes certain particle sizes. The theory of the design of filters is highly evolved, and we can only touch on the subject here; we shall have to content ourselves with showing what the idea is and how to evaluate certain filters. Those interested in *designing*, as contrasted with *evaluating*, filters for specific situations are referred to the standard courses in circuit and filter design.

In this chapter, we shall study the role of filters in two typical situations, that of smoothing data and that of differentiating data. Both topics have been either carefully avoided or mentioned briefly in Part II; both are difficult and delicate topics to treat, and the polynomial approach provides no insight. Further examples will be given in Chap. 25.

The typical polynomial approach to differentiating some data is to pass an exact-matching polynomial through the sample points, differentiate the polynomial, and then evaluate the derivative at a sample point. A glance at the error term of the exact-matching polynomial [Eq. (8.6-1)]

$$\frac{(x - x_1)(x - x_2) \cdots (x - x_{n+1})f^{(n+1)}(\theta)}{(n + 1)!}$$

shows that almost certainly at the sample points the approximating polynomial is crossing the function; hence a sample point is about the worst possible place to evaluate the derivative.

Occasionally the least-squares approach is used to obtain a smoothed polynomial before differentiating. This process minimizes the sum of the squares of the residuals of the original data at the sample points but says nothing about the behavior between the sample points. Thus, while we can hope that the derivative is estimated accurately, we are not in a very good position to be sure that the least-squares polynomial does not have some wiggles between the sample points which will greatly influence the estimate of the derivative.

The approach to the topics via band-limited functions, while not answering all questions, does give some insight into the situation.

24.2 AN EXAMPLE OF A SIMPLE SMOOTHING FILTER

Suppose that we have a time-varying function $f(t)$ and for one reason or another decide to smooth our equally spaced samples f_k by the linear formula

$$\frac{f_k + f_{k+1} + f_{k+2}}{3} \equiv g_k \qquad (t_k = hk) \tag{24.2-1}$$

What do we do to the spectrum of the function?

Suppose first that $f(t)$ is a pure sinusoid

$$f(t) = e^{2\pi i\sigma t} \qquad (\sigma = \text{real number})$$

It is convenient to introduce an abbreviation

$$2\pi\sigma = \omega$$

σ is measured in cycles per unit time, and ω in radians (of angle) per unit time. Thus we write

$$f(t) = e^{i\omega t}$$

Let us take $h = \Delta t = 1$ for convenience, then using (24.2-1)

$$\begin{aligned} f_k &= e^{i\omega k} \\ g_k &= \frac{e^{i\omega(k+1)}}{3}(e^{-i\omega} + 1 + e^{i\omega}) \\ &= \frac{e^{i\omega(k+1)}}{3}(1 + 2\cos\omega) \end{aligned}$$

Thus the values of f have picked up a multiplicative factor *independent of* k and of size

$$\omega = \left|\frac{1 + 2\cos\omega}{3}\right|$$

It is important to note that some multiplicative factor independent of k will occur because of the linear form of g_k. If g_k were a nonlinear function of the data f_k, then the result would no longer be true.

In the spectrum, the factor becomes

$$\left(\frac{1 + 2\cos\omega}{3}\right)^2 \tag{24.2-2}$$

which is 0 when $\omega = 2\pi/3$, or $\sigma = \frac{1}{3}$. When we plot this curve (Fig. 24.2-1) as a function of the frequency σ, we see that the simple smoothing formula (24.2-1) has the drastic effect of suppressing most of the frequencies in the upper half of the spectrum. Thus the band-limited-function approach sheds some light on the effect that the simple linear smoothing process (filter) [Eq. (24.2-1)] has on the original signal; it weights the frequencies according to the curve of Fig. 24.2-1.

We asked a question about the effect of the linear filter on a single frequency, but our answer really gives us the effect on every frequency;

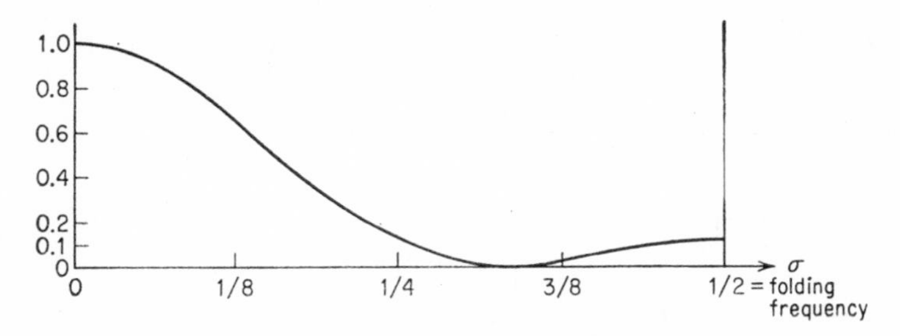

FIG. 24.2-1. The filter $\frac{1}{3}(f_k + f_{k+1} + f_{k+2})$.

hence by using a Fourier-transform representation we can handle arbitrary functions composed of all frequencies.

24.3 AN EXAMPLE OF THE DESIGN OF A FILTER

We have already learned that the sampling frequency is directly connected with the folding frequency in the spectrum. Suppose that we have a signal that is band-limited and has a maximum frequency of 10 cycles per second. This forces us to 20 samples per second to avoid aliasing.

But suppose that we are interested only in frequencies less than 5 cycles per second. We dare not sample at 10 per second, since then the part of the spectrum above 5 cycles per second would be aliased down into the interval of interest.

The filter in Sec. 24.2 suggests a way of eliminating the upper half of the spectrum. As the filter stands, it reduces the frequencies by at least a factor of $\frac{1}{9}$.

We can do better by following the first filter with a second smoothing filter

$$\frac{g_k + g_{k+1} + g_{k+2} + g_{k+3}}{4} \equiv h_k \tag{24.3-1}$$

Then h_k has the effect on $e^{2\pi i\sigma t} = e^{i\omega t}$:

$$\begin{aligned} h_k &= \frac{e^{i\omega k}}{4}(1 + e^{i\omega} + e^{2i\omega} + e^{3i\omega}) \\ &= \frac{e^{i\omega k}}{4}\frac{1 - e^{4i\omega}}{1 - e^{i\omega}} = \frac{e^{i\omega(k+2)}}{4e^{i\omega/2}}\frac{e^{2i\omega} - e^{-2i\omega}}{e^{i\omega/2} - e^{-i\omega/2}} \\ &= \frac{e^{i(k+3/2)}}{4}\frac{\sin 2\omega}{\sin \omega/2} \end{aligned}$$

Hence the spectrum is affected in magnitude by

$$\left(\frac{\sin 2\omega}{\sin \omega/2}\right)^2 = \left(\frac{\sin 4\pi\sigma}{4 \sin \pi\sigma}\right)^2 \tag{24.3-2}$$

which has zeros at $\sigma = 1/4, 1/2, \ldots$.

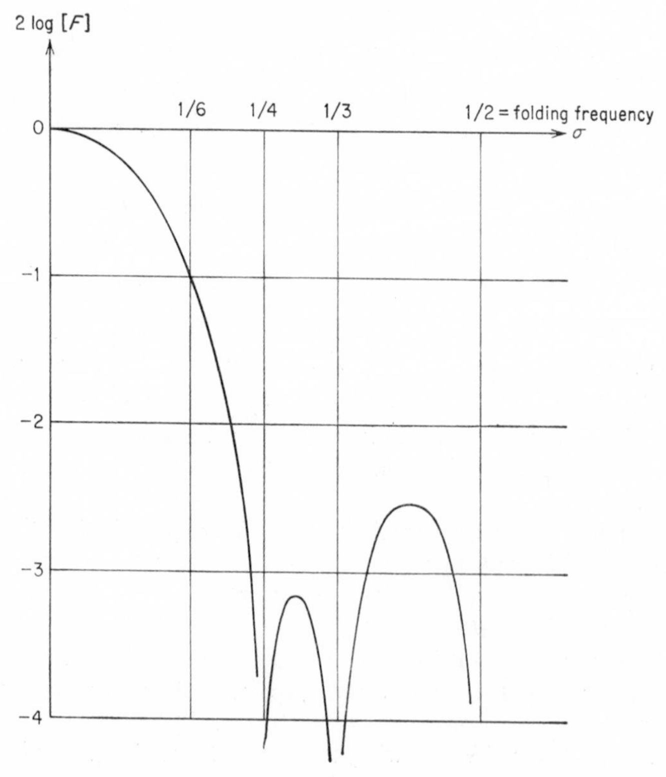

FIG. 24.3-1. Response of the filter

$$h_k = \frac{f_k + 2f_{k+1} + 3f_{k+2} + 3f_{k+3} + 2f_{k+4} + f_{k+5}}{12}$$

The result of the two filters (24.2-1) and (24.3-1), the smoothing by 3's followed by smoothing by 4's, is the filter

$$h_k = \frac{f_k + 2f_{k+1} + 3f_{k+2} + 3f_{k+3} + 2f_{k+4} + f_{k+5}}{12} \tag{24.3-3}$$

and its ability to reject the upper half of the spectrum is very good. See Fig. 24.3-1 where we have plotted the log of the spectrum versus σ.

Having smoothed, we can safely drop alternate terms h_k, keeping, say, h_{2k}, and analyze the result without any serious fear of the resulting aliasing (the folding frequency is now at 5 cycles per second) unless there was a very large amount of "power" in the upper half of the spectrum in relation to that in the lower half. When we finally use the results, we must make due allowance for the effect of filtering on the frequencies up to 5 cycles per second. Thus the filtering action of (24.3-3) allows us to subsample the function without having aliasing troubles. The accuracy for frequencies between ⅓ and ½ of the original folding frequency is not very good.

EXERCISE

24.3-1. Develop the effect on the spectrum of the signal of smoothing by averaging m consecutive terms.

Ans. $(\sin \pi m\sigma / m \sin \pi\sigma)^2$

24.4 FILTERS IN GENERAL

It should now be clear that when we take any linear combination of our equally spaced samples, then the result can be viewed as "filtering" the function in some manner. It has happened many times in the past, before the theory was well understood, that some innocent-looking operation performed at an early stage of a computation radically affected the result obtained, and *the result was interpreted as a physical effect* rather than an effect of the data processing.

As an example, if we have white noise (meaning, just as in white light, all frequencies in a band present to the same degree but at random phases to each other) and happen to process the data to remove most of the frequencies except those in some small region, then the subsequent analysis of the data would show the presence of a dominant frequency, and this same frequency would appear *regardless* of the source of the white noise, perhaps stockmarket prices, perhaps sun-spot data. Such data may have other than a flat spectrum, but the analysis of data contaminated with noise requires considerable judgment.

On the other hand, as was shown in Sec. 24.3, by the proper use of filters very useful effects can be realized. The field is vast and the usual courses in network theory are probably the best source of further knowledge, as the concepts have rarely filtered into computing circles. A further reference, authoritative but hard to read, is Blackman and Tukey.[1]

24.5 ANALYSIS OF A SIMPLE FORMULA FOR DIFFERENTIATING

Perhaps the simplest and most widely used formula for computing a derivative from equally spaced samples is

$$f'_k = \frac{f_{k+1} - f_{k-1}}{2h} \qquad (h = \text{spacing}) \tag{24.5-1}$$

Let us analyze how this treats various frequencies. Suppose that

$$f(t) = e^{2\pi i\sigma t} = e^{i\omega t} \tag{24.5-2}$$

At the kth sample point, let $t_k = hk$. Then

$$f_k = e^{2\pi i\sigma kh} = e^{i\omega kh}$$

Differentiating $f(t)$ gives

$$\left.\frac{df(t)}{dt}\right|_{t=hk} = \frac{df_k}{dt} = i\omega e^{i\omega kh} \tag{24.5-3}$$

On the other hand, (24.5-1) gives the estimate

$$f'_k = e^{i\omega kh}\,\frac{e^{i\omega h} - e^{-i\omega h}}{2h} = i\omega e^{i\omega kh}\,\frac{\sin \omega h}{\omega h} \tag{25.5-4}$$

The ratio of the computed answer (24.5-4) to the true answer (24.5-3) is again independent of k and has the value

$$\frac{\sin \omega h}{\omega h}$$

(see Fig. 23.4-2). Thus formula (24.5-1) clearly underestimates all frequencies except $\omega = 0$ (often called d.c., from direct current). As one would expect when $\omega = \pi/h$, that is, $\sigma = 1/2h$, we get an estimate of zero for the derivative [$\sigma = 1/2h$ is folding frequency].

There is a similar effect for the second derivative

$$f''_k = \frac{f_{k+1} - 2f_k + f_{k-1}}{h^2}$$

[1] Reference 2 or *Bell System Tech. J.*, January and March, 1958.

Using (24.5-2), the estimate is

$$\hat{f}_k'' = e^{i\omega kh} \frac{e^{i\omega h} - 2 + e^{-i\omega h}}{h^2} = -\omega^2 e^{i\omega hk} \frac{2(1 - \cos \omega h)}{\omega^2 h^2}$$

$$= -\omega^2 e^{i\omega hk} \frac{\sin^2 (\omega h/2)}{(\omega h/2)^2}$$

The correct answer is

$$f'' = -\omega^2 e^{i\omega hk}$$

so that the ratio of the results is

$$\frac{\sin^2 (\omega h/2)}{(\omega h/2)^2} \sim 1 - \frac{\omega^2 h^2}{12} + \cdots$$

which again underestimates the quantity except at d.c. At the folding frequency $h\omega = \pi$ we find the estimated value is $-(4/h^2)e^{i\pi k}$.

24.6 AVOIDING ESTIMATION OF DERIVATIVES

Many times the problem of estimating a derivative can be avoided by proper analysis. As an example, suppose that we have data f_k and we know that theoretically $f(t)$ satisfies a second-order differential equation of the form

$$f'' = H(f,t)$$

Using this, we can go from values of $f(t)$ to values of $f''(t)$ and can integrate the latter to obtain $f'(t)$. We frequently prefer integration to differentiation, provided that we are not operating on so long a run of data that a slow steady drift in the integration process would ultimately produce a large error.

Numerous other tricks have been found in special cases, but there appears to be no general theory of when and how derivatives can be avoided.

24.7 FILON'S METHOD

The purpose of this section is to show that the same simple ideas that we have been using can produce insight into what a formula really does without going through the detailed analysis that we have been making in the smoothing and differentiating filter. At the same time, we shall introduce a pair of useful formulas.

The problem of computing integrals of the form

$$I(k) = \int_a^b f(t) \cos kt \, dt \tag{24.7-1}$$

occurs frequently. Filon[1] introduced a method which, technically speaking, belongs in Part II of this book, since it is based on the polynomial approximation of $f(t)$. In particular, the interval (a,b) is divided into $2N$ intervals, and in each double interval $f(t)$ is approximated by a quadratic. Thus it resembles Simpson's composite formula, except for the presence of the extra factor $\cos kt$.

The method of derivation is straightforward but messy,[2] and we shall give only the results. Let

$$h = \frac{b-a}{2N} \tag{24.7-2}$$

be the spacing, let C_{2n} be the sum of all the even-numbered ordinates of $f(t) \cos kt$ *except* that we take one-half of the first and last ordinates,

$$C_{2n} = \tfrac{1}{2}f(a) \cos ka + f(a+2h) \cos k(a+2h) + f(a+4h) \cos k(a+4h) + \cdots + \tfrac{1}{2}f(b) \cos kb \tag{24.7-3}$$

and let C_{2n-1} be the sum of all the odd-numbered ordinates

$$C_{2n-1} = f(a+h) \cos k(a+h) + f(a+3h) \cos k(a+3h) + \cdots + f(b-h) \cos k(b-h) \tag{24.7-4}$$

Then

$$\int_a^b f(t) \cos kt\, dt = h\{\alpha[f(b) \sin kb - f(a) \sin ka] + \beta C_{2n} + \gamma C_{2n-1}\} \tag{24.7-5}$$

where

$$\begin{aligned}
\alpha &= \frac{\theta^2 + \theta \sin\theta \cos\theta - 2\sin^2\theta}{\theta^3} = \frac{2\theta^3}{45} - \frac{2\theta^5}{315} + \frac{2\theta^7}{4{,}725} + \cdots \\
\beta &= \frac{2[\theta(1+\cos^2\theta) - 2\sin\theta\cos\theta]}{\theta^3} = \frac{2}{3} + \frac{2\theta^2}{15} - \frac{4\theta^4}{105} + \frac{2\theta^6}{567} + \cdots \\
\gamma &= \frac{4(\sin\theta - \theta\cos\theta)}{\theta^3} = \frac{4}{3} - \frac{2\theta^2}{15} + \frac{\theta^4}{210} - \frac{\theta^6}{11{,}340} + \cdots
\end{aligned} \tag{24.7-6}$$

and

$$\theta = kh = \frac{k(b-a)}{2N} \tag{24.7-7}$$

A similar formula applies to the integral

$$\int_a^b f(t) \sin kt\, dt = h\{-\alpha[f(b) \cos kb - f(a) \cos ka] + \beta S_{2n} + \gamma S_{2n-1}\} \tag{24.7-8}$$

where S_{2n} and S_{2n-1} are the corresponding sums for $f(t) \sin kt$.

[1] L. N. G. Filon, *Proc. Roy. Soc. Edinburgh*, vol. 49, pp. 38–47, 1928–1929.
[2] See, for example, Ref. 40, pp. 67–72.

Let us try to understand these formulas by the simple use of the ideas that we have so far developed.

The quantities that we are calculating are clearly the Fourier coefficients in a slight disguise; that is, we are computing the frequency content of $f(t)$ at the radian frequency k. The quantity h is the distance between samples, and the quantity

$$\theta = kh$$

scales this to the frequency that we are examining. The sampling theorem indicates that we cannot expect to do this reliably unless we sample at a rate of at least two samples per cycle of length $2\pi/k$, and even then we would require an infinitely long set of samples. Because of the finite range (a,b), we expect θ to be a good deal less than π. (We shall discuss this point more in the next chapter.)

Tables of the coefficients α, β, γ are given in Tranter [40] (for θ in radians) up to $\theta = 1.50$, and in Kopal [20, p. 539] (for θ in degrees) up to 45°. In both cases, and in Filon's original paper, there is no discussion as to *why* the tables break off where they do—why, if $f(t)$ is locally a polynomial, cannot we go as far as we please and make k and hence θ quite large? The integration is done *analytically* once we have made the "analytic substitution" of the polynomial approximation in place of the function. What is to stop us? Evidently it is the sampling theorem in the background that is determining the size θ that we may use.

We also expect that as $\theta \to 0$ we shall approach Simpson's formula. Equations (24.7-6) with (24.7-5) and (24.7-8) clearly show this, the deviation being of order θ^2.

In principle, we could do a detailed analysis and find exactly what Filon's method does to each frequency, but we shall not do it here. The method is adequate, but the process is messy to carry out in a textbook.

EXERCISES

24.7-1. Derive Eq. (24.7-5).
24.7-2. Derive Eq. (24.7-8).

24.8 CLOSING REMARKS

We have not treated (nor do we intend to do so) smoothing and differentiating[1] with full details; as we said, both are delicate topics. However, the main lines of an adequate treatment should be clear. We are given a function (signal) contaminated with noise, perhaps a signal from a distant planet or a function from a computer contaminated by roundoff

[1] The reader is reminded of Lanczos' σ method of Sec. 22.9.

noise. Ideally we have theoretical estimates of the spectra of the function and the noise. We then face the difficult statistical question of shaping a filter to do what we want to the signal while removing as well as possible the unwanted noise.

In practice we often have to estimate the separate spectra from the spectrum of the data that we have. Sometimes the measured spectrum looks something like Fig. 24.8-1.

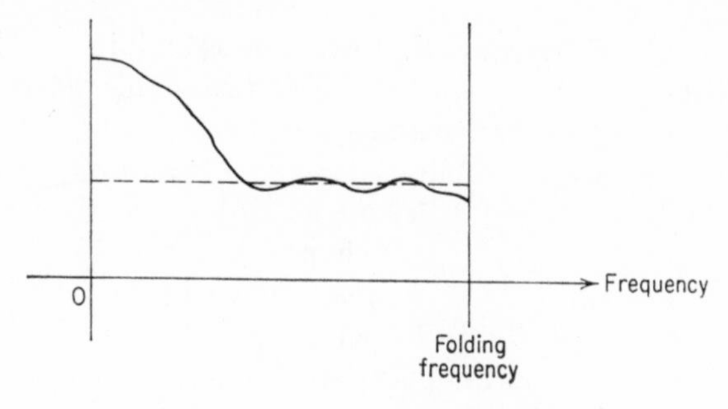

FIG. 24.8-1. A typical spectrum.

This suggests a rectangle[1] of white noise (probably heavily aliased because of sampling) plus a signal in the lower frequencies. We might, therefore, plan a cutoff filter to remove the upper part of the spectrum (along with some of the signal) and combine it with a second filter to get the result that we want from the signal.

It should be clear that adequate estimates of the spectra involved are necessary before designing the filters. Unfortunately, the topic of the measurement of power spectra (Sec. 24.4) lies outside a beginning text, and we must therefore drop these two important topics after only slight treatment.

[1] Spectral analyses of some random numbers of Chap. 32 have given flat, white spectra to within sampling fluctuations but adequate studies of the spectra of round-off noise remain to be done.

CHAPTER 25

Integrals and Differential Equations

25.1 OUTLINE OF CHAPTER

In this chapter we first study the evaluation of a few integration formulas from the point of view of how they treat various frequencies. We shall also examine formulas for integrating differential equations. This naturally leads to the problem of the design of integration methods based on the band-limited-function approach. In particular, we shall look at the Chebyshev criterion.

We have found that the complex notation $e^{2\pi i\sigma t}$ is useful in manipulating various expressions; it also provides a useful criterion for measuring formulas. This latter property has been used implicitly several times, and before continuing we propose to examine it more critically.

Suppose that we are operating on the sinusoid $A \sin 2\pi\sigma t$. If we measure the quality of the formula by the size of the error, we need to know whether $A = 1$ or $A = 1{,}000$ before stating how serious a given error is. A more reasonable measure is the relative error. But if we are near a crossing of the axis, then the relative error can be very large, and we prefer the absolute size of the error.

When we use the complex sinusoid $Ae^{2\pi i\sigma t}$, then, for real σ and t, the modulus is fixed in size

$$|Ae^{2\pi i\sigma t}| = |A|$$

and the relative error provides a satisfactory measure of accuracy.

One further property is that, if we have

$$Ae^{2\pi i\sigma t+i\varphi}$$

and compute the modulus, we find that the phase angle φ drops out. Thus when we try the function

$$Ae^{2\pi i\sigma t}$$

we can often neglect the phase angle φ in our discussion. Alternatively, we can put the phase angle in the coefficient A (which can be a complex number). But we shall see that the *relative* phase of various functions cannot be neglected.

25.2 THE TRANSFER-FUNCTION APPROACH TO INTEGRATION FORMULAS

In Chap. 13 we used a general approach to a class of formulas for computing indefinite integrals. The basic form (13.3-1) was

$$y_{n+1} = a_0 y_n + a_1 y_{n-1} + a_2 y_{n-2} + h(b_{-1} y'_{n+1} + b_0 y'_n + b_1 y'_{n-1} + b_2 y'_{n-2}) \tag{25.2-1}$$

where the y'_k are the integrand values and the y_k are the computed answers to

$$y(t) = y(0) + \int_0^t y'(x)\,dx$$

We propose to reexamine these formulas in the light of our present approach. Suppose, first, that the integrand (the input) is a pure sinusoid

$$y'(t) = A_I e^{2\pi i \sigma t} = A_I e^{i\omega t} \qquad (\omega = 2\pi\sigma) \tag{25.2-2}$$

(The subscript I refers to input.) The true answer is

$$y(t) = -\frac{iA_I}{\omega} e^{i\omega t} + C, \qquad C = y(0) + \frac{iA_I}{\omega} \tag{25.2-3}$$

Since the basic formula (25.2-1) is linear, we expect the computed values (the output) to have the same frequency but perhaps a different phase and amplitude. Let A_O be the output amplitude, where we allow A_O to be complex to include the phase angle. Thus we can assume that the computed $y(t)$, except for roundoff, are of the form

$$y(t) = A_O e^{i\omega t} \tag{25.2-4}$$

We now put these two functions (25.2-2) and (25.2-4) into the basic form (25.2-1) and solve for the ratio A_O/A_I:

$$\frac{A_O}{A_I} = \frac{h(b_{-1} + b_0 e^{-i\omega h} + b_1 e^{-2i\omega h} + b_2 e^{-3i\omega h})}{1 - a_0 e^{-i\omega h} - a_1 e^{-2i\omega h} - a_2 e^{-3i\omega h}} \tag{25.2-5}$$

This ratio is called "the transfer function" and is the multiplicative factor by which an input function at frequency ω is transformed to obtain the output function at the same frequency. This idea of a transfer function is a more formal statement of what we did frequently in Chap. 24 when we examined what happened to a single frequency.

Let us consider in more detail the simple case of the trapezoid rule (12.2-1)

$$y_{n+1} = y_n + \frac{h}{2}(y'_{n+1} + y'_n)$$

Clearly, in (25.2-1),

$$a_0 = 1 \qquad a_1 = a_2 = 0$$
$$b_{-1} = b_0 = \tfrac{1}{2} \qquad b_1 = b_2 = 0$$

and the transfer function is

$$\frac{A_O}{A_I} = \frac{h}{2}\frac{1 + e^{-i\omega h}}{1 - e^{-i\omega h}} = \frac{h \cos(\omega h/2)}{2i \sin(\omega h/2)} = -i\left(\frac{h}{2}\cot\frac{\omega h}{2}\right) \qquad (25.2\text{-}6)$$

When we compare the phase angle of this result with the true answer (25.2-3) we find that both are purely imaginary and we have exactly the correct phase for all ω. When we compare the amplitudes (we set aside the initial conditions as irrelevant), we have to compare

$$\frac{h}{2}\cot\frac{\omega h}{2} \qquad \text{with} \qquad \frac{1}{\omega}$$

For small ωh we have

$$\frac{h}{2}\cot\frac{\omega h}{2} = \frac{h}{2}\left[\frac{2}{\omega h} - \frac{\omega h}{6} - \frac{(\omega h)^3}{360} - \cdots\right] = \frac{1}{\omega} - \frac{\omega h^2}{12} - \frac{\omega^3 h^4}{720} - \cdots$$

Hence the *ratio* of the computed answer to the true answer

$$1 - \frac{\omega^2 h^2}{12} - \frac{\omega^4 h^4}{720} - \cdots$$

deviates from 1 by an order of magnitude

$$-\frac{\omega^2 h^2}{12}$$

Both the transfer function and the ratio of the computed to the true answer are functions of the frequency ω and the spacing h. For the ratio, only the product ωh occurs and shows how the choice of spacing affects the frequency if we try to maintain the same error. The quantity h is the reciprocal of the sampling rate.

Next let us examine Simpson's formula (12.2-2):

$$y_{n+1} = y_{n-1} + \frac{h}{3}(y'_{n+1} + 4y'_n + y'_{n-1})$$

Here we have for (25.2-1)

$$\begin{array}{llll} a_0 = 0 & a_1 = 1 & a_2 = 0 & \\ b_{-1} = 1/3 & b_0 = 4/3 & b_1 = 1/3 & b_2 = 0 \end{array}$$

and the transfer function is

$$\frac{A_O}{A_I} = \frac{h(1 + 4ei^{-i\omega h} + e^{-2i\omega h})}{3(1 - e^{-2i\omega h})} = -\frac{ih}{3}\frac{\cos\omega h + 2}{\sin\omega h} \qquad (25.2\text{-}7)$$

which again has exactly the correct phase for all ω (the factor $-i$) and for small ωh has an amplitude like $1/\omega$.

Finally, let us consider the three-eighths rule [Eq. (12.2-3)]. The transfer function is

$$\frac{3h}{8}\frac{1+3e^{-i\omega h}+3e^{-2i\omega h}+e^{-3i\omega h}}{1-e^{-3i\omega h}} = (-i)\frac{3h}{8}\frac{\cos(3\omega h/2)+3\cos(\omega h/2)}{\sin(3\omega h/2)} \tag{25.2-8}$$

In order to standardize these formulas, let us use the *ratio of the computed to the true answers* and scale ω so that $h = 1$. We have

$$\begin{aligned}
&\text{Trapezoid rule:} && H_1(\omega) = \frac{\omega}{2}\cot\frac{\omega}{2}\\
&\text{Simpson's formula:} && H_2(\omega) = \frac{\omega}{3}\frac{2+\cos\omega}{\sin\omega}\\
&\text{Three-eighths rule:} && H_3(\omega) = \frac{3\omega}{8}\frac{\cos(3\omega/2)+3\cos(\omega/2)}{\sin(3\omega/2)}
\end{aligned} \tag{25.2-9}$$

These three formulas are plotted in Fig. 25.2-1. We have used the $\log_{10}$ of the square of the ratio as the measure, since being twice too large is about as serious as being twice too small and this makes logarithms the natural scale to use. We have shown these formulas as plotted against both $\sigma = \omega/2\pi$ *and* the sampling rate. It is easy to see that Simpson's formula is the best of the three (for reasonably high sampling rates) as far as the size of the error is concerned. The table in Fig. 25.2-1 shows some values which are too small to be read conveniently from the graph but which are important for applications.

Figure 25.2-1 also shows the curve for the formula

$$y_{n+1} = y_{n-1} + h(0.3584y'_{n+1} + 1.2832y'_n + 0.3584y'_{n-1})$$

The coefficients were determined experimentally by Leo Tick so that in the interval $0 \le \omega \le \pi/2$ the error would have the Chebyshev form (see Chap. 19) and also would be exact for $\omega = 0$ (d.c.). He is clearly using twice the Nyquist sampling rate.

The significance of the trapezoid rule's going down while Simpson's formula goes up requires some explanation. Simpson's formula increases the amplitude of the higher frequencies whereas the trapezoid tends to smother them. Although not enough is actually known about roundoff effects, we know that sudden jumps in a function due to roundoff tend to produce some high frequencies; hence Simpson's formula amplifies these whereas the trapezoid rule tends to smother them. If we examine Simpson's formula as a function of the sampling rate, we see that at five samples per cycle the error per step is about 1.5 per cent, which would almost always be too big. At seven samples per cycle we approach crude but sometimes useful accuracy, and at 10 we are committing an error of less than 1 part per 1,000, which in many cases is good enough.

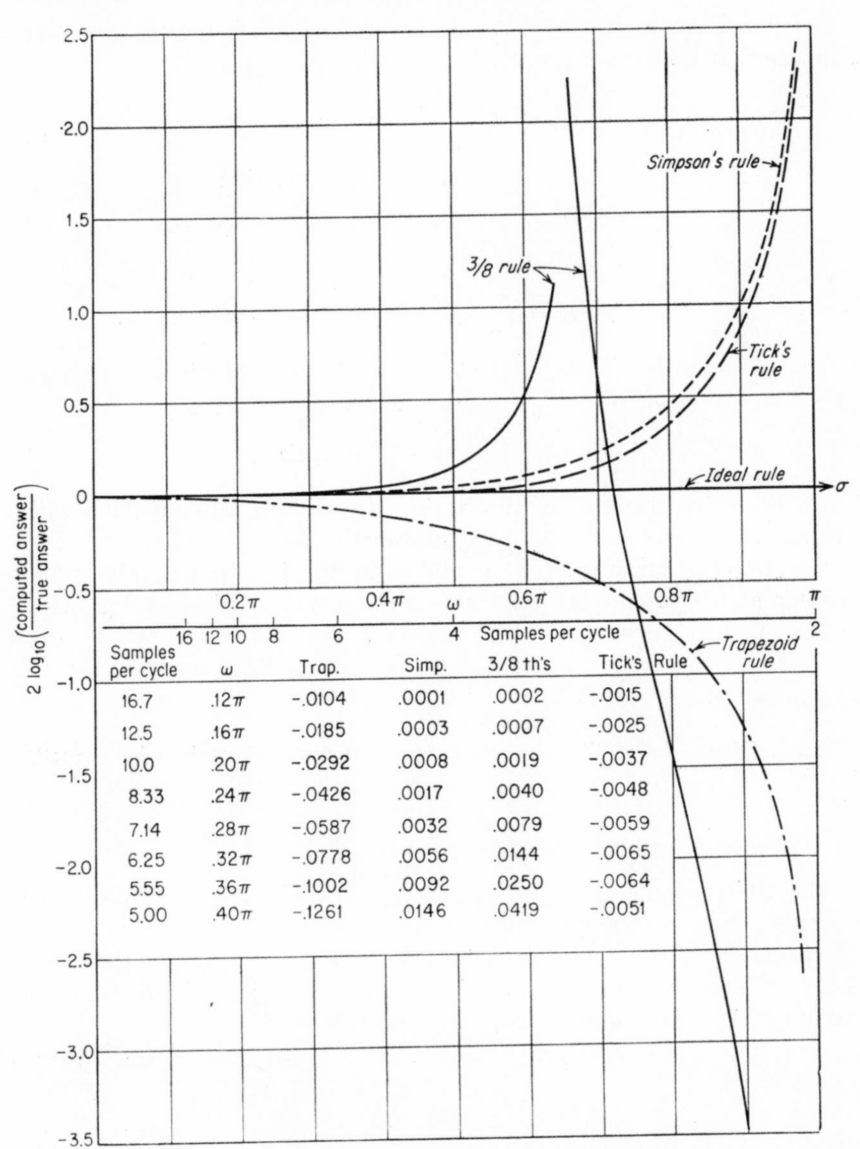

Samples per cycle	ω	Trap.	Simp.	3/8 th's	Tick's Rule
16.7	.12π	−.0104	.0001	.0002	−.0015
12.5	.16π	−.0185	.0003	.0007	−.0025
10.0	.20π	−.0292	.0008	.0019	−.0037
8.33	.24π	−.0426	.0017	.0040	−.0048
7.14	.28π	−.0587	.0032	.0079	−.0059
6.25	.32π	−.0778	.0056	.0144	−.0065
5.55	.36π	−.1002	.0092	.0250	−.0064
5.00	.40π	−.1261	.0146	.0419	−.0051

FIG. 25.2-1. Frequency response to some integration formulas.

So far we have, in principle, been examining a single frequency $\omega = 2\pi\sigma$. Suppose that we have a general function (input)

$$f_I(t) = \int_{-\infty}^{\infty} F(\sigma)e^{2\pi i\sigma t}\,d\sigma$$

Each term $e^{2\pi i\sigma t}$ is transformed by the transfer function

$$G(\omega) = G(2\pi\sigma) = G_1(\sigma)$$

and we obtain for the output

$$f_O(t) = \int_{-\infty}^{\infty} F(\sigma)G_1(\sigma)e^{2\pi i\sigma t}\,d\sigma$$

We have on the right-hand side the transform of a product, and applying the convolution theorem (Sec. 23.6) we have

$$f_O(t) = \int_{-\infty}^{\infty} f_I(s)g_1(t-s)\,ds \tag{25.2-10}$$

Thus, the transform $g_1(t)$ of the transfer function $G_1(\sigma)$ convolved with the input function $f_I(t)$ gives the output function $f_O(t)$.

The transfer function $G_1(\sigma)$ contains all the information about the formula and in a sense is equivalent to the formula of integration being used.

EXERCISE

25.2-1. Discuss the conditions necessary for an integration formula to have exactly the right phase $(-i)$.

25.3 GENERAL INTEGRATION FORMULAS

The transfer-function approach gave the exact phase for the three formulas that we examined in Sec. 25.2 as well as those examined in Chap. 24. When we examine the general form (25.2-1) in the cases listed in Table 13.7-1, matters do not go so smoothly. For example, using the Adams-Bashforth method, we get $(h = 1)$

$$\frac{9 + 19e^{-i\omega} - 5e^{-2i\omega} + e^{-3i\omega}}{24(1 - e^{-i\omega})} = -i\,\frac{[9e^{i\omega/2} + 19e^{-i\omega/2} - 5e^{-3i\omega/2} + e^{-5i\omega/2}]}{48\sin(\omega/2)}$$

and the numerator is not purely imaginary since the imaginary part of the brackets is, for small ω,

$$\left[9\sin\frac{\omega}{2} - 19\sin\frac{\omega}{2} + 5\sin\frac{3\omega}{2} - \sin\frac{5\omega}{2}\right] \sim \omega^5$$

The real part of the brackets is

$$\left[9\cos\frac{\omega}{2} + 19\cos\frac{\omega}{2} - 5\cos\frac{3\omega}{2} + \cos\frac{5\omega}{2}\right] \sim 24$$

so that the deviation from a purely imaginary transfer number is quite small for small ω.

This incorrect phase should not be taken too seriously. While both the amplitude and phase can be studied separately, we are concerned with the failure to have the value 1, and so we now consider

$$\left| \frac{i\omega A_O}{A_I} - 1 \right| \tag{25.3-1}$$

which gives a good single measure of the error.[1] The ratio we used before, $|\omega A_O/A_I|$, can be close to 1 and still have a large phase error; this did not occur in the cases studied previously, since the phase was exactly correct.

Figure 25.3-1 shows this quantity as a function of frequency.

Table 25.3-1 gives the values of the quantity (25.3-1) for the usual range of sampling used. An examination of the table shows that of the stable methods (which excludes Simpson) the two-thirds has the least error for low and moderate sampling rates.

TABLE 25.3-1. $\left| \frac{i\omega A_O}{A_I} - 1 \right|$ FOR VARIOUS INTEGRATION METHODS

Sampling rate	Angle	Trapezoid	Adams-Bashforth	Simpson	Three-eighths
25.00	0.08π	5.27×10^{-3}	1.05×10^{-4}	2.23×10^{-5}	5.06×10^{-5}
16.67	0.12π	1.19×10^{-2}	5.29×10^{-4}	1.14×10^{-4}	2.61×10^{-4}
12.50	0.16π	2.11×10^{-2}	1.66×10^{-3}	3.66×10^{-4}	8.49×10^{-4}
10.00	0.20π	3.31×10^{-2}	4.02×10^{-3}	9.08×10^{-4}	2.15×10^{-3}
8.33	0.24π	4.78×10^{-2}	8.25×10^{-3}	1.92×10^{-3}	4.67×10^{-3}
7.14	0.28π	6.53×10^{-2}	1.51×10^{-2}	3.66×10^{-3}	9.16×10^{-3}
6.25	0.32π	8.57×10^{-2}	2.54×10^{-2}	6.44×10^{-3}	1.68×10^{-2}

Sampling rate	Angle	One-third	One-half	Two-thirds	"S"
25.00	0.08π	5.05×10^{-5}	5.00×10^{-5}	3.44×10^{-5}	1.28×10^{-4}
16.67	0.12π	2.60×10^{-4}	2.55×10^{-4}	1.76×10^{-4}	6.24×10^{-4}
12.50	0.16π	8.39×10^{-4}	8.10×10^{-4}	5.66×10^{-4}	1.88×10^{-3}
10.00	0.20π	2.11×10^{-3}	2.00×10^{-3}	1.41×10^{-3}	4.34×10^{-3}
8.33	0.24π	4.53×10^{-3}	4.18×10^{-3}	3.02×10^{-3}	8.51×10^{-3}
7.14	0.28π	8.75×10^{-3}	7.85×10^{-3}	5.78×10^{-3}	1.49×10^{-2}
6.25	0.32π	1.57×10^{-2}	1.36×10^{-2}	1.03×10^{-2}	2.41×10^{-2}

[1] In some problems, such as in many acoustical ones, phase errors are of minor importance; in some problems such as in the estimation of cospectra, they can be of prime importance.

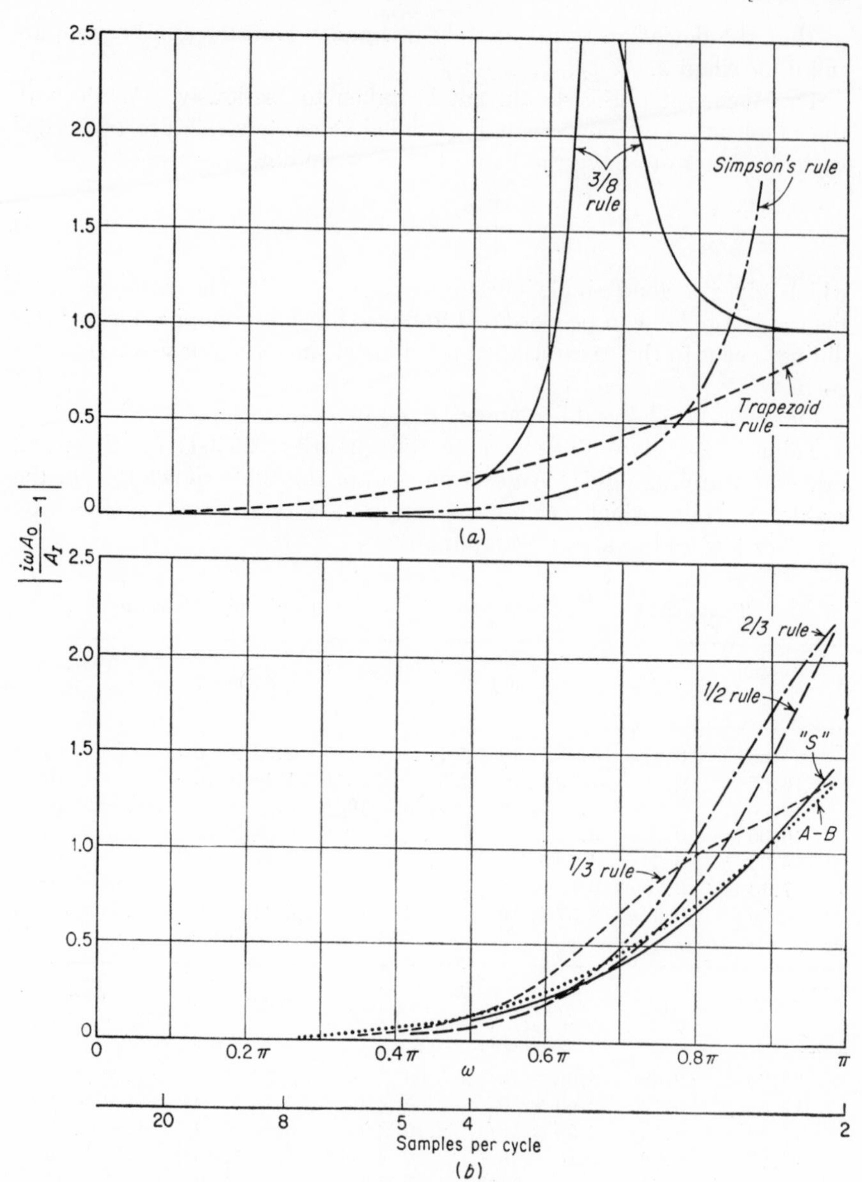
2.5
2.0
1.5
1.0
0.5
0
3/8 rule
Simpson's rule
Trapezoid rule
(a)
$\left|\frac{i\omega A_O}{A_I}-1\right|$
2/3 rule
1/2 rule
"S"
A–B
1/3 rule
0
0.2π
0.4π
0.6π
0.8π
π
ω
20
8
5
4
2
Samples per cycle
(b)

FIG. 25.3-1

25.4 DIFFERENTIAL EQUATIONS

There is an essential difference between computing an indefinite integral and solving an ordinary differential equation. For the purpose of clarity, suppose that the differential equation is

$$y' = Ay + f(t)$$

and the indefinite integral is

$$y' = f(t)$$

We have for the integral the schematic diagram

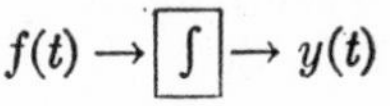

$$f(t) \rightarrow \boxed{\int} \rightarrow y(t)$$

and for the differential equation

$$f(t) \rightarrow \boxed{\int} \rightarrow y(t)$$
$$\boxed{.A}$$

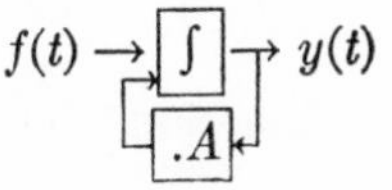

where we have a feedback loop.

This difference makes a different approach more suitable, although the transfer-function approach can be used if desired. We shall use the approach of studying the size of the error in a step, rather than the ratio of output to input, since it is usually of more interest for differential equations. We are, in fact, giving different answers to question 4, "What accuracy?", since different situations require the accuracy criterion to be applied in different ways.

We adopt the general form (15.2-1) (which looks the same, as far as notation is concerned, as Secs. 25.2 and 25.3 but is different in meaning since the y' values now come from the y values via the differential equation):

$$\begin{aligned} y_{n+1} = a_0 y_n + a_1 y_{n-1} + a_2 y_{n-2} \\ + h(b_{-1} y'_{n+1} + b_0 y'_n + b_1 y'_{n-1} + b_2 y'_{n-2}) \end{aligned} \tag{25.4-1}$$

In order to study the frequency response of (25.4-1), we substitute the function

$$y(t) = e^{2\pi i \sigma t} = e^{i\omega t} \qquad (\omega = 2\pi\sigma)$$

and compute the error as a function of frequency.

$$\begin{aligned} G(\sigma) = G_1(\omega) = e^{i\omega(t+h)} - a_0 e^{i\omega t} - a_1 e^{i\omega(t-h)} - a_2 e^{i\omega(t-2h)} \\ - hi\omega[b_1 e^{i\omega(t+h)} + b_0 e^{i\omega t} + b_1 e^{i\omega t(t-h)} + b_2 e^{i\omega(t-2h)}] \end{aligned} \tag{25.4-2}$$

$$\begin{aligned} |G_1(\omega)| = |1 - a_0 e^{-i\omega h} - a_1 e^{-2i\omega h} - a_2 e^{-3i\omega h} \\ - ih\omega(b_{-1} + b_0 e^{-i\omega h} + b e^{-2i\omega h} + b_2 e^{-3i\omega h})| \end{aligned}$$

If now we use a general function

$$y(t) = f(t) = \int_{-\infty}^{\infty} F(\sigma)e^{2\pi i\sigma t}\, d\sigma$$

we find that the error is

$$R[f(t)] = \int_{-\infty}^{\infty} F(\sigma)G(\sigma)e^{2\pi i\sigma t}\, d\sigma \qquad (25.4\text{-}3)$$

Using the convolution theorem, we get

$$R[f(t)] = \int_{-\infty}^{\infty} f(s)g(t-s)\, ds = \int_{-\infty}^{\infty} g(t-s)y(s)\, ds \qquad (25.4\text{-}4)$$

In this form the error may be compared with the error form of Chap. 11. In particular, we had [Eq. (11.3-8)]

$$R[f(t)] = \int_{A}^{B} f^{(m)}(s)G(s)\, ds$$

where $G(s)$ was the influence function and was composed of terms of the form

$$(x_k - s)_+^j$$

We again see a significant difference between the two theories: The polynomial approach leads us to express the error as an integral of the mth *derivative* of the function times an influence function, whereas the Fourier-integral approach leads us to use *the function itself* or, alternatively, its transform rather than the mth derivative and, of course, a different influence function. [See also (25.2-10).] However, the form (25.4-4) is deceptive; if the function $g(t - s)$ is examined closely, then it is seen to be merely (25.4-1) in disguise.

25.5 CHEBYSHEV DESIGN OF FILTERS

The availability of a method of evaluation leads promptly to the first stages of design. So many different possibilities arise in practice that we cannot hope to cover them all; rather we shall content ourselves with sketching an approach to Chebyshev design, as this type is becoming increasingly important. One of the aims of this book has been gradually to bring the reader to the position where he is able to use the principles given in the text to design his own formulas, and we believe that this point has now been reached.

Suppose that we wish to design a method for integrating differential equations such that the error in the frequency domain has a Chebyshev form rather than the form that we have had so far—very good at the origin and increasingly poor as the frequency increases. For some purposes the Chebyshev error form is clearly preferable to the power-series error form.

We begin by transforming our range of frequencies in ω so that they lie between -1 and $+1$; hence we can now use the standard Chebyshev polynomials. If we adopt the general form of the last section [Eq. (25.4-1)], we are led to the G function for the error at frequency ω [Eq. (25.5-2)]

$$\begin{aligned} G(\sigma) = e^{i\omega(t+h)} &- a_0 e^{i\omega t} - a_1 e^{i\omega(t-h)} - a_2 e^{i\omega(t-2h)} \\ &- ih\omega[b_{-1}e^{i\omega(t+h)} + b_0 e^{i\omega t} + b_1 e^{i\omega(t-h)} + b_2 e^{i\omega(t-2h)}] \end{aligned} \qquad (25.5\text{-}1)$$

We now recall the observation in Chap. 19 on Chebyshev polynomials to the effect that if we wish to have the error in the form of a Chebyshev polynomial then we should consider getting the whole expression in Chebyshev polynomials. For this we need to be able to express $e^{i\omega z}$ in terms of Chebyshev polynomials. A search of Watson's "Bessel Functions" [41, p. 14] reveals the generating function

$$e^{(z/2)(t-1/t)} = \sum_{n=-\infty}^{\infty} t^n J^n(z)$$

Set $t = ie^{i\theta}$ and use

$$J_n(z) = (-1)^n J_n(z)$$

which is the same as

$$(i)^{-n} J_{-n}(z) = (i)^n J_n(z)$$

to get

$$e^{iz\cos\theta} = J_0(z) + 2\sum_{n=1}^{\infty} (i)^n J_n(z) \cos n\theta$$

We set $\cos\theta = \omega$, and we have

$$e^{iz\omega} = J_0(z) + 2\sum_{n=1}^{\infty} (i)^n J_n(z) T_n(\omega) \qquad (-1 \le \omega \le 1) \quad (25.5\text{-}2)$$

In (25.5-1) we factor out the term $e^{i\omega t}$, and we have

$$\begin{aligned} |G(\sigma)| = |e^{i\omega h} &- a_0 - a_1 e^{-i\omega h} - a_2 e^{-2i\omega h} \\ &- ih\omega(b_{-1}e^{i\omega h} + b_0 + b_1 e^{-i\omega h} + b_2 e^{-2i\omega h})| \end{aligned}$$

Using (25.5-2) ($z = h, 0, -h, -2h$ in turn) and the appropriate identities involving $\omega T_k(\omega)$, we can arrange everything in an expansion in Chebyshev polynomials with complex coefficients (involving Bessel functions).

If we now start equating coefficients to zero, we get one equation for each complex coefficient, and we could proceed as far as we wished, but we would have to save a parameter or two for stability and other desirable properties. The result would be an error curve that did not pass through zero at zero (d.c.) frequency, and this would probably, though not always, cause us some embarrassment; the simple equation

$$y' = 0 \qquad y(0) = A \neq 0$$

would have a solution that was not a constant. To remedy this, we could start by requiring that zero frequency fit exactly; that is,

$$1 = a_0 + a_1 + a_2$$

and then begin equating coefficients to zero. Thus we would be sure that, within roundoff, d.c. would be computed correctly. Note that, if we keep the same frequency band and change the sampling rate by changing h, then *all* the coefficients change, rather than merely changing a factor h as in the polynomial case.

Thus we see the main outlines of the Chebyshev approach. We see how to find the formula desired, provided that we are willing to do a little work to obtain it. And this is true over a wide range of formulas—the general approach frequently works provided that we are willing to do some hard work and at times use a little imagination to overcome a few messy spots. The art of finding formulas has been practically reduced to a science; the methods developed so far suffice to provide more formulas than there is room to list in a book many times thicker than this one, since there are many combinations of ideas that can be used in different circumstances. It is hoped that the reader will now feel that he can, with some labor, design formulas to fit the situation rather than force the situation to fit the classical formulas; it is also hoped that he will do so.

25.6 SOME DETAILS OF CHEBYSHEV DESIGN

Since the Chebyshev design methods are not readily available in the literature, we shall indicate some of the details for integration formulas. Let the integration formula be of the standard form

$$y_{n+1} = a_0 y_n + a_1 y_{n-1} + a_2 y_{n-2} + h(b_{-1} y'_{n+1} + b_0 y'_n + b_1 y'_{n-1} + b_2 y'_{n-2}) \tag{25.6-1}$$

We assume, as in the Fourier approach, that

$$y(t) = e^{i\omega t}$$

and using Eq. (25.2-2) we have

$$y(t) = J_0(t) + 2 \sum_{n=1}^{\infty} (i)^n J_n(t) T_n(\omega) \tag{25.6-2}$$

The derivative is

$$y'(t) = J'_0(t) + 2 \sum_{n=1}^{\infty} (i)^n J'_n(t) T_n(\omega) \tag{25.6-3}$$

We now substitute these two expressions into (25.6-1) using the appropriate values of t, namely, $t = h,\ 0,\ -h,\ -2h$, and factor out the term $e^{i\omega nh}$.

$$
\begin{aligned}
J_0(h) + 2 \sum_{n=1}^{\infty} (i)^n J_n(h) T_n(\omega) = {} & a_0 J_0(0) \\
& + a_1 \left[J_0(-h) + 2 \sum_{n=1}^{\infty} (i)^n J_n(-h) T_n(\omega) \right] \\
& + a_2 \left[J_0(-2h) + 2 \sum_{n=1}^{\infty} (i)^n J_n(-2h) T_n(\omega) \right] \\
& + b_{-1} h \left[J_0'(h) + 2 \sum_{n=1}^{\infty} (i)^n J_n'(h) T_n(\omega) \right] \\
& + b_0 h [2i J_1'(0) T_1(\omega)] \\
& + b_1 h \left[J_0'(-h) + 2 \sum_{n=1}^{\infty} (i)^n J_n'(-h) T_n(\omega) \right] \\
& + b_2 h \left[J_0'(-2h) + 2 \sum_{n=1}^{\infty} (i)^n J_n'(-2h) T_n(\omega) \right]
\end{aligned}
$$

We next rearrange this to have the form of an expansion in Chebyshev polynomials, and we equate coefficients of T_n.

$T_0(\omega)$:

$$J_0(h) = a_0 + a_1 J_0(h) + a_2 J_0(2h) + h[b_{-1} J_0'(h) - b_1 J_0'(h) - b_2 J_0'(2h)]$$

$\dfrac{T_1(\omega)}{2i}$:

$$
\begin{aligned}
J_1(h) = -a_1 J_1(h) - a_2 J_1(2h) + h[b_{-1} J_1'(h) + b_0 J_1'(0) + b_1 J_1'(h) \\
+ b_2 J_1'(2h)] \qquad (25.6\text{-}4)
\end{aligned}
$$

$\dfrac{T_k(\omega)}{2(i)^k}$:

$$
\begin{aligned}
J_k(h) = (-1)^k [a_1 J_k(h) + a_2 J_k(2h)] + h[b_{-1} J_k'(h) - (-1)^k b_1 J_k'(h) \\
- (-1)^k b_2 J_k'(2h)]
\end{aligned}
$$

The first (and degenerate) case we consider is the equivalent of the trapezoid rule; that is, we set $a_1 = a_2 = b_1 = b_2 = 0$. We shall also require the condition that (25.6-1) be exact for $y \equiv 1$ (the d.c. case), which amounts to setting

$$a_0 = 1$$

We now take the coefficients of $T_0(\omega)$ and $T_1(\omega)$

$$
\begin{aligned}
J_0(h) &= 1 + h b_{-1} J_0'(n) \\
J_1(h) &= h[b_{-1} J_1'(n) + b_0 J_1'(0)]
\end{aligned}
$$

From these we get [where $J_0'(h) = -J_1(h)$]

$$b_{-1} = \frac{1 - J_0(h)}{hJ_1(h)}$$
$$b_0 = 2\frac{J_1(h) - 2hb_{-1}J_1'(h)}{h} \qquad (25.6\text{-}5)$$

In order to understand and check these equations we examine what happens as h approaches zero. We know that

$$J_n(z) = \frac{1}{n!}\left(\frac{z}{2}\right)^n\left[1 - \frac{(z/2)^2}{n+1} + \frac{(z/2)^4}{2!(n+1)(n+2)} - \frac{(z/2)^6}{3!(n+1)(n+2)(n+3)} + \cdots\right]$$

so that

$$J_0(h) = 1 - \frac{h^2}{4} + \frac{h^4}{64} - \cdots$$
$$J_1(h) = \frac{h}{2} - \frac{h^3}{16} + \frac{h^5}{384} - \cdots$$
$$J_2(h) = \frac{h^2}{8} - \frac{h^4}{96} + \frac{h^6}{3{,}072} - \cdots$$

and

$$b_{-1} = \frac{h^2/4 - h^4/64 + \cdots}{h^2/2 - h^4/16 + \cdots} \simeq \frac{1 - h^2/16}{2(1 - h^2/8)} \to \frac{1}{2}$$
$$b_0 = \frac{h - h^3/8 + \cdots - 2b_{-1}h(\frac{1}{2} - 3h^2/16 + \cdots)}{h} \to \frac{1}{2}$$

At first it may seem surprising that the two coefficients b_{-1} and b_0 are not equal, but a little thought shows that there was nothing else to be expected. The difficulty lies in the fact that we made the error curve go through zero at $\omega = 0$ and then tried to make the dominant term in the error expansion proportional to $T_2(\omega)$, which clearly does not have a zero at $\omega = 0$. The case is really too trivial to be meaningful.

The next case to be considered is that for which we use two old values of the function and the derivative, setting only $a_2 = b_2 = 0$. In this case we expect to save one parameter for stability. Thus we have the equations

$$\begin{aligned}
&\text{d.c.:} & 1 &= a_0 + a_1\\
&T_0(\omega): & J_0(h) &= a_0 + a_1J_0(h) + h[b_{-1}J_0'(h) - b_1J_0'(h)]\\
&T_1(\omega): & J_1(h) &= -a_1J_1(h) + h[b_{-1}J_1'(h) + b_0J_1'(0) + b_1J_1'(h)]\\
&T_2(\omega): & J_2(h) &= a_1J_2(h) + h[b_{-1}J_2'(h) - b_1J_2'(h)]
\end{aligned}$$

The second equation can be written, eliminating a_0, as

$$0 = (1 - J_0)(1 - a_1) + hJ_0'(b_{-1} - b_1)$$

and the fourth equation as

$$0 = -J_2(1 - a_1) + hJ_2'(b_{-1} - b_1)$$

from which we see that

$$a_1 = 1$$
$$b_{-1} = b_1$$

and we are on the edge of stability (Ex. 13.7-1). From the first equation

$$a_0 = 0$$

The third equation is now [where $J_1'(0) = \frac{1}{2}$]

$$2J_1 = 2hb_{-1}J_1' + \frac{hb_0}{2} \tag{25.6-6}$$

As $h \to 0$ this approaches

$$2b_{-1} + b_0 = 2 \tag{25.6-7}$$

as it should.

These results remind us of Tick's method (Sec. 25.2) which used four samples per cycle, $h = \pi/2$. In his method he added the condition (25.6-7) that the method of integration be exact when $y(t)$ is a straight line. Using Tick's method we have to solve (25.6-6) and (25.6-7), which gives us

$$b_{-1} = \frac{2J_1(h) - h}{2J_1'(h) - 1} = 0.35785 \qquad (h = \pi/2)$$

and agrees remarkably well with Tick's 0.3584. We have, of course, the solution for any sampling rate. In this example the higher-order terms in the Chebyshev expansion apparently contribute very little.

An alternative approach would have been to make the formula exact for the frequencies that correspond to the zeros of $T_3(\omega)$, namely, $\omega = 0$ and $\pm\sqrt{3}/2$; thus the error curve would have the form of $T_3(\omega)$.

The above examples show some of the elements of the Chebyshev design method. In particular cases, however, there are often special details that influence how we attack the problem of finding a formula to fit a situation. The variety of these details precludes developing the method further in an elementary book.

CHAPTER 26

Exponential Approximation

26.1 INTRODUCTION

Of the three classes of functions which we studied in Sec. 7.4 and which are invariant under a translation of the independent variable axis—polynomials, sines and cosines, and exponentials—we have examined polynomials in Chaps. 7 through 20, sines and cosines in Chaps. 21 through 25, and we now come to one short chapter on exponentials. Why this disparity in the treatment? Several reasons can be given. First, the class of polynomials is also invariant under scale changes and hence is more important in some respects. Second, much of the treatment of polynomials was devoted to developing techniques which can be used on the other two classes. Third, the polynomial often has fewer parameters than has the exponential form, which tends to make them both easier to discuss and less effective to use in many situations. It is also true that exponentials can be regarded as sines and cosines of pure imaginary argument, and we shall again find that some of the details have already been developed. Yet another reason is that both the literature and the history of mathematics give the class of exponentials a minor role as compared with the other two classes. Finally, it is part of the plan of this book to pass from specific details to general ideas, and we therefore now deliberately omit many details.

The problems of exponential approximation can be divided roughly into two classes: those in which the exponents are known and those in which they are not. We first examine the case where they are known.

26.2 ON FINDING FORMULAS USING EXPONENTIALS WHEN THE EXPONENTS ARE KNOWN

In Chap. 10 we developed a rather uniform approach to finding formulas and applied it to the polynomial case; we use the method again.

Suppose that we have a linear operator $L(f)$, such as integration, differentiation, interpolation, etc. (see Sec. 10.3), and we want to represent the answer as a linear combination of weighted sample points (we exclude derivatives for convenience *only*).

$$L(f) = w_0 f(x_0) + w_1 f(x_1) + \cdots + w_{n-1} f(x_{n-1}) \qquad (26.2\text{-}1)$$

We have n parameters, $w_0, w_1, \ldots, w_{n-1}$, and hence can make the formula exact for n functions (we hope), $e^{\alpha_0 x}, e^{\alpha_1 x}, \ldots, e^{\alpha_{n-1} x}$. Often $\alpha_0 = 0$, so that we have a constant term.

The defining equations corresponding to (10.3-1) are

$$m_k = w_0 e^{\alpha_k x_0} + w_1 e^{\alpha_k x_1} + \cdots + w_{n-1} e^{\alpha_k x_{n-1}} \qquad (k = 0, 1, \ldots, n-1) \qquad (26.2\text{-}2)$$

where, of course, now

$$m_k = L(e^{\alpha_k x}) \qquad (26.2\text{-}3)$$

We therefore ask if we can invert the system of equations; this depends on the value of the determinant

$$|e^{\alpha_k x_j}| \qquad (k = 0, 1, \ldots, n-1) \quad (j = 0, 1, \ldots, n-1) \qquad (26.2\text{-}4)$$

Suppose that the α_k are equally spaced, in particular, that $\alpha_k = k$. Then we set

$$e^{x_j} = u_j \qquad (26.2\text{-}5)$$

and the determinant is

$$|u_j^k| \qquad (26.2\text{-}6)$$

which is the Vandermonde determinant (Sec. 8.2) and is not zero unless $x_i = x_j$ for some $i \neq j$. Indeed, in this case the change of variables (26.2-5) reduces the problem to the polynomial case.

We now consider the general case of equally spaced values of α_k and set

$$\alpha_k = a + bk$$

Then (26.2-2) becomes

$$m_k = \sum_{j=0}^{n-1} w_j e^{\alpha_k x_j}$$

$$= \sum_{j=0}^{n-1} (w_j e^{a x_j})(e^{b x_j})^k$$

If we set

$$w_j e^{a x_j} = \bar{w}_j$$

$$e^{b x_j} = u_j$$

we have

$$m_k = \bar{w}_j u_j^k$$

which is again the polynomial case. Thus the case of equally spaced exponents is simply the polynomial case in disguise.

When we solve for the w_k (or some multiple of them, $\bar{w}_k$), we have found the formula for which we were looking.

When the exponents are known but not equally spaced, then the proof that the determinant (26.2-4) is not zero is not so easy, and we shall not discuss it further here. In all cases, the solution of the defining equations (26.2-2) gives the formula that is being sought.

EXERCISE

26.2-1. Find, by two different methods, the formula

$$\int_0^1 f(x)\,dx = a_0 f(0) + a_{1/2} f(1/2) + a_1 f(1)$$

which is exact for $f(x) = 1$, e^{-x}, and e^{-2x}.

26.3 UNKNOWN EXPONENTS

The case of interpolation, or representation of a function, using sums of exponentials with unknown exponents is of importance, as it is the basis of the analytic-substitution approach. Prony gave a simple method for finding the exponents when the data are equally spaced. We have

$$f(x) = A_0 e^{\alpha_0 x} + A_1 e^{\alpha_1 x} + \cdots + A_{k-1} e^{\alpha_{k-1} x} \tag{26.3-1}$$

for some set of values $x = x_j$ $(j = 1, \ldots, n)$ which are equally spaced. It is no real restraint to assume $x_j = j$.

Prony observed that each of the

$$e^{\alpha_i x} \qquad (i = 0, 1, \ldots, k-1)$$

satisfies a kth-order difference equation with constant coefficients whose characteristic roots are

$$\rho = e^{\alpha_i}$$

Hence $f(x)$ also satisfies this difference equation. Let this difference equation be

$$f(j) + C_1 f(j+1) + \cdots + C_k f(j+k) = 0 \qquad (j = 0, 1, 2, \ldots) \tag{26.3-2}$$

We now consider two cases. If we have exactly as many equations (26.3-2) as unknowns C_m $(m = 1, 2, \ldots, k)$, then we simply have to examine the persymmetric determinant (see Sec. 10.6)

$$|f(j+n)|$$

If this is not zero, then we can solve for the C_j values. From the C_j, we find the characteristic equation

$$\rho^k + C_1 \rho^{k-1} + \cdots + C_k = 0$$

and from its roots, we find the α_i. Note how this resembles the method used in solving the system for Gauss quadrature (Sec. 10.6). When we know the α_i, we can solve the first k equations for the A_i. Thus $2k$ of equally spaced samples of $f(x)$ determine the $2k$ unknowns α_i and A_i.

In the second case, we have more than $2k$ samples, and we can use least squares (Chaps. 17 and 18) if we wish and find the *normal equations* corresponding to (26.3-2), from which we find in turn the α_i and A_i.

Once we have the interpolating function, we can use analytic substitution and find any formulas that we wish.

If we wish to go directly to the answer without passing through the interpolating function, we have merely to substitute the moments in place of the function values in (26.3-1); we of course use as many moments as there are unknowns—$2k$ in all. Again the general method suffices to find us the formula that we want. However, we must be careful which part of the problem is being approximated by a sum of exponentials since different places for the approximation may give different answers.

EXERCISE

26.3-1. Fit the function $y = A_1e^{-\alpha_1 x} + A_2e^{-\alpha_2 x}$ to the data

x	0	1	2	3
y	5.5	5.0	5.1	4.3

26.4 WARNINGS

While, in principle, we have shown how to find the exponents, things do not always go so well in practice; sometimes it happens that the number of terms to use is not known but is to be found. An illustration of this is radioactive decay where the terms correspond to various half-lives in the decay chain being investigated.

Let us examine the simple case of trying to distinguish between

$$Ae^{-\alpha t} \text{ and } \frac{A}{2}e^{-(\alpha+\epsilon)t} + \frac{A}{2}e^{-(\alpha-\epsilon)t} = Ae^{-\alpha t}\left[\frac{e^{-\epsilon t} + e^{\epsilon t}}{2}\right]$$

The expression in brackets is

$$1 + \frac{\epsilon^2 t^2}{2} + \frac{\epsilon^4 t^4}{24} + \cdots$$

The difference depends on ϵ^2, and only for large t can we hope to detect this amid the background noise of measurement, but, for large t, $e^{-\alpha t}$ is small! A similar situation applies to Laplace transforms

$$f(t) = \int_0^\infty F(\sigma)e^{-\sigma t}\,d\sigma$$

Given $F(\sigma)$, it is easy to compute $f(t)$, but given $f(t)$ in the form of data the problem of finding $F(\sigma)$ is more difficult. Lanczos states [23, p. 287]: ". . . and thus it is demonstrated that physical *observations* of the Laplace transform can never solve the problem of restoring the indicial functions with any degree of accuracy." One of the difficulties is that it is the values of $F(\sigma)$ for large σ that determine the values of $f(t)$ for small t, and vice versa. With proper care and known limitations on $f(t)$ or some extra information about $F(\sigma)$, the inverse transform can sometimes be found.

The reader should not give up when faced with an inverse Laplace transform, since isolated results have been successfully found, but generally satisfactory methods are not presently known to the author.

26.5 EXPONENTIALS AND POLYNOMIALS

When the over-all behavior of a problem is exponential in character, the use of a polynomial times a suitable exponential often is more manageable than is a sum of exponentials. Gauss-Laguerre integration (Sec. 12.5) and the method used in the latter part of Sec. 16.5 illustrate this point. Since the general techniques necessary to carry out this suggestion have been developed in Part II, we do not need to discuss them further.

26.6 ERROR TERMS

When the formulas have been found, it is natural to ask for the error terms. The approach in Chap. 11 of Part II started with an expansion [Eq. (11.3-2)] of the arbitrary function $f(x)$ in terms of the particular functions $1, x, \ldots, x^{m-1}$ for which the formula was made exact. This expansion has been generalized to arbitrary sets of functions by Petersson, and a treatment of it can be found in A. S. Householder's excellent (but difficult to read) book "Principles of Numerical Analysis" [16].

In practice the usefulness of such error terms is open to debate, and so far they have seldom been used. The error terms corresponding to that developed for band-limited functions have apparently not been investigated.

CHAPTER 27

Singularities

27.1 INTRODUCTION

In Sec. 7.4 we discussed the fact that there were essentially only three classes of functions which, as classes, were invariant under translation along the axis of the independent variable. These three classes have been treated in Part II, Chaps. 21 to 25, and Chap. 26 of Part III devoted, respectively, to polynomials, the Fourier functions ($\sin nx$, $\cos nx$), and exponentials.

The importance of the invariance under translation along the independent variable axis fades when the problem contains a singularity, since the location of the singularity provides a natural origin. It is also true that except for rational functions we could not approximate singularities where the values of the function go to infinity for finite values of x.

We have skirted the treatment of singularities several times and indicated in Sec. 16.6 that using a multiplicative factor to handle a singularity is usually better than using an additive term. We also used a change of variable to eliminate a singularity in Sec. 1.9.

The proper use of the special knowledge of the behavior of the function near the singularity greatly aids the computation. Most of this art lies in the field of mathematics and hence outside the scope of this book. We therefore content ourselves with two illustrative examples followed by some general remarks. However, once the class of functions is determined, the process of finding a particular member of the class follows the same general lines as we have used in the past and need not be discussed again.

27.2 AN EXAMPLE OF AN INTEGRAL WITH A SINGULARITY AT INFINITY

Suppose that we consider the integral

$$f(x) = \int_0^x e^{x^2}\,dx$$

It is known that the asymptotic expansion is

$$f(x) \simeq \frac{e^{x^2}}{2x} + \cdots$$

but this is not useful near $x = 0$. Hence we write

$$f(x) = \int_0^x e^{x^2}\,dx = e^{x^2}D(x)$$

and try to determine $D(x)$. We differentiate this to get the differential equation

$$\begin{aligned} D'(x) + 2xD(x) &= 1 \\ D(0) &= 0 \end{aligned}$$

We now have a simple differential equation to integrate numerically, and we know that $D(x) \to 1/(2x)$ as $x \to \infty$.

The convenience of a single function $D(x)$ (for the whole range $0 \leq x \leq \infty$) suggests that Dawson's integral

$$D(x) = e^{-x^2} \int_0^x e^{x^2}\,dx$$

is the better form in which to study the function than the original one.

From this example one can see how some integrals with singularities can be computed easily. It is also true for cases where the function has some awkward behavior which remains finite; thus if there is a singularity in the complex plane near the real axis, similar tricks can be used. In our particular example the result amounts to merely introducing a special factor to compensate for the singular behavior and resorting to solving a differential equation to do the integration.

27.3 A SINGULARITY IN A LINEAR DIFFERENTIAL EQUATION

It occasionally happens that a numerical solution is required for a linear differential equation which has a singularity[1] in the range of integration. One way of doing this, which was first suggested by Prof. J. W. Tukey, is the following; it again illustrates the technique of using the structure of the singularity in a multiplicative fashion. We study

$$y'' + P(x)y = 0 \qquad \begin{cases} y(x_0) = A \\ y'(x_0) = B \end{cases} \tag{27.3-1}$$

where $P(x)$ has a singularity in the range of integration, say at $x = 0$.

We choose a comparison equation

$$u'' + Q(x)u = 0 \tag{27.3-2}$$

having known solutions $u_1(x)$ and $u_2(x)$ *and* having the same kind of singularity in $Q(x)$ at the same place as $P(x)$. We shall later examine how this may be done.

[1] By singularity we mean that the function does not behave as a power series about the point.

We next assume that

$$y(x) = \alpha(x)u_1(x) + \beta(x)u_2(x) \tag{27.3-3}$$

where $\alpha(x)$ and $\beta(x)$ are functions to be determined but which we expect will vary smoothly. The method now goes along the same lines as the classical method of "variation of parameters." Having introduced two unknown functions $\alpha(x)$ and $\beta(x)$, we may require both (27.3-1) *and* one more condition, which we choose as

$$\alpha' u_1 + \beta' u_2 = 0 \tag{27.3-4}$$

Thus we have

$$\begin{aligned} y &= \alpha u_1 + \beta u_2 \\ y' &= \alpha u_1' + \beta u_2' \\ y'' &= \alpha u_1'' + \alpha' u_1' + \beta u_2'' + \beta' u_2' \end{aligned}$$

Putting these in (27.3-1), we get

$$\alpha(u_1'' + Pu_1) + \beta(u_2'' + Pu_2) + \alpha' u_1' + \beta' u_2' = 0 \tag{27.3-5}$$

But using the fact that u_1 and u_2 are solutions of (27.3-2), we have

$$\begin{aligned} u_1'' + Pu_1 &\equiv (P - Q)u_1 \\ u_2'' + Pu_2 &\equiv (P - Q)u_2 \end{aligned}$$

Thus we have, from (27.3-5),

$$\alpha' u_1' + \beta' u_2' = (Q - P)(\alpha u_1 + \beta u_2) = (Q - P)y$$

Solving this with (27.3-4), we get

$$\begin{aligned} \alpha'(u_1 u_2' - u_2 u_1') &= -u_2(Q - P)y \\ \beta'(u_1 u_2' - u_2 u_1') &= u_1(Q - P)y \end{aligned}$$

But $u_1 u_2' - u_2 u_1'$ is the Wronskian of u_1 and u_2, and for this case (no y' term) it is a constant W_0, where

$$W_0 = u_1(x_0)u_2'(x_0) - u_2(x_0)u_1'(x_0)$$

W_0 is readily determined from the known solutions $u_1(x)$ and $u_2(x)$.

We thus have to solve

$$\begin{aligned} \alpha' &= -\frac{u_2(Q - P)y}{W_0} \\ \beta' &= \frac{u_1(Q - P)y}{W_0} \qquad \begin{cases} y(x_0) = A \\ y'(x_0) = B \end{cases} \\ y &= \alpha u_1 + \beta u_2 \end{aligned} \tag{27.3-6}$$

The problem is to choose a $Q(x)$ so that the factor $Q - P$ vanishes sufficiently strongly at the singularity to cover the peculiar behavior of the products u_2y and u_1y (or, equivalently, the products u_2^2, u_1u_2, and u_1^2). If this can be done, then the equations determining $\alpha(x)$ and $\beta(x)$ will have a much less peculiar behavior at the point where there was a singularity.

We now show how $u_1(x)$ and $u_2(x)$ may be chosen in one particular example. Suppose that

$$P(x) = \frac{a_{-1}}{x} + a_0 + a_1x + \cdots \qquad (a_{-1} \neq 0) \tag{27.3-7}$$

Thus we know that, if $y(0) \neq 0$, then $y''(0)$ is infinite, and it is apparent that we could not approximate the solution by a polynomial with any fidelity near $x = 0$. We try

$$u_1 = \frac{x}{1 + ax + bx^2} \equiv \frac{x}{D(x)} \equiv xD^{-1}(x) \tag{27.3-8}$$

Then
$$u_1'' = [2(D^{-1})' + x(D^{-1})''] \left(\frac{D}{x} u_1\right), \left(\frac{D}{x} u_1 \equiv 1\right)$$
$$= \frac{x[2(D')^2 - DD''] - 2DD'}{xD^2} u_1$$

But
$$\begin{aligned} D &= 1 + ax + bx^2 \\ D' &= a + 2bx \\ D'' &= 2b \end{aligned}$$

Hence for small x

$$u_1'' = \frac{-2a - 6bx + 2b^2x^2}{xD^2} u_1 = \frac{-2a - 6bx + 2b^2x^2}{x[1 + 2ax + (a^2 + 2b)x^2 + \cdots]} u_1$$
$$= -2\left[\frac{a}{x} + (3b - 2a^2) + \cdots\right] u_1$$

Thus

$$Q = -\frac{2a}{x} + (-6b + 4a^2) + (\cdots)x + \cdots$$

and we want to choose (comparing Q and P)

$$-2a = a_{-1} \qquad \qquad a = \frac{a_{-1}}{2}$$
$$\text{or}$$
$$-6b + 4a^2 = a_0 \qquad \qquad b = \frac{a_{-1}^2 - a_0}{6}$$

With this choice of a and b in (27.3-8), we have [using (27.3-7)]

$$Q - P \sim Cx + \cdots$$

and $Q - P$ vanishes sufficiently rapidly.

We now determine u_2 in the usual fashion:

$$\begin{aligned} u_2(x) &= u_1(x) \int^x \frac{d\theta}{u_1^2(\theta)} \\ &= u_1(x) \int^x \frac{1 + 2a\theta + (a^2 + 2b)\theta^2 + 2ab\theta^2 + b^2\theta^4}{\theta^2}\, d\theta \\ &= u_1(x)\left[-\frac{1}{x} + 2a \ln |x| + (a^2 + 2b)x + abx^2 + \frac{b^2x^3}{3} \right] \end{aligned}$$

and the rest is easy. We have $u_1(x)$ with no singularity at $x = 0$, and $u_2(0) = -1$. We also have a $u_2(x)$ whose second derivative is infinite at $x = 0$. The products

$$(Q - P)u_1y \qquad \text{and} \qquad (Q - P)u_2y$$

vanish at $x = 0$, allowing the smooth integration of Eqs. (27.3-6) to determine $\alpha(x)$ and $\beta(x)$, since the peculiar behavior is contained in the u_2 and is partially masked by the zero of $Q - P$.

In this example we again see the advantage of handling the singularities by first examining the dominant behavior of the solution near the singularity and then using a known function to approximate this behavior.

27.4 GENERAL REMARKS

We have given a pair of examples of how to handle singularities in two specific cases. We have no detailed rules for the general case, especially for singularities in nonlinear problems. Much depends on the mathematical ability of the person planning the solution.

With modern floating-point machines it is sometimes possible to integrate close enough to the singularity so that an analytic approximation can be fitted and used to get over the singularity. Once past the singularity, numerical values can be computed from the analytic expression in order to resume the computation. However, this should be used only when better methods fail, as the accuracy is often hard to estimate and control.

When the singularity is a so-called "movable singularity," which can occur in nonlinear problems, then it is particularly important to be careful how the numerical data are used to fit the function approximating the singularity. Yet such things can be and have been done successfully; it is a matter of courage and careful computation.

PART IV

Algorithms and Heuristics

CHAPTER 28

On Finding Zeros

28.1 ALGORITHMS AND HEURISTICS

The formulas that we have been discussing so far have all (possibly excluding predictor-corrector methods) given the desired quantity in an explicit form. Many times, such as in the case of finding a zero of a function or solving a system of linear algebraic equations, the quantities to be computed are given only implicitly. When we have a definite process for computing the unknowns in these implicit situations, we say that we have an *algorithm*.

In the sense that many different special methods are known, we can say that the field of algorithms is highly developed, but in the sense that there are general principles which are regularly translated into everyday practice, the field can be said to be in a more primitive state. The purpose of this book is to give the general techniques and ideas used in the computing field, and since there appear to be very few general ideas in the area of algorithms, we shall have to be content with describing for a few selected algorithms what kinds of things can and cannot be expected. It is also true that most of the standard situations which call for the use of an algorithm are usually coded in a library form for the occasional user, and the details of the method need not concern him, *provided* that he understands what kinds of results he is obtaining.

There are many situations in which there is no known algorithm that will produce the result or else the known method is one of exhaustion—exhaustion of the problem, machine time, and the proposer of the problem. In such a situation we have to resort to whatever methods look as if they might be useful. Such situations typically occur in problems of finding maxima or minima of functions of many variables. It is simply too expensive to explore, say, a 14-dimensional space for all the local minima. We are then driven to trying hunches, or good ideas, or reasonable processes for searching for the desired quantities. Such processes are called *heuristics* and will be examined briefly in Chap. 31. They lie on the exciting frontier of modern research in artificial intelligence and in pattern recognition and of necessity will be treated only briefly. Thus this book goes from explicit formulas to implicit formulas for which there are known methods of solution (algorithms) and ends with suggestions on what to do with implicit formulas for which we do not know a practical algorithm (heuristics).

28.2 THE BISECTION METHOD FOR FINDING A REAL ZERO

Given a continuous real function of a real variable x which has a negative value at $x = a$ and a positive value at $x = b$, then we know that there exists a place between a and b where the function takes on the value zero. If we bisect the interval and note at this point whether the function is positive or negative, then we have found a subinterval in which there is a change in sign (or else we have found the zero). By repeating this bisection process, we can come arbitrarily near to the zero, in principle. Since each step halves the interval in which the zero lies, 10 steps will reduce the interval by a factor of about 1,000; 20 steps, 1,000,000; etc. Thus the method, which assumes *only* continuity and the ability to evaluate the function at any point, is fairly effective.

We said "in principle" because there is, of course, noise in the computation. Let us examine what this does to the process. Suppose that, in computing some value of the function, we get the wrong sign, because of roundoff noise. The result will be that we shall choose the wrong half to explore during the next step. However, if no other noise produces further errors in determining the sign of the function, the zero will be located in the subinterval that has one end at the place where the sign was computed incorrectly. And this is a quite reasonable result; we have determined the location within our ability to compute the function itself. The small error has not led us very far astray.

The vulnerability to noise should be examined in every process that is proposed. Many of the more rapid processes go badly astray when they reach noise level—and remember that, by chance, we may be near the zero at any stage of the computation and hence subject to errors due to noise. It should again be remarked that this bisection method requires only the ability to compute the functional values and does not suppose that the routine for the derivative has also been produced and coded.

EXERCISE

28.2-1. Make a flow diagram of the logic of the program for the bisection process.

28.3 LINEAR INTERPOLATION

The first idea which is apt to occur to the reader is that the bisection method can be greatly improved by using the value of x where linear interpolation between the two known values of opposite sign yields zero as the place to choose the next sample rather than the mid-point.

Figure 28.3-1 shows a function for which this process of linear interpolation would be definitely slower than the bisection method. Of course we can argue that for a reasonable population of functions the interpolation process will, on the average, require fewer steps of evaluation of the function. However, this points up one of the virtues of the bisection method, namely, that we know *in advance* how long it will take to obtain a given accuracy in the position x of the zero; although other methods may, on the average, be faster, still we cannot know that the particular case on hand is faster. Indeed, we see from Fig. 28.3-1 how a function can be found which will take more than any given number of steps; the value at $x = b$ need only be increased sufficiently.

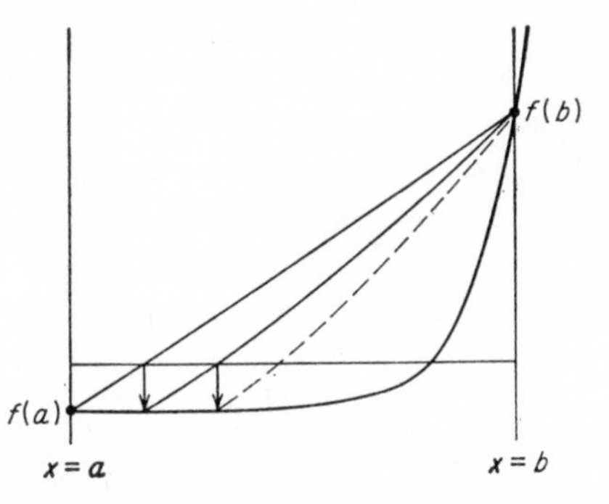

FIG. 28.3-1. An example of slow convergence.

An alternative method of linear interpolation which uses the two most recently computed values runs the risk of diverging from the zero on successive steps if the curve is sufficiently peculiar.

EXERCISE

28.3-1. Make a flow diagram of the linear-interpolation process.

28.4 PARABOLIC INTERPOLATION

If linear interpolation is better than bisection, then perhaps the process of using three points to determine a parabola would be even better; certainly we need to examine it a bit.

We start as in the bisection method and arbitrarily bisect the interval to get the third point. (Other strategies might be preferable, but we shall not discuss them.) We now fit a parabola, using divided differences and, say, Newton's formula (Sec. 8.4). The zero of this quadratic which falls in the interval gives us a fourth point. At this stage there are at least two possibilities: We can take the two points closest to the estimate of the zero that we just found, or we can be careful to keep two close points which *also* have a change in sign. The author favors the latter on the grounds that he likes to have the zero for which he is looking boxed in within a known interval and can therefore be sure that the parabola has a real zero in the interval. On the first step both possibilities give the same result, but at subsequent stages the two can give different results.

We leave it to the reader to examine the noise effects of the quadratic process as it closes in on the zero.

These last two methods illustrate a vague, general principle: the fancier the method and the better it is supposed to be, the worse it can behave when things go wrong for some functions; it may, in fact, be worse than simpler methods and quite likely is more vulnerable to noise.

We can also gain some perspective on what we have been doing by referring back to the four basic questions of Sec. 7.1.

1. What samples?
2. What class of functions?
3. What criterion of goodness of fit?
4. What error?

We have been choosing our samples as we go along, using polynomials for our analytic substitution, using exact matching, and have inferentially used the error in x as the measure of success; it would require very small changes to shift to the error in the value of y. Evidently other classes of functions could be used—and should be if they are appropriate to the nature of the function being studied.

EXERCISES

28.4-1. Draw a flow diagram of the quadratic process, keeping the change-of-sign approach.

28.4-2. Examine the noise effects of the quadratic process.

28.4-3. Discuss how you would provide the choice of error control in either x or y.

28.5 SOME GENERAL REMARKS

We have used linear and quadratic interpolation to approximate the zeros of a function. Evidently we could go further and try higher-order polynomials. The best strategy depends on how hard it is to compute the function. If it is very difficult, requiring perhaps hours of machine time per point, we would probably resort to high-order processes, raising the degree every time we got another point. On the other hand, if the function is easy to compute, we would prefer some simple, reliable method such as the bisection method.

It is customary, when using polynomials, to find the zeros of a function

$$y = y(x)$$

to approximate the data as a polynomial in x and then use inverse interpolation for finding the zero. Frequently it is just as reasonable to assume that the x is a polynomial in y and use direct interpolation; it certainly is a lot less labor.

We also discussed Newton's method (Sec. 7.1) and in Exercise 7.1-1 indicated an extension which made use of the second derivative. Ostrowski, in his excellent book "Solution of Equations and Systems of Equations" [34], assumes that evaluating the function be taken as one unit of work and that each derivative computed be considered another unit of work. From this he rates the linear interpolation of the two most recent values (also called *regula falsi*) as worth 1.618 . . . , meaning that one step multiplies the number of correct digits by 1.618 . . . , while Newton's method rates as 1.414 per unit of work. Thus one is apt to conclude that the linear-interpolation method is more efficient. But as we have repeatedly stressed, by the time the typical function is computed, one has most of the parts for computing the derivative, and it requires very little more work to obtain it; thus if only machine time during the running of the problem is considered, it is quite likely that Newton's method compares favorably with, or is better than, linear interpolation. On the other hand, Newton's method requires the analytical finding of the derivative (possibly with a human algebraic error), coding, and loading time for the extra coding of the problem on the machine.

Another point should be made clear. It is customary to rate various methods according to their final rate of convergence to the true solution (in a mathematical sense) and neglect the question of how many steps are involved in the process before the final rate takes over. Again experience shows that some of the most highly rated methods require a very long time to get started. The examples in the texts usually show 10 to 20 decimal places, but in practice all that is usually sought is three to five places. Thus the examples are often misleading.

Finally, it has been tacitly assumed that the zero is isolated before we start, and the reader is likely to want to know how to do this essential first step. Unfortunately, the author has nothing really useful to say on this topic. Often the problem itself, when studied carefully, will give some information, and there are numerous mathematical theorems dealing with the subject, but it is an art to isolate the various zeros.

28.6 BAIRSTOW'S METHOD FOR FINDING ZEROS OF A POLYNOMIAL

The problem of finding the zeros of a polynomial arises frequently enough to justify studying the special properties of this problem and developing special methods to handle it. We shall consider only real polynomials, since they are the ones usually encountered in practice. A fundamental property of real polynomials is that they may be factored into real linear and real quadratic factors. We assume that the real linear factors are removed first. It is only after the real linear factors have been found that any attention should be paid to the complex zeros.

A whole book could be written on the various known methods for finding the real linear and real quadratic factors of a real polynomial. The fact that there are so many methods indicates that there is apparently no completely satisfactory method; each person has his own favorite method or methods. It appears that, on the average, Bairstow's method is more widely used than any other comparable method. It is not infallible; occasionally it takes very long to converge on a quadratic factor or even fails; but, on the average, it seems to be better than any other single method.[1] It is likely that the local computing library has some routine already coded and in running order, and probably you would use that one rather than attempt to program your own, but it is perhaps worth discussing one method which is rather like many others.

Let the polynomial be

$$P(x) = a_0 + a_1 x + a_2 x^2 + \cdots + a_n x^n \qquad (28.6\text{-}1)$$

and let us assume that we have a guess at a quadratic factor

$$x^2 + px + q \qquad (28.6\text{-}2)$$

(Initially, we can choose $p = q = 0$, which will simplify the first step.) Using synthetic division (see Sec. 1.6), we divide the polynomial by the quadratic factor to get a quotient and a remainder, e.g.,

$$\begin{aligned} a_n x^n + a_{n-1}x^{n-1} + \cdots + a_0 = (x^2 + px + q)(b_n x^{n-2} \\ + b_{n-1}x^{n-3} + \cdots + b_2) + b_1 x + b_0 \end{aligned} \qquad (28.6\text{-}3)$$

The reason for the peculiar subscripts on the b's will become apparent as we go on; it makes notation easier. In a skeleton synthetic form, we have

$$\begin{array}{c|cccccccc} 1 \quad p \quad q & a_n & a_{n-1} & a_{n-2} & a_{n-3} & \cdots & a_2 & a_1 & a_0 \\ & - & - & qb_n & qb_{n-1} & \cdots & qb_4 & qb_3 & qb_2 \\ & - & pb_n & qb_{n-1} & & & pb_3 & pb_2 & - \\ \hline & b_n & b_{n-1} & b_{n-2} & & \cdots & b_2 & b_1 & b_0 \end{array}$$

where the remainder is

$$b_1 x + b_0$$

[1] A method due to D. Muller is also highly regarded. Some people favor Lin's method.

The algebraic relations between the coefficients are

$$
\begin{aligned}
b_n &= a_n \\
b_{n-1} &= a_{n-1} - pb_n \\
b_{n-2} &= a_{n-2} - pb_{n-1} - qb_n \\
&\cdots\cdots\cdots\cdots \\
b_{n-k} &= a_{n-k} - pb_{n-k+1} - qb_{n-k+2} \qquad (k = 2, 3, \ldots, n-1) \\
&\cdots\cdots\cdots\cdots\cdots\cdots\cdots\cdots \\
b_0 &= a_0 - qb_2
\end{aligned}
\tag{28.6-4}
$$

We would have the desired quadratic factor if, and only if, the remainder were identically zero, that is,

$$b_1 = b_0 = 0$$

Let us consider these coefficients as functions of p and q:

$$
\begin{aligned}
b_1 &= b_1(p,q) \\
b_0 &= b_0(p,q)
\end{aligned}
$$

We now use the two-dimensional analog of Newton's method (Sec. 7.1) and expand b_1, b_0 in a Taylor series about the present guess (p,q). Writing p^* and q^* as the desired solution, we have

$$
\begin{aligned}
b_1(p^*,q^*) &= 0 = b_1(p,q) + \frac{\partial b_1}{\partial p}\Delta p + \frac{\partial b_1}{\partial q}\Delta q + \cdots \\
b_0(p^*,q^*) &= 0 = b_0(p,q) + \frac{\partial b_0}{\partial p}\Delta p + \frac{\partial b_0}{\partial q}\Delta q + \cdots
\end{aligned}
\tag{28.6-5}
$$

where

$$
\begin{aligned}
\Delta p &= p^* - p \\
\Delta q &= q^* - q
\end{aligned}
\tag{28.6-6}
$$

are the errors to be corrected (approximately) for our next guess. Neglecting all but the linear terms in (28.6-5), we have a pair of linear equations for the changes to be made in p and q.

The problem is how to find the partial derivatives which are the coefficients of the unknowns Δp and Δq. We could make a small change in p and note the changes in b_1 and b_0 and do the same using a small change in q. We prefer, instead, to find them in a more analytical fashion. We differentiate Eqs. (28.6-4) with respect to p.

$$
\begin{aligned}
\frac{\partial b_n}{\partial p} &= 0 \\
\frac{\partial b_{n-1}}{\partial p} &= -b_n - p\frac{\partial b_n}{\partial p} \\
\frac{\partial b_{n-2}}{\partial p} &= -b_{n-1} - p\frac{\partial b_{n-1}}{\partial p} - q\frac{\partial b_n}{\partial p} \\
&\cdots\cdots\cdots\cdots\cdots\cdots \\
\frac{\partial b_{n-k}}{\partial p} &= -b_{n-k+1} - p\frac{\partial b_{n-k+1}}{\partial p} - q\frac{\partial b_{n-k+2}}{\partial p} \\
&\cdots\cdots\cdots\cdots\cdots\cdots \\
\frac{\partial b_0}{\partial p} &= \qquad\qquad - q\frac{\partial b_2}{\partial p}
\end{aligned}
$$

If we now write

$$\frac{\partial b_k}{\partial p} = -C_k^* \tag{28.6-7}$$

then we have

$$\begin{aligned}
C_n^* &= 0 \\
C_{n-1}^* &= b_n - pC_n^* \\
C_{n-2}^* &= b_{n-1} - pC_{n-1}^* - qC_n^* \\
&\cdots\cdots\cdots\cdots\cdots\cdots \\
C_{n-k}^* &= b_{n-k+1} - pC_{n-k+1}^* - qC_{n-k+2}^* \\
&\cdots\cdots\cdots\cdots\cdots\cdots \\
C_0^* &= \qquad\qquad - qC_2^*
\end{aligned} \tag{28.6-8}$$

These equations are of the same form as (28.6-4), provided that we note that $C_n^* = 0$ means the following:

C_{n-k}^* is to be identified with b_{n-k+1}.
b_k is to be identified with a_{k-1}.
Also the last equation is not quite correct.

These observations suggest repeating the process of synthetic division, using the same quadratic factor $x^2 + px + q$, on the b's (instead of the a's) to obtain coefficients C_k. This we now do:

$1 \quad p \quad q$	b_n	b_{n-1}	b_{n-2}	b_{n-3}	$\cdots$	b_2	b_1	b_0
			qC_n	qC_{n-1}	$\cdots$	qC_4	qC_3	qC_2
		pC_n	pC_{n-1}		$\cdots$	pC_3	pC_2	
	C_n	C_{n-1}	C_{n-2}		$\cdots$	C_2	C_1	C_0

Thus we get

$$\begin{aligned}
C_n &= b_n \\
C_{n-1} &= b_{n-1} - pC_n \\
C_{n-2} &= b_{n-2} - pC_{n-1} - qC_n \\
&\cdots\cdots\cdots\cdots\cdots\cdots \\
C_{n-k} &= b_{n-k} - pC_{n-k+1} - qC_{n-k+2} \\
&\cdots\cdots\cdots\cdots\cdots\cdots \\
C_1 &= b_1 - pC_2 - qC_3 \\
C_0 &= b_0 \qquad\quad - qC_2
\end{aligned} \tag{28.6-9}$$

The partial derivatives for which we are looking in (28.6-5) are [using (28.6-7)]

$$\frac{\partial b_1}{\partial p} = -C_1^* \qquad \frac{\partial b_0}{\partial p} = -C_0^*$$

Comparing (28.6-8) and (28.6-9), we see that

$$\begin{aligned}
C_{k-1}^* &= C_k \qquad (k = n, n-1, \ldots, 3, 2) \\
C_0^* &= C_1 - b_1 + pC_2
\end{aligned}$$

Hence $$\frac{\partial b_1}{\partial p} = -C_2 \qquad \frac{\partial b_0}{\partial p} = -(C_1 - b_1 + pC_2) \tag{28.6-10}$$

We now examine the process for the partial derivatives with respect to q. Again differentiating (28.6-4),

$$\begin{aligned}
\frac{\partial b_{n-1}}{\partial q} &= \qquad\qquad - p\frac{\partial b_n}{\partial q} = 0\\
\frac{\partial b_{n-2}}{\partial q} &= -b_n \qquad - p\frac{\partial b_{n-1}}{\partial q} - q\frac{\partial b_n}{\partial q}\\
&\cdots\cdots\cdots\cdots\cdots\\
\frac{\partial b_{n-k}}{\partial q} &= -b_{n-k+2} - p\frac{\partial b_{n-k+1}}{\partial q} - q\frac{\partial b_n}{\partial q}\\
&\cdots\cdots\cdots\cdots\cdots\\
\frac{\partial b_0}{\partial q} &= -b_2 \qquad\qquad\qquad - q\frac{\partial b_2}{\partial q}
\end{aligned}$$

We now set
$$\frac{\partial b_k}{\partial q} = -C_k^{**} \tag{28.6-11}$$

to get
$$\begin{aligned}
C_{n-1}^{**} &= 0\\
C_{n-2}^{**} &= b_n \qquad - pC_{n-1}^{**} - qC_n^{**}\\
&\cdots\cdots\cdots\cdots\cdots\\
C_{n-k}^{**} &= b_{n-k+2} - pC_{n-k+1}^{**} - qC_{n-k+2}^{**}\\
&\cdots\cdots\cdots\cdots\cdots\\
C_0^{**} &= b_2 \qquad\qquad\qquad - qC_2^{**}
\end{aligned} \tag{28.6-12}$$

Since $C_n^{**} = C_{n-1}^{**} = 0$, we see that we need to identify

$$\begin{aligned}
C_{k-2}^{**} &= C_k \qquad (k = n, n-1, \ldots, 3)\\
C_0^{**} &= C_2 - pC_3
\end{aligned}$$

if we are to compare (28.6-9) and (28.6-12).

The partial derivatives that we want for (28.6-5) are

$$\frac{\partial b_1}{\partial q} = -C_1^{**} = -C_3 \qquad \frac{\partial b_0}{\partial q} = -C_0^{**} = -(C_2 + pC_3)$$

Thus we have for (28.6-5)

$$\begin{aligned}
b_1(p,q) &= C_2\,\Delta p + C_3\,\Delta q\\
b_0(p,q) &= (C_1 - b_1 + pC_2)\,\Delta p + (C_2 + pC_3)\,\Delta q
\end{aligned} \tag{28.6-13}$$

The solution of these two equations produces the amounts to change our guess at the quadratic factor $x^2 + px + q$. The convergence, when it works, is quadratic; that is, the errors, when small, are approximately squared each step.

Thus we have an iterative process which we hope will converge to a quadratic factor of the original polynomial. When we find it, we can factor it out (the first division step) and use the quotient as a new polynomial to be examined by the same process.

CHAPTER 29

Simultaneous Linear Algebraic Equations

29.1 INTRODUCTION

The problem of solving a system of simultaneous linear algebraic equations occurs quite frequently and has received much attention from many people. As a result, there are many methods available, and, just as in the case of finding the zeros of a polynomial, a whole book could be written on the topic. We shall confine our attention to two widely used methods, which are typical of two large classes of methods. The method that is best suited to a given machine is a complex question.[1]

To say that a system of linear equations is ill-conditioned means, loosely speaking, that they are almost linearly dependent. A great deal of effort has gone into the study of how to solve ill-conditioned systems, but it may be questioned whether or not one wishes to solve such a system in a practical situation. In what physical circumstances would the answers be useful when they depend so sensitively on the coefficients of the system? It is usually true that, in place of an answer, what is desired are the labels of a minimum set of equations which are almost linearly dependent. From this information the problem can be better understood and usually reformulated in a more satisfactory way. It is quite likely to be true that the ill-conditioned equations, except for the errors of roundoff and measurement, are linearly dependent and hence do not represent an adequate formulation of the situation.

29.2 GAUSS ELIMINATION METHODS

Of the methods for solving systems of linear equations in a direct way, as contrasted with iterative methods, the method due to Gauss is perhaps the most widely used. Gauss' method is the same as the one used to solve systems of linear equations by eliminating one variable at a time. There are many variations on the method, and we shall discuss

[1] See chap. 2 by A. Orden in Ref. 37.

one of them later. Since the method is so basic, let us examine a very simple case first. Suppose that we have

$$\begin{aligned} x + 2y + 3z &= 10 \\ x + 3y - 2z &= 7 \\ 2x - y + z &= 5 \end{aligned}$$

We first eliminate x by subtracting suitable multiples of the first equation from the second and third equations. This gives us

$$\begin{aligned} x + 2y + 3z &= 10 \\ y - 5z &= -3 \\ -5y - 5z &= -15 \end{aligned}$$

We next eliminate the second unknown y by subtracting -5 times the second equation from the third:

$$\begin{aligned} x + 2y + 3z &= 10 \\ y - 5z &= -3 \\ -30z &= -30 \end{aligned}$$

We now solve these equations, the so-called "back solution," by working from the bottom up. The last equation gives

$$z = 1$$

Putting this in the second equation, we get

$$y = 5 - 3 = 2$$

Putting these two in the first equation, we get

$$x = 10 - 4 - 3 = 3$$

and we have finished.

Now suppose that we have n equations in n unknowns $x_1, x_2, \ldots, x_n$.

$$\begin{aligned} a_{11}x_1 + a_{12}x_2 + \cdots + a_{1n}x_n &= b_1 \\ a_{21}x_1 + a_{22}x_2 + \cdots + a_{2n}x_n &= b_2 \\ \cdots\cdots\cdots\cdots\cdots\cdots\cdots& \\ a_{n1}x_1 + a_{n2}x_2 + \cdots + a_{nn}x_n &= b_n \end{aligned}$$

We begin by "scaling" the problem; we multiply the various equations by some power of 10 so that the largest a_{ij} in each equation lies between 1 and 10 in size. If we find that the largest coefficient in each equation is always the coefficient of the same variable, say x_k, then clearly we should also scale the x_k to bring them to a reasonable range. We may, in fact, want to scale several of the x's to keep a reasonable balance in the coefficients. This scaling provides a reasonable meaning for large and small and does not produce any roundoff errors if we are computing in floating point (we would clearly use powers of 2 on a binary machine).

Since a_{11} may be zero, we cannot be sure that we can use the first coefficient. Instead we search for the largest coefficient of x_1, say a_{k1}, which is called the *pivot*. We use this kth equation to eliminate x_1 from the other equations. Since the labeling of the equations is arbitrary, we can suppose for the sake of talking that it is the first equation. (It is not very expensive to shift the equations in storage to achieve this label, although one may prefer merely to control the logic of the problem to achieve the same result.)

We now scan the coefficients of x_2 in the remaining equations and use the equation with the largest such coefficient to eliminate x_2 from the other equations. And so we proceed, unknown by unknown.

Some remarks are in order. In principle, if the system is large enough, our initial scaling can be completely lost, and an occasional rescaling may be called for, but we shall not discuss such large systems. If the systems have 50 or more equations, it would be better to consult an expert *before* wasting machine time. Indeed, when systems pass 10 or 20 in size, new effects and dangers frequently begin to appear which we shall not go into here.

When we choose the equation that we are going to use to eliminate the next variable, we can first divide through the equation by the leading coefficient so that the first coefficient is now 1. Thus when we come to the back solution we have no divisions to do. But if division is expensive of machine time, we can calculate the reciprocal of the chosen pivot and then form the multipliers (all ≤ 1 in size) to use on the equation to eliminate the unknown.

It may be thought that we should also search for which x_k to use next, rather than proceeding $x_1, x_2, \ldots, x_n$, by selecting the largest coefficient of all those that we have left, but this additional refinement does not seem to add much accuracy.

29.3 VARIANTS OF THE GAUSS METHOD

There are many variants of the Gauss method. Perhaps the most significant one is that called "complete elimination," or Gauss-Jordan. In this method, when an x_k is eliminated, it is eliminated from *all* the other equations, including those solved for the earlier variables. In this way there is no need of a "back solution." It is sometimes claimed that this has an advantage in accuracy.

In the case of symmetric coefficients, if the pivots are picked on the main diagonal, the labor can be cut almost in half, but there may be some loss in accuracy owing to a poor choice of pivots. If the matrix is positive definite, there is little loss in accuracy.

Most of the variants appear to be minor changes made to fit the characteristics of a particular machine (although they can have a noticeable effect on the results obtained); hence they are not suitable for discussion in a first course. The book "Modern Computing Methods" [29][1] is a good reference, and it in turn has an extensive bibliography on many aspects of computing, including simultaneous equations.

29.4 GAUSS-SEIDEL METHOD

In contrast to the direct methods of solution such as the Gauss elimination method, there are iterative methods such as the Gauss-Seidel method. Again we first examine a simple case. We rearrange the equations of the example in Sec. 29.2 so that the largest coefficient of the first equation is x, the largest of the second is y, and the largest of the third is z.

$$\begin{aligned} 2x - y + z &= 5 \\ x + 3y - 2z &= 7 \\ x + 2y + 3z &= 10 \end{aligned}$$

We start with a guessed solution, say $x = y = z = 0$. We now use the first equation to solve for the new x value:

$$x = \frac{5 + y - z}{2} = \frac{5}{2}$$

Using $x = 5/2$, $z = 0$, we solve the second equation for the new y:

$$y = \frac{7 - x + 2z}{3} = \frac{3}{2}$$

Finally, using these calculated values, we solve the third equation for the new z:

$$z = \frac{10 - x - 2y}{3} = \frac{3}{2}$$

These three values provide a new guess, and we repeat the cycle:

$$\begin{aligned} x &= 5/2 \\ y &= 5/2 \\ z &= 5/6 \end{aligned}$$

Another repetition gives

$$\begin{aligned} x &= 10/3 = 3.33 \rightarrow 3 \\ y &= 16/9 = 1.780 \rightarrow 2 \\ z &= 28/27 = 1.04 \rightarrow 1 \end{aligned}$$

We note that the convergence is slow. Also, if an error occurs it can affect the number of steps but does not affect the final answer (in prin-

[1] See also Ref. 16.

ciple). It is customary to iterate until the changes are sufficiently small, but what this may mean with respect to the answer is another question.

The general case is much the same. The equations are arranged so that the large terms are "on the main diagonal." If these main-diagonal terms are sufficiently larger than the other terms in the equations, then the convergence is assured; if not, then not. One sufficient condition is

$$|a_{ii}| > |a_{i1}| + |a_{i2}| + \cdots + |a_{ii-1}| + |a_{i,i+1}| + \cdots + |a_{in}|$$

for all i (assuming that the system is indecomposable into two independent systems of equations).

When there are many zero coefficients,[1] a_{ij}, the method is very favorable as compared with the elimination method. The Gauss-Seidel method takes advantage of the zeros, whereas the elimination method does not.

There are many variants of the iteration method. If the convergence is slow, then we try to guess by various techniques where it is headed and to jump there in one step. The so-called under-and-over relaxation methods are of this type. Again the reader is referred to "Modern Computing Methods" [29] for details.

29.5 INCREASED ACCURACY

If we have any method of solving a system of equations and we want more accuracy, then the following method can be used. Let us substitute into the equations the values of x that we have found and then calculate the residuals, using double-precision arithmetic. We then solve the system of equations again, except with the residuals of the equations as the right-hand sides. The new solution plus the old solution gives a more accurate solution. It is easy to see why this is so. The first set of x values $x^{(1)}$ satisfy the equations with the right-hand side equal to the given values b_i minus the residuals r_i in the ith equation, while the second solution $x^{(2)}$ satisfies, approximately, the equations with the residuals on the right-hand side. Hence $x^{(1)} + x^{(2)}$ satisfies the same equations but with the sum of the two right-hand sides $(b_i - r_i) + r_i = b_i$.

29.6 GENERAL REMARKS

It is evident that, in the scheme of Chap. 7, we are not here using any class of functions as approximations in the process of analytic substitution. The only relevant question is the fourth: What accuracy?

[1] Sometimes called "sparse equations."

Do we want the x_i accurate? Do we want the residuals to be small? Do we want some system of equations for which this computed solution is the exact solution to be close to the original system? The latter condition is confusing because there are $n(n + 1)$ coefficients and only n answers whose n differences from the true solution can be distributed among the $n(n + 1)$ coefficients in many, many ways, the residuals being but one example. Thus the question in this case differs significantly from the same question for the zeros of a polynomial, where we have n errors in the zeros to be distributed among essentially n coefficients. It seems that the accuracy question is more difficult to answer in this case than in most other cases. Frequently small residuals—small in relation to the given right-hand sides—are the best answer to the accuracy question.

CHAPTER 30

Inversion of Matrices and Eigenvalues

30.1 INTRODUCTION

The study of matrices has been diligently pursued for a long time, and a great deal is known about them. They are also found to occur in many different kinds of physical problems. As a result, the literature is vast, and numerous methods are associated with various problems. We shall examine only a few examples and refer the reader to Ralston and Wilf [37], "Modern Computing Methods" [29], and Householder [16] for further material, especially the bibliography of the latter two. We assume that the reader is familiar with the elements of matrix theory.

30.2 MATRIX INVERSION BY GAUSS ELIMINATION

Frequently it is necessary to invert a square matrix

$$A = (a_{ij})$$

Let us imagine that we have the rectangular matrix

$$\begin{pmatrix} a_{11} & a_{12} & \cdots & a_{1n} & 1 & 0 & \cdots & 0 \\ a_{21} & a_{22} & \cdots & a_{2n} & 0 & 1 & \cdots & 0 \\ \cdots & \cdots & \cdots & \cdots & \cdots & \cdots & \cdots & \cdots \\ a_{n1} & a_{n2} & \cdots & a_{nn} & 0 & 0 & \cdots & 1 \end{pmatrix}$$

which is really A followed by the identity matrix I.

We now apply the Gauss elimination method, letting the n right-hand columns behave as they may. When we finish the Gauss-Jordan complete elimination, we have

$$\begin{pmatrix} 1 & 0 & \cdots & 0 & b_{11} & b_{12} & \cdots & b_{1n} \\ 0 & 1 & \cdots & 0 & b_{21} & b_{22} & \cdots & b_{2n} \\ \cdots & \cdots & \cdots & \cdots & \cdots & \cdots & \cdots & \cdots \\ 0 & 0 & \cdots & 1 & b_{n1} & b_{n2} & \cdots & b_{nn} \end{pmatrix}$$

We claim that the matrix B

$$B = \begin{pmatrix} b_{11} & b_{12} & \cdots & b_{1n} \\ b_{21} & b_{22} & \cdots & b_{2n} \\ \cdots & \cdots & \cdots & \cdots \\ b_{n1} & b_{n2} & \cdots & b_{nn} \end{pmatrix}$$

is the inverse of A. To see this, we need to recognize that each step of the elimination process is equivalent to multiplying on the left by a matrix. The product of all these left-hand matrices is clearly A^{-1} since it reduced A to the identity matrix. But when applied to the n right-hand columns it produced from the identity matrix the matrix B:

$$A^{-1} \cdot I = B$$

Hence B is the inverse of A.

30.3 THE EIGENVALUE PROBLEM

Many problems lead to the system of equations

$$\begin{aligned} a_{11}x_1 + a_{12}x_2 + \cdots + a_{1n}x_n &= \lambda x_1 \\ a_{21}x_1 + a_{22}x_2 + \cdots + 2_{2n}x_n &= \lambda x_2 \\ \cdots\cdots\cdots\cdots\cdots\cdots & \cdots\cdots \\ a_{n1}x_1 + a_{n2}x_2 + \cdots + a_{nn}x_n &= \lambda x_n \end{aligned}$$

or

$$Ax = \lambda x$$

where λ is unknown. These equations are consistent if, and only if,

$$\Delta = \begin{vmatrix} a_{11} - \lambda & a_{12} & \cdots & a_{1n} \\ a_{21} & a_{22} - \lambda & \cdots & a_{2n} \\ \cdots & \cdots & \cdots & \cdots \\ a_{n1} & a_{n2} & & a_{nn} - \lambda \end{vmatrix} = 0$$

Now this is a polynomial in λ of degree n. In principle, if we were to evaluate the determinant for $n + 1$ values of λ, we could use these $n + 1$ values to determine the polynomial

$$\Delta(\lambda) = 0$$

Thus we would have to find the zeros of the polynomial $\Delta(\lambda)$. While it is easy to understand, the roundoff errors and amount of computation to be done have generally caused this method to be passed over in favor of other methods.

Sometimes all that is wanted is the largest λ in absolute value. Assume that λ is real. In such cases we can proceed as follows: We pick an arbitrary vector

$$y = \begin{pmatrix} y_1 \\ y_2 \\ \cdots \\ y_n \end{pmatrix}$$

It is known that if the eigenvalues are distinct then the corresponding eigenvectors (the solutions of the equation for the corresponding eigenvalue) form a complete system, that is, for any vector y

$$y = c_1y_1 + c_2y_2 + \cdots + c_ny_n$$

where the y_i's are the eigenvectors and the c's are constants. Let us multiply both sides by A.

$$Ay = \sum_{i=1}^{n} c_iAy_i = \sum_{i=1}^{n} c_i\lambda_iy_i$$

and in general

$$A^ky = \sum_{i=1}^{n} c_i\lambda_i^ky_i$$

Now if $|\lambda_1| > |\lambda_i|$ for $i = 2, 3, \ldots, n$, then, as $k \to \infty$,

$$A^ky \sim c_1\lambda_1^ky_1$$

Consider the ratio

$$\frac{A^{k+1}y}{A^ky} \sim \frac{c_1\lambda_1^{k+1}y_1}{c_1\lambda_1^ky_1} = \lambda_1$$

In the limit, we would expect that each component of our vector would be multiplied by the same factor λ_1.

As a matter of practice, it is advisable to scale the resultant vector at each stage so that the largest element is equal to 1. At the last step this scale factor gives the value of the largest (in absolute value) eigenvalue. The eigenvector is also obtained as a by-product.

This process clearly can be speeded up if we form some powers of A first. We could first compute

$$A^2, \quad A^4, \quad A^8, \quad \ldots, \quad A^{2k}$$

Then applying

$$A^{2k}y = \sum_{i=1}^{n} c_i\lambda_i^{2k}y_i$$

we would obtain a much more rapid approach to the complete dominance of the largest eigenvector. The (2^k)th root of the ratio of successive steps would give λ_1.

The question of what happens if by chance $c_1 = 0$ is not serious. Even if it were true at some early step, chance roundoff would usually bring it in at a later step, and it would ultimately become dominant.

This points up the fact that a good guess at the structure of the eigenvector of the largest eigenvalue speeds up the process in the early stages.

The idea that, once we know the largest eigenvalue and vector, we can carefully subtract it at each stage and thus give the second largest (in absolute value) eigenvalue a chance to grow occurs to almost everyone. It can be done, but it is not as accurate as one could wish. Indeed, several of the larger eigenvalues can be found before the process breaks up into "noise," and each one is less and less accurately determined than the preceding ones.

30.4 THE SMALLEST EIGENVALUES

A simple observation suffices to find the algebraically smallest eigenvalues by the same method. Let y_i be an eigenvalue, that is,

$$Ay_i = \lambda y_i$$

Then $(A - pI)y_i = (\lambda - p)y_i$.

By a suitable choice of p the smallest eigenvalue can be converted to the largest (in size). Suppose that the eigenvalues are approximately

$$1, \quad 2, \quad 3, \quad 4, \quad 5, \quad 6, \quad 7, \quad 8, \quad 9, \quad 10$$

If we choose $p = -10$, then $\lambda - p$ is

$$-9, \quad -8, \quad -7, \quad -6, \quad -5, \quad \ldots, \quad 0$$

Once we know the approximate size of the largest eigenvalue, we can use p as the negative of this value, and the smallest becomes the largest (in size).

30.5 A FEW REMARKS

The theory of how to find all the eigenvalues at one time is quite complex, and we cannot examine the topic adequately in an introductory text. At present, for symmetric matrices, Householder's modification of Givens' method seems to be about the best direct method.

While there are direct methods for the nonsymmetric matrix, they require care because of the inherent instability of the problem, and if the matrix is at all large it is best to consult an expert *before* wasting machine time.

The reader is referred to the three books mentioned in Sec. 30.1 for further details.

CHAPTER 31

Some Examples of the Simulation of Situations and Processes

31.1 INTRODUCTION

The idea of simulation of a situation or a process is intuitively clear, but apparently no satisfactory definition exists. As a result we shall have to use examples to develop and extend the intuitive idea. For both situations and processes a simulation is usually closely related to an optimization. While the simulation part is usually straightforward, there is often no known algorithm which is practical for solving the optimization part. The purpose of the following sequence of examples is gradually to develop new ideas of how to attack situations in which an algorithm for solving the problem is not known. This lack of a definite usable algorithm leads to the popular field of *artificial intelligence* in which games of chess, theorem-proving routines, and such have attracted much attention. We shall not, however, enter very far into this important and active frontier of current research.

Simulations occupy a very large part of the time on many digital computers, and consequently the field cannot be ignored merely because there are no general methods or deep results. There is some evidence that any fool can say "Let us simulate the process exactly," and it is much harder for an expert to show that many times an exact simulation is not what is wanted and is a great waste of machine time (and hence money). In the absence of general methods of what, when, and how to simulate, and when not to, we shall offer vague philosophic comments for general guidance. As we go through the various examples, which are slightly altered actual case histories, these remarks will not be found to lead to a consistent, complete philosophy.

There is one crude and useful classification of simulations as either *discrete* or *continuous*. These correspond to situations in which the quantities of interest are basically discrete or continuous, but unfortunately the division is not always sharp. From a practical point of view, we shall adopt the division as being based on whether or not there is a serious problem due to truncation error. We shall first take up only those problems in which there is no trouble with truncation errors. Later we examine briefly some of the details of simulations which involve truncation problems, but most of the relevant material has already been developed, especially in Chap. 25. In the next chapter we shall examine the important problems of random numbers and random processes which are often needed in simulation problems.

31.2 A SIMPLE EXAMPLE OF A DISCRETE SIMULATION

We begin with an example of a conceptually simple simulation. Suppose that we are concerned with the construction of a digital computer or other large piece of electronic equipment. Suppose further that the basic circuits are on printed circuit boards which plug into a large panel supplying the mechanical support. Lastly, suppose that we are concerned merely with the physical layout of the equipment and we are supplied with a list of the interconnections of the various pins of the printed circuit boards. We are then in the position of the equipment layout engineer.

Depending on how we place the individual circuit boards on the panel, we shall get different wiring patterns. Most importantly, these will differ in the *amount* of wire used (although the number of wires will be the same). It is desirable to be reasonably near to the minimum wire length for these reasons:

1. The weight of the wire may be a serious factor when transporting the equipment.
2. The cost of the wire, though not great in any one case, can become considerable in mass production.
3. Circuit troubles in the form of shorts between wires, or stray capacitances, may be increased by the use of extra wire lengths.

One final remark is perhaps necessary for those not familiar with modern wiring practice. It is conventional to run the wires along channels in N-S or E-W directions rather than to go "as the crow flies" between the various pins of the circuit boards.

Suppose that we begin with what we hope is a reasonable assignment of the circuit boards to the various locations on the panel. Figure 31.2-1 shows a small part of such a situation. It is not hard, although it may be tedious, to write a program for a digital computer which will take the circuit-board assignments on the panel and, using the information of the interconnections (say one wire per Hollerith punched card which may have come from the output of the design stage), compute the wire placement and the corresponding length of wire for each interconnection, as well as the total wire length. Let this length be labeled L_0.

We next try interchanging a pair of circuit boards. Will this reduce the total length of wire? We could recompute the whole process, but it is faster to have a second program which removes the old wires and

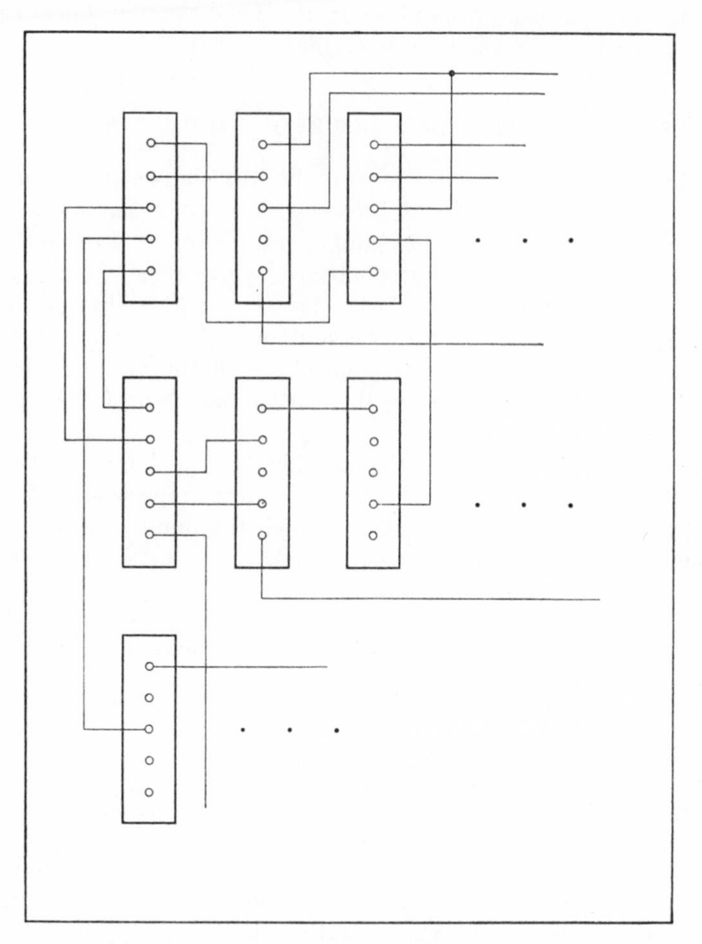

FIG. 31.2-1. Panel-board wiring.

puts in new ones. Thus we get the change $\Delta L_0 = L_1 - L_0$, where L_1 is the new length. If the total length is decreased, we naturally keep the change and try another; if not, we do not make the change. Thus the equipment layout engineer can use the computer to simulate the wiring of the panel. Every time he tries a change in the layout he can obtain an answer to the question "Was the change good or bad?"

This is about as simple a simulation as there is. There is no time dependence, no truncation error worth mentioning, no use of random processes, and the conception is immediately clear—we are calculating what wire length we would find were we to use the proposed assignment of the circuit boards on the panel. Furthermore, the output can be put in a form suitable to guide a mechanical wire wrap gun and hence fits into a total plan of mechanization.

We have solved our simulation problem but we now propose to explore further the optimization process which used human intervention at each step and which is very wasteful of machine capacity if the machine is at all fast; we naturally ask why the machine itself cannot select and make the trials. It may (or may not) be true that the layout engineer can arrive at an answer in fewer guesses, but the time and cost, as well as the result, may be more favorable if we let the machine make many more trials than the engineer will have the time or patience to make.

We now face the problem of telling the machine how to make changes in the circuit-board-location assignment—changes which we hope will decrease the total wire length. We could, of course, interchange adjacent boards, first the N-S neighbors and then the E-W neighbors, and after each trial either keep it or reject it, depending on whether or not the total wire length was reduced. Repeated sweeps of trials over the whole panel until there is no more change will gradually reduce the total length, but we have no assurance that in the end we would have, or be near, a minimum. In any case, the crude "interchange of adjacent neighbors" process seems to be very tedious and wasteful of machine time.

We therefore ask ourselves how we would attack the problem if we had to do it by hand. One heuristic that is bound to come to mind after a little introspection and trial is the following: Take an arbitrary board and count how many wires go east and how many go west. If there are more to the east, then probably we should consider moving the circuit board in that direction. How far? Well, it is not hard to see that we should move it to a location where as many wires go east as go west. We then repeat the process for the N-S direction. Thus we have located a place, or at least a general area on the panel, where we would like to put this particular circuit board. We therefore hunt in this general area for a second circuit board which has the property that it should be moved to the general area of the first one. If we find such a board, then we make the interchange. We continue until there are no good interchanges left.

One can vaguely describe the model in these words. There is a general tendency to force the individual circuit board toward the center so that the wires departing from it will be evenly balanced in each direction, but there is also a population pressure which forces some of the boards to be on the edges of the panel. The minimum wiring is a compromise between these two forces.

Returning to the above plan of operation, which appears to require more coding but less machine execution time than the first crude plan, we still have no assurance that we shall get a minimum or that we shall find the same solution as we found by the slow "interchange of adjacent neighbors" process. Indeed, experiences with such plans of operation have shown that, starting with different layouts, we get significantly different amounts for the total wire length. At first this suggests starting with several random layouts and processing each to see which gives us the least. This idea leaves much to be desired, and we cast about for other possibilities.

The idea that maybe we should blindly try changing three at a time usually fills the proposer with horror at the number of trials if the panel has a considerable number of boards on it. Blind trials can be replaced by a more reasonable process such as that used for speeding up the interchange of two boards, but again it is not clear that we shall be able to find triples. Well then, perhaps we should try our method of relocating the first one, then taking *that* board (or a nearby one that does not have a balance of wires) and finding where to relocate that one, etc., hoping that some board in the chain will fall near some previous one (not necessarily the first board) and thus provide a cycle of shifts to be made.

What is it that we need? What are we looking for? At the end of the interchange-of-pairs process there is a local minimum, and we want to break out of the local minimum without paying too high a price in extra wire. Once well out of the valley of that minimum, we can resume the simple interchange of pairs and go down the next valley. Thus we want to "shake up" the arrangement without too great a loss in the wire-length minimization. If we find such an alternative process, we can use it for a while, but after a few such uses, we shall probably find ourselves in a local minimum which is lower than any of the nearby local minima. Thus we need a stronger "shaking" process if we are to go on. But before we get too involved in this interesting problem, we need to ask how much we really want to find the minimum and for what deviation from the absolute minimum we might be willing to settle to save machine (and real) time. Unfortunately, with the analysis that we have so far produced, we have no real idea of how far we are from the minimum, but after a little experience we can *guess* how much we are going to be able to decrease the total wire length by a given amount of computing. This is the real question being asked most of the time. We know that with sufficient machine time we can find the minimum by exhaustively trying all combinations, but we do not care to pay the price since the gain is not worth it. How far we are willing to go is a matter of judgment in each individual case, but at this point in the history of using computers for optimization it is probably better to overstudy the theory of finding algorithms for reducing the total length than to understudy it, because we know that we are going to face the same type of problem many, many times in the future and probably anything that we can learn from this case will help us with the next ones.

It is not the purpose of this book to go into this field in any depth but rather to survey a few of the possible methods that have come to mind during various simulations. The problem of finding minima is a real, important, and often very difficult problem and is one which will undoubtedly receive much more attention in the future. Already we have such fields of knowledge as *linear programming* and *dynamic programming* which have grown out of the search for the best, or at least better, solutions.

There is, however, one remark to be made. In a situation which is too large to optimize in a precise fashion there is a great temptation to try to optimize small blocks of it and to hope that this will improve matters. In the above example we might try to optimize small blocks of boards on the panel. Experience seems to indicate that generally this is a bad policy; optimization of small parts causes very great losses when the parts are finally put together. There are undoubtedly many exceptions to this rule, but it seems to be a moderately well-known principle in systems engineering: Do not try too hard to optimize the small pieces of a tightly interrelated system because it will cost you more than you gained when you put the parts together.

31.3 AN EXAMPLE OF AN INVENTORY SIMULATION

The simulation to be described in this section falls more in what is usually considered the field of business applications than in the field of scientific computing, but it has close relations to many problems that arise in normal scientific work. There are a number of special morals to be drawn from this case history, and they are part of the reason for including this particular example.

In an early study of the Western Electric inventory practice, the inventory was simulated on a moderate-sized computer. The main elements under study were as follows:

1. When to reorder, that is, when to place an order for more stock
2. How much to reorder
3. The cost of delays in filling the order

To some extent all three are under the control of the company. The third factor may be influenced in many ways: more rapid processing of orders, use of the telephone to place orders rather than the usual letter, more rapid handling of received goods, *especially* through the stages of quality inspection and delivery into the live inventory bins.

There are two conflicting desires in keeping an inventory, and it is this clash that makes the problem interesting. First, with a small inventory, it is possible to save expenses (and taxes). Second, it is desirable to avoid an "outage" when some item is needed but is not in stock; hence we want a large inventory. Evidently some common measure of cost, such as the dollar value of avoiding an outage, will enable us to compare these two so that a balance between the two opposing forces can be decided upon.

In principle, the simulation was very simple. We read into a storage register of the computer the amount of stock on hand at the beginning of the period under study. We then processed the *actually observed withdrawals and returns* to stock that occurred during an 18-month period. After each withdrawal we examined the amount left to see if we had reached a reorder point; if so, we decided how much to order, placed the order, and computed the time when it would arrive. This latter point involved a random number from a Poisson distribution but need not bother us here as we shall take it up in the next chapter. When the time came for the order to be filled, we added the amount to the inventory stock by adding the proper number to the register, which gave the number of items on hand. We also had to make a test each time we tried to make a withdrawal to see if there was any left, and if there was not we recorded an outage and watched how long it ran until it was filled.

Finally, at regular intervals we printed out the amount on hand, along with a list of the outages and their durations.

By running the simulation with different reorder formulas, both for "when" and for "how much," and also varying the mean delay time for filling the orders, we could compare the results of different policies, and in principle we could say what would have happened over the 18-month period had we followed the various practices being tested. There is one flaw in this argument, however. During the actual 18-month period occasional outages were recorded, and then no new demands for withdrawals were recorded until the stock came in. Thus the simulation was not entirely accurate.

A second fault occurs in the study when we try to ask not what would have happened but what will happen in the future. Because we did not construct a model of the withdrawals and returns to stock but used actual observations, we are in no position to project ourselves into typical future situations. The particular situation of the 18 months would undoubtedly, by chance, favor one formula over another at times when their relative merits were in the opposite direction. A much more theoretical study might have produced much better answers, but it was not done.

How, then, can the study be justified? To a great extent it is justified by the consumers of the results. They fancied themselves as (and were, in fact) hard-headed practical engineers, and they had not too much faith in theory. The statements of what would have happened during those 18 months had they followed a different policy (even when the flaw was pointed out) carried more weight in their minds than a more theoretical treatment. We shall not question which approach would have been better, since both have their faults, but we rest the case on the obvious point that it is of little use to compute results in a form which the customer is not prepared to use. What you propose to compute must be reasonably well matched to the user.

A second point that came up in this simulation should be mentioned. Although we were assured that the records were complete and accurate, nevertheless when we came to use them we found inconsistencies. This points to a general observation that it is very difficult to find accurate records of the past, especially when they have been kept by hand, and you should not take anyone else's word for their quality; you should include adequate testing of the data to assure yourself that they are in fact what they were advertised to be. Large amounts of accurate data are very rare.

31.4 THREE-DIMENSIONAL TIC-TAC-TOE

We now take up a slightly different type of simulation, and although it is stated in the form of a game, the principles are very practical. There is a version of tic-tac-toe that is played in three dimensions in a 4 by 4 by 4 cube. The object is to get four men in a straight line before the opponent gets four of his men in a straight line. (We assume that the reader is familiar with the two-dimensional game played on a 3 by 3 board.)

In the two-dimensional version a definite strategy for playing the game is known, but apparently this is not true for three dimensions. Thus we face a problem of deciding what move to make when we cannot be sure of the right move—we have to work out a practical way of choosing our next move. In principle, all possible games could be tried, but this approach is exhaustive, to say the least. In this respect the game resembles many practical situations. In principle, there is the possibility of exhaustively trying all possible combinations, but practically this approach is beyond reach, and we have to develop an alternative approach to the problem. The advantage of studying a game rather than a practical situation is that the game is widely known and is simply and clearly stated, while many practical situations have small details which are difficult to discuss and which add nothing to the understanding of how to attack the problem.

Let us begin by assigning numbers to the positions in the cube. (See Fig. 31.4-1.) A little study will show that if four men are in a line the numbers of their squares form an arithmetic progression. The converse is not true.

1	2	3	4
5	6	7	8
9	10	11	12
13	14	15	16

Top layer

17	18	19	20
21	22	23	24
25	26	27	28
29	30	31	32

Second layer

33	34	35	36
37	38	39	40
41	42	43	44
45	46	47	48

Third layer

49	50	51	52
53	54	55	56
57	58	59	60
61	62	63	64

Bottom layer

FIG. 31.4-1

While the players make a move by placing a man in a given cell of the cube, it is not the positions but rather the lines that are significant when it comes to studying the possibilities of winning. A brief study will show that there are 76 lines. These may be considered as being the following:

Ten lines on each horizontal plane, two being diagonal—total 40 lines
Sixteen vertical lines
Sixteen slant lines crossing the four horizontal planes (two from each corner cell and one from each side cell)
Four body diagonals starting in an upper corner and going through the cube to end on the diagonally opposite corner

We need to keep in the storage of the computer both a map of the board and a list of the state of the lines. After each play we have to change the lines involved. The value that we associate with each line is equal to the sum of the number of men on the line of the first player if only his men are on the line, the negative of the sum if only the second player has men on the line, and zero if there are men of both players. We also need a way of recording that the line is entirely empty.

We now ask ourselves how to go about playing the game. The following rules are clearly a start toward a strategy. When it is your move:

1. If you have three men on a line, play the fourth and win.
2. If the opponent has three men on a line, block his win by playing the fourth cell.

3. If there is a cell which has at least two lines crossing it, both of which have only two of your men on them, play the "fork" and you will win on the next move.

4. If the opponent has a similar fork, you had better block it or he will win.

But suppose that none of these apply, then what? We need to study the game a bit further. We first note that 16 of the cells, 1, 4, 13, 16, 22, 23, 26, 27, 38, 39, 42, 43, 49, 52, 61, 64, have seven lines through them, while all the other cells have only four lines through them. This suggests that, other things being the same, playing in such places gives a greater chance of later finding a winning combination.

The situation in which we now find ourselves is characteristic of many problems in this area. We have definite algorithms to cover some of the configurations which may arise, and we must revert to heuristics to cover the rest.

While the main object of this example is to illustrate this principle, the general interest in this game perhaps justifies a few further remarks, although we do not have a complete strategy.

Further study of the game shows that there is a transformation, an inversion of the cube into itself, which leaves all lines invariant but which interchanges the eight outer vertices with the eight in the center of the cube (see Fig. 31.4-2).

22	21	24	23
18	17	20	19
30	29	32	31
26	25	28	27

6	5	8	7
2	1	4	3
14	13	16	15
10	9	12	11

54	53	56	55
50	49	52	51
62	61	64	63
58	57	60	59

38	37	40	39
34	33	36	35
46	45	48	47
42	41	44	43

FIG. 31.4-2

Thus any strategy which involves the center is equivalent to one which involves the vertices.

At this point, it is probably necessary to play a reasonable number of games to become familiar with the progress of the game. In the author's experience with this game, the play proceeds somewhat as follows: Both players start playing mainly on the seven-line cells, but soon one of them begins a sequence of forcing moves by putting three men on one line, to which the other can only reply by blocking in the obvious place. The first player hopes to build up a fork against which there is no defense, and as he chooses his forcing moves he has this partly in mind. He also must watch that a forced defensive move of his opponent does not suddenly give the opponent the initiative by providing the opportunity for a forcing move. If the first attacker fails to push through a win and has to give up the initiative, then usually the second player attacks and carries through a win. It also seems to be impossible to play a purely defensive game for any length of time against an aggressive player and not lose.

If we were to try to convert these vague statements into a strategy, we would find that different stages of the game require quite different strategies. In the opening play, it is mainly a matter of trying to fill the seven-line cells, protecting against a sneak attack, and also preventing the opponent from getting too many men on any plane without some of our men there to block the potential combinations that he could develop.

We would then have to develop a measure of the appropriate time to shift from the opening play into an attack, remembering that if we attack too soon we shall probably have to give up the initiative and end by losing, while if we are too late the opponent may attack and be able to push through to a win.

It is clear that we would include in any forcing move an analysis of both the opponent's reply and the situation at that point to see if that lost us the initiative. We would also tend to favor forcing moves in which we played a seven-line cell and the opponent was forced to play a three-line cell. In any deep study it would be advantageous to study not only the cell configurations and the line configurations but also to develop a strategy for planes.

Experience with the game played by human beings suggests that a computer with such a program could probably play a fairly good game because of its ability to examine the algorithmic part with care each time and not overlook certain combinations, as the human player usually does. On the other hand, the long sequences of combinations seem to be hard to program for the machine.

The following exercises give some combinatorial results that have been found by careful analysis and are characteristic of the kinds of partial results that can be found in complex problems. Methods for solving them are not given in the text.

EXERCISES

31.4-1. Show that 16 men properly placed can block 73 lines (excluding three body diagonals). (One answer: 1, 8, 10, 15, 19, 22, 28, 29, 36, 37, 43, 46, 50, 55, 57, 64.)

31.4-2. Show that, if the rules are modified to include the rule that the first eight plays must fall in the 2 by 2 by 2 center cube (or in the eight vertices), the first player has a win.

31.5 GENERAL REMARKS ON DISCRETE SIMULATIONS

These few examples indicate that the hard part of a discrete simulation is not usually finding the way to arrange a simulation, although it may in practice be a tedious exercise in coding for a particular machine, but rather finding an algorithm for solving a closely related optimization problem. Usually in discrete cases there is the known method of exhaustively examining every case which is prohibitively expensive to use. Thus one is driven to the examination of heuristic methods and the general fields of *pattern recognition* and *artificial intelligence*.

When something along these lines is tried in a specific problem, it usually turns out that in a few situations there are definite algorithms, such as we found in the tic-tac-toe example, but that most situations in the problem have a less specific treatment. When no specific move is given, either by the algorithmic rules or the heuristic rules, still, as in the tic-tac-toe case, some move must be chosen. For this we appeal to the idea of a *random move* or a random move among a given class. The rule of using the first move, when the moves are arranged in some order, does not meet a number of conditions as satisfactorily as does the random move, for several reasons, among which are the following:

1. If a specific move is always chosen, the opponent can gradually explore the game, uncover a weakness, and then exploit it, while the use of random moves makes this very difficult.
2. The systematic choice, although we do not know why, may have, in the long run, a bad effect, and we feel intuitively that a random move will, on an average, do us less harm.

Thus are we led to the examination of the idea of "*random*" which is taken up in the next chapter.

A few words of caution should perhaps be included at this point. Frequently an examination will show that there are various alternative formulations of the situation being explored and that some are far better to use for machine simulation than others. Thus the reader should pause a long time before plunging into all the details of a simulation; especially should he assure himself that the proposed simulation will actually answer the questions which *should* have been asked.

31.6 CONTINUOUS SIMULATIONS

Most continuous simulations involve the solution of one or more differential equations. Of these simulations, many involve "band-limited functions," since the electronic circuitry being simulated usually limits the range of frequencies that can be transmitted. Thus we find Chaps. 21 to 25, especially Chap. 25, particularly relevant.

However, the details of various simulations are often quite different, and it is necessary to study each proposed simulation carefully before starting to program a numerical solution. Failure to take a sufficiently small step size, thus getting involved in unwanted aliasing of frequencies, is but one example of how haste in the planning stage can affect the answers obtained.

A large simulation needs to be approached carefully and slowly. When the cost of the total machine time involved is considered, it becomes apparent that common sense dictates a large expenditure of time and effort in the planning stages.

One of the main aims of this book has been to develop the ideas and feeling for a procedure to find the kinds of formulas that a specific situation requires. It is also hoped that the book has developed the ability and confidence necessary to carry the reader through what may be long months of detailed development and testing of the appropriate formulas for the simulations.

Continuous simulations often involve the simulation of "noise," and the methods for generating random numbers given in the next chapter are useful to start the generation of "noise" having the required properties of spectral distribution, sample-to-sample correlation, etc.

So much depends on the specific situation that a particular set of formulas cannot be recommended, and we are forced to be vague. About the only general rule is that it is necessary to grasp the total plan of the simulation, and what relation it bears to the total project, *before* plunging into details. Along this line Chap. $N + 1$ is especially relevant.

CHAPTER 32

Random Numbers and Monte Carlo Methods

32.1 THE IDEA OF A RANDOM NUMBER

In mathematics and statistics there is a well-defined concept of a random process, but the idea of a random number is not so simple. Random numbers come from random processes, but when such a sequence is written down, is it still random? Once written down, it is a perfectly predictable sequence. It is evident that we need to do some thinking on this matter before plunging into the use of random numbers.

By a random integer between 0 and 9 we generally mean a digit drawn from a population, with each digit having an equal probability of being chosen. But a random number between 0 and 1 is another matter. For almost all such numbers it would take forever to recite the digits. If we propose to generate a random number in a machine, it must mean that we are not proposing to produce a random number in a mathematical sense. What do we mean? Well, we might mean that of the whole population of possible 8-, 10-, or 12-digit numbers we propose to select one in some "equally likely" manner. Actually, we find that we mean to compromise even further; we propose to sample from a part of the whole population of numbers.

When we propose to produce a sequence of random numbers, what do we mean? Usually we mean going through a part of the whole population of possible numbers that could be found in the machine, taking one at a time, and not repeating until we exhaust them. Thus we mean to sample in a rather strange way; sampling without replacement, it might be called.

But when we propose to find a whole sequence of numbers, then further tests in addition to equal probability are involved. What sequence are we going to get? If the numbers were monotone increasing in size, we would not like to consider that sequence random, although in a sense it is as random as any other specific sequence. Evidently, we want some further properties. At first we say that we want to produce a sequence in a random manner, but what does that mean? Any order is as random as any other one. What we are forced to say is that we do not want to see a pattern in the way in which they are selected. One of the patterns for which we might check is that of correlation of one number to the next; can we predict the size of the next number from the size of the present one? Thus if the x_i are the random numbers, then we want

$$\sum_i (x_i - 1/2)(x_{i+1} - 1/2) = 0$$

where we have taken out the mean $\bar{x} = 1/2$. Given enough time, we can imagine a large number of possible tests—so many, in fact, that we would not care to apply them all.

It is necessary to become practical about the matter and say that, if we can see no pattern with reference to the particular application, the numbers are random for that application, and we must be content with very little more. Indeed, when viewed from the practical point of view, it may be that we really do not want a truly random sequence; we may prefer our much too flat distribution which exhausts the subpopulation before repeating any number. The way in which we are going to use a random sequence in many applications is to try to estimate the statistics of a large population of events from a small sample. Thus we would like to get an "over-typical" sample so that we would have a large sample stability for a small sample in actual use. We would want a flatter distribution than usual without including too many of the extreme peculiarities—just enough extremes to protect us from the opposite error of too homogeneous a population. We want a homogenized, pasteurized, and certified random sequence of numbers if we can get it. These remarks are not true, for example, if we wish to estimate the variance of some quantity.

32.2 GENERATION OF RANDOM NUMBERS ON A BINARY MACHINE

The most commonly used method of generating random numbers is the following routine:

$$x_{n+1} \equiv \rho x_n \qquad \text{(modulo the word length of the machine)}$$

Thus we use one multiplication per number and take the low-order digits of the product for the next number. The main problems are what to choose for ρ and how to select the first number x_0.

The answer to these questions and a proof that this method produces long chains of random numbers before repeating depend on some results in number theory. For ease in derivation[1] (the purpose of which is to

[1] Borrowed from Milton Levy.

demonstrate procedure and results) we shall use the problem of generating random numbers on a binary machine and leave the results for decimal machines to be stated in the next section.

The standard notation in number theory

$$x \equiv a \pmod{m}$$

means that $x - a$ is divisible by m.

If we have a k-digit binary machine and take the last k digits of the double-length product of ρx_n, then we want

$$\begin{aligned} x_0 &= a \\ x_{n+1} &\equiv \rho x_n \pmod{2^k} \end{aligned} \qquad (k \geq 3) \qquad (32.2\text{-}1)$$

Now if 2 divides a, then the sequence will be equivalent to

$$\begin{aligned} y_0 &= \frac{a}{2} \\ y_{n+1} &\equiv \rho y_n \pmod{2^{k-1}} \end{aligned}$$

which really is a shorter word length, and we are wasting machine capacity. Hence we take a to be an odd number.

Similarly, ρ should be odd, since if it were even

$$x_{k+1} = \rho^{k+1}a = 0$$

and we would have a trivial sequence of all zeros past this point.

Now all odd numbers ρ can be written in one of the forms

$$8t - 3 \quad 8t - 1 \quad 8t + 1 \quad 8t + 3$$

for some t.

Theorem 1. *If* $\rho = 8t \pm 1$, *then*

$$\rho^{2^{k-3}} \equiv 1 \pmod{2^k} \qquad (32.2\text{-}2)$$

(*that is, the order of* ρ *is not greater than* 2^{k-3} *and is a divisor of* 2^{k-3}).

Proof. This is equivalent to saying

$$2^k \text{ divides } \rho^{2^{k-3}} - 1$$

Since in general

$$a^2 - 1 = (a + 1)(a - 1)$$

then $$\rho^{2^{k-3}} - 1 = (\rho^{2^{k-4}} + 1)(\rho^{2^{k-5}} + 1) \cdots (\rho + 1)(\rho - 1) \qquad (32.2\text{-}3)$$

For each term $(i \geq 1)$

$$\begin{aligned} \rho^{2^i} + 1 &= (8t \pm 1)^{2^i} + 1 = 1 + (1 \pm 8t)^{2^i} \\ &= (1 + 1) + \sum_{k=1}^{2^i} (-1)^k C(2^i,k)t^k \cdot 8^k \end{aligned}$$

Hence 2 divides $\rho^{2^i} + 1$, and, from (32.2-3),

$$2^{k-4} \text{ divides } (\rho^{2^{k-4}} + 1)(\rho^{2^{k-5}} + 1) \cdots (\rho^2 + 1) \qquad (32.2\text{-}4)$$

We have also

$$(\rho + 1)(\rho - 1) = \rho^2 - 1 = (8t \pm 1)^2 - 1 = 16(4t^2 \pm t)$$

Hence $$2^4 \text{ divides } (\rho + 1)(\rho - 1)$$

and, using (32.2-4) in (32.2-3),

$$2^4 \cdot 2^{k-4} = 2^k \text{ divides } (\rho^{2^{k-3}} - 1)$$

Theorem 2. *If* $\rho = 8t \pm 3$, *then*

$$\rho^{2^{k-3}} \not\equiv 1 \pmod{2^k}$$
$$\rho^{2^{k-2}} \equiv 1 \pmod{2^k}$$

(*that is, the order of* ρ *is* 2^{k-2}).

Proof. Again using (32.2-3), we have, for this case,

$$\rho^{2^i} + 1 = 1 + (3 + 8t)^{2^i}$$
$$= 1 + 3^{2^i} + \sum_{k=1}^{2^i} (\pm 1)^k C(2^i,k) t^k \cdot 8^k \cdot 3^{(2^i-k)}$$

But $$1 + 3^{2^i} = 1 + (4 - 1)^{2^i} = 1 + (1 - 4)^{2^i}$$
$$= 1 + 1 + \sum_{k=1}^{2^i} C(2^i,k)(-4)^k$$

Hence $$\rho^{2^i} + 1 = 2 + \sum_{k=1}^{2^i} C(2^i,h)[3^{2^i-k} t^k \cdot 2^k + (-1)^k] 4^k$$

and $$2 \text{ divides } \rho^{2^i} + 1$$

but $$4 \text{ does not divide } \rho^{2^i} + 1$$

We therefore have from (32.2-3)

$$2^{k-4} \text{ divides } (\rho^{2^{k-4}} + 1)(\rho^{2^{k-5}} + 1) \cdots (\rho^2 + 1)$$

but $$2^{k-3} \text{ does not divide it}$$

Observe that

$$(\rho + 1)(\rho - 1) = 8(8t^2 \pm 6t + 1) = 8(\text{an odd number})$$

Thus, in either case,

$$2^3 \text{ divides } (\rho + 1)(\rho - 1)$$

and $$2^4 \text{ does not}$$

Therefore $$2^{k-1} \text{ divides } \rho^{2^{k-3}} - 1$$

but $$2^k \text{ does not}$$

Or, in other words,

$$\rho^{2^{k-3}} \not\equiv 1 \pmod{2^k}$$

We now write

$$\rho^{2^{k-2}} - 1 = (\rho^{2^{k-3}} + 1)(\rho^{2^{k-3}} - 1)$$

and

$$2 \text{ divides } \rho^{2^{k-3}} + 1$$

$$2^{k-1} \text{ divides } \rho^{2^{k-3}} - 1$$

so that

$$2^k \text{ divides their product}$$

and

$$\rho^{2^{k-2}} \equiv 1 \pmod{2^k}$$

Theorem 3. *If* $\rho = 8t - 3$, *then the sequence* $x_0, x_1, x_2, \ldots, x_{(2^{k-2}-1)}$ *generated by* (32.2-1) *is some permutation of*

$$1,\ 5,\ 9,\ \ldots,\ (2^k - 3) \qquad \text{if } a \equiv 1 \pmod 4$$

or

$$3,\ 7,\ 11,\ \ldots,\ (2^k - 1) \qquad \text{if } a \equiv 3 \pmod 4$$

Proof. Consider the values $(a\rho^n)$, $n = 0, 1, \ldots, 2^{2^{k-2}} - 1$. The difference between consecutive terms

$$a\rho^{n+1} - a\rho^n = a\rho^n(\rho - 1)$$
$$= a\rho^n(8t - 4)$$

is divisible by 4. But we know (Theorem 2) that the order of ρ is 2^{k-2}; hence we have 2^{k-2} distinct terms whose differences are divisible by 4, and the theorem follows.

Thus we see that, depending on how we select a, provided that we pick $\rho = 8t - 3$ (for any value t), we can get a permutation of one of the sequences in Theorem 3. While we have regarded the x_n as integers in the derivation, when we come to use them, we place the binary point on the extreme left; then we use

$$x_n \cdot 2^{-h}$$

But it is not satisfactory to select just any value for t. For example, $t = 1$ gives $\rho = 5$, and whenever a small value of x_n occurs, say $1 \cdot 10^{-k}$, then there will be a long sequence of gradually increasing numbers following it. This suggests that perhaps we want to pick a t so that $\rho \cdot 2^{-k}$ has leading digits either $0.01 \cdots$ or $0.10 \cdots$ to avoid these troubles. There has apparently been no careful study[1] of the advantages of choosing different ρ values to obtain different permutations of the same numbers.

For a given $\rho = 8t - 3$ we have two cycles, each of which exhausts one-fourth of the available numbers. The remaining half of the numbers occur in smaller cycles, depending on the choice of a.

If, in the sequences of Theorem 3, the last two binary digits are dropped, then the resulting sequence is a permutation of $0, 1, 2, \ldots, 2^{k-2} - 1$ having a full cycle of length 2^{k-2}. If we do not drop the last

[1] See R. R. Coveyou, Serial Correlation in the Generation of Pseudo-random Numbers, *J. Assoc. Computing Machinery*, vol. 7, pp. 72–74, 1960.

two digits, then the last digit is always a 1 and certainly is not random. As we progress from right to left and examine the various trailing digits, we find less and less structure. Thus it is customary not to depend on the last few digits as being random.

32.3 GENERATION OF RANDOM NUMBERS ON A DECIMAL MACHINE

The basic results in this area are in Moshman's paper,[1] which gives the formula for an s-digit decimal machine ($s > 4$)

$$\rho = 7^{4k+1}$$

which has a period[2] of length $5 \cdot 10^{s-3}$, as well as the results of testing it on an 11-digit machine. Another choice[3] of ρ

$$\rho = 76{,}768{,}779{,}754{,}638{,}671{,}877 \text{ reduced mod } 10^s$$

gives maximal periods.

Subsequent investigations using more elaborate formulas have been made. For a binary machine

$$x_{n+1} = (2^a + 1)x_n + C \pmod{2^{35}}$$

can be found in a paper by Rotenberg.[4]

Also studied are formulas of the form[5]

$$x_{n+1} = \alpha x_n + \beta x_{n-k}$$

The problem is evidently in a great state of flux and cannot be settled here at this time. We leave the matter for future investigation.

32.4 OTHER DISTRIBUTIONS

When we know how to obtain a flat distribution of random numbers, we are then in a position to generate other distributions. Many tricks are known, and we shall investigate only a few of them.

[1] Jack Moshman, The Generation of Pseudo-random Numbers on a Decimal Machine, *J. Assoc. Computing Machinery,* vol. 1, pp. 88–91, 1954.

[2] E. Bofinger and V. J. Bofinger, On a Periodic Property of Pseudo-random Numbers, *J. Assoc. Computing Machinery,* vol. 5, pp. 261–265, 1958, amend Moshman's results slightly.

[3] Due to E. N. Gilbert.

[4] A. Rotenberg, A New Pseudo-random Number Generator, *J. Assoc. Computing Machinery,* vol. 7, pp. 75–77, 1960. See also J. Certain, On Sequences of Pseudo-random Numbers of Maximal Length, *J. Assoc. Computing Machinery,* vol. 5, pp. 353–356, 1958.

[5] B. Green, J. Smith, and L. Klem, Empirical Tests of an Additive Random Number Generator, *J. Assoc. Computing Machinery,* vol. 6, pp. 527–537, 1959.

Suppose that we want a distribution $f(y)$; how can we get it from a flat distribution? One way is to equate the two cumulative distributions, the flat one in the variable x and the desired one in y,

$$\int_0^x 1 \cdot dx = \int_0^y f(y)\, dy = F(y) = x$$

Thus if we can find the inverse function of $F(y)$, we have

$$y = F^{-1}F(y) = F^{-1}(x)$$

In principle, this is all that is required.

A simple case of this is the exponential distribution

$$f(y) = e^{-y} \tag{32.4-1}$$

$$x = \int_0^y e^{-y}\, dy = 1 - e^{-y}$$

or

$$e^{-y} = 1 - x$$

In practice, we can replace the random number $1 - x_i$ by x_i so that we have $e^{-y_i} = x_i$, or

$$y_i = -\ln x_i \tag{32.4-2}$$

This has been used with excellent success.

If this method is tried on the normal distribution

$$\frac{1}{\sqrt{2\pi}} e^{-y^2}$$

the inverse function, erf ( ), has to be approximated (say by a Hastings [13, p. 191] approximation). But experience shows that in the tails of the distribution the approximation sometimes leads to serious errors.

Instead of the inverse transformation, we may make use of the well-known result that the sum of comparatively few random numbers from a distribution usually gives a very good approximation to a normal distribution. In the case of a flat distribution, it requires the sum of about 10 numbers to get a very close approximation to the normal distribution. In the case of a decimal machine, it is customary to use 12 numbers to avoid any possible interaction between the generation mechanism and the addition process. The sum does not have mean zero, and so we have to subtract the necessary amount: 5 if we add 10 numbers and 6 if we add 12. Using the results of Sec. 2.9, we have the variance

$$\sigma^2 = \frac{n}{12} \qquad (n = 10, 12)$$

Thus for the choice $n = 12$ we have unit variance, which provides a second reason for using $n = 12$.

The following device has been used with success in the case of a Poisson distribution. We calculate the k such that if the x_i are from a flat distribution

$$x_1 x_2 \cdots x_k \geq e^{-m}$$
$$x_1 x_2 \cdots x_{k+1} < e^{-m}$$

where m is the mean of the Poisson distribution. The integer k is then the desired number.

If m is small, this is an economical routine, but if m is fairly large, it is very slow and not very good. But if m is fairly large, then the Poisson distribution is accurately approximated by a normal distribution, and we can obtain the value that way.

32.5 MONTE CARLO METHODS

We have discussed a few of the many different known tricks for producing random numbers from various distributions, and we now turn to a discussion of the possible use of them.

The name Monte Carlo for a computation which systematically uses random numbers goes back to von Neumann and Ulam during the late years of World War II when they used random numbers to simulate the behavior of neutrons. The use of a combination of human beings and machines to simulate the random process of traffic in a telephone central office goes back at least to 1926, although in those days it was (and in some places still is) called a "throw down," from the act of throwing dice to get the random numbers. But the idea itself goes back at least to the French naturalist Buffon[1] who observed that, if a needle were thrown down at random on a ruled surface, then the probability of the needle crossing a line depended on the number π.

Let us derive this fact. Suppose that the family of parallel, equally spaced, straight lines have a spacing of unit length. Let the needle have length $l < 1$. When we throw it down, the center of the needle can be at any distance from a line from 0 to ½. Let this be the variable x. Next consider the angle φ at which the needle lies (see Fig. 32.5-1a). We consider the variables x and φ as random and independent.

If we are to have an intersection of the needle with the line, then

$$x < \frac{l}{2}\cos\varphi \qquad \left(-\frac{\pi}{2} < \varphi < \frac{\pi}{2}\right)$$

The shaded area in Fig. 32.5-1b, where the bounding curve is $(l/2)\cos\varphi$, is the area in which x and φ must lie. The ratio of this area to the total

[1] *Proc. Paris Acad. Soc.*, 1773.

area is the probability of a crossing. This ratio is

$$p = \frac{\int_{-\pi/2}^{\pi/2} (l/2) \cos d\varphi}{\frac{1}{2} \cdot \pi} = \frac{l}{\pi} \int_{-\pi/2}^{\pi/2} \cos \varphi \, d\varphi = \frac{2l}{\pi}$$

Thus we can consider Buffon's observation as being a Monte Carlo method of determining the value of π. Now if the value of π were only known to be between 1 and 10, this would be an excellent way to find that π was around 3. With some labor and many trials, we might get a value around 3.1. But to try to get greater accuracy would tax our ability to make the lines accurate, to have the needle just the right length, to count the questionable crossings, etc. And this is a general observation. Monte Carlo methods are perhaps best in the preliminary stages

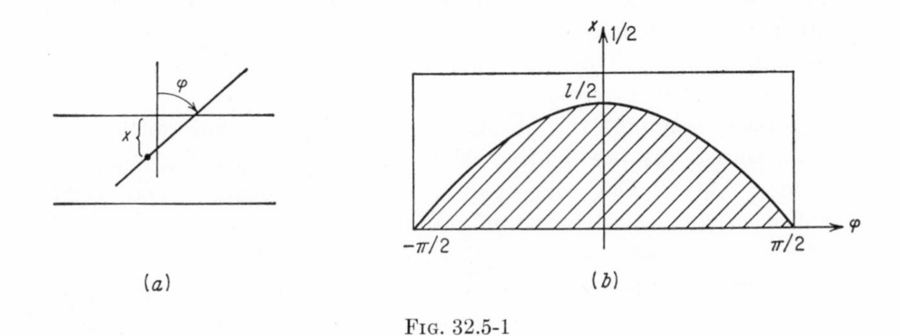

FIG. 32.5-1

where they help to give a general idea of a situation, but they are of less value if accurate results are desired.

As another example of this remark, consider the following case history. A problem was proposed in the early days of computing machines which in the original analytic form would have taxed even the fastest of modern machines, but a little probing uncovered the fact that it was really an ion accelerated by an electric field and trickling through a gas. A Monte Carlo simulation using 10,000 collisions produced a distribution plot of the velocity across and along the field. After the physicist stopped complaining about the low level of accuracy, he said something like, "Hmm. That suggests an elliptical maxwellian distribution, but a bit off center. Hmm." And with that he had the clue to how to attack the problem in an analytical fashion, and that was the end of the numbers found by the simulation.

It is unfortunately true that most of the time the Monte Carlo approach is used only after other methods have been exhausted, and in these circumstances it loses some of its advantages.

EXERCISE

32.5-1. Describe a program to simulate Buffon's needle problem on a digital computer.

32.6 ANOTHER ILLUSTRATION OF MONTE CARLO METHODS

Usually the basic process itself has a random element in it which suggests the use of random numbers to simulate the process, but sometimes, as in the Buffon needle example, it is necessary to find an alternative formulation of the problem which does contain a random element. (Some people would restrict the name Monte Carlo to these cases only.) The following example illustrates another use of the random approach.

In a psychological experiment a team of five people were seated around a table and given a problem to solve as a group. The main part of the study concerned their behavior as the channels of communication among them were altered. The results showed that certain organizations of the channels gave better results than others.

But there remained the question, "How well did they do?" What was measured was the relative relationship of various communication systems. This finally suggested asking the question "If they had behaved in an entirely random fashion, what would have been the results?" In a sense, the random study gave an absolute measure of their success at organizing their work. The random case also reflected more clearly the structural effects due to the various channels available.

Thus sometimes the random approach, although in no way a part of the model of the system being studied, can nevertheless contribute to an understanding of the working of the model. We did not believe that the people acted in a random fashion, but the question "What would have happened if they had?" sheds some light on what was observed.

32.7 SWINDLES

In view of the popularity of the colorful words "Monte Carlo," we have adopted the colorful word "swindle"[1] to describe the use of correlated variables in Monte Carlo calculations. The proper use of correlated (especially negatively correlated) variables can greatly reduce the amount of computing to be done in a Monte Carlo simulation.

As has so often happened in discussing various topics in computing, we soon find ourselves deep in statistics, and we have to say that the topic lies outside the domain of an introduction to computing. But the importance

[1] Apparently due to Prof. J. W. Tukey and used to describe *any* method for speeding up a computation by going to an equivalent problem.

of swindles, when they can be made to work, is so great that a few words are necessary. As a single illustration, Hammersley and Morton[1] describe the use of negatively correlated variables in the Buffon experiment. They first use two needles rigidly fixed in the form of a cross and show that, allowing the labor to toss and record the intersections of the cross with the ruled lines to be considered as twice the labor of tossing and recording for a single needle, the efficiency gain in labor is 12.2; that is, to get the same accuracy we can expect to do 1/12.2 times as much work. When they extend this to three needles crossing at equal angles and allow three times the work, they find a gain of 44.3, and for four, a gain of 107.2. If the three needles form an equilateral triangle, similar results can be obtained. They go on to show that even greater gains can be obtained by using known devices of statistics. Thus, before starting a Monte Carlo computation, it is advisable to consult a good statistician to see how he can help.

[1] J. M. Hammersley and K. W. Morton, A New Monte Carlo Technique: Antithetic Variates, *Proc. Cambridge Phil. Soc.*, vol. 52, pt. 3, pp. 449–457, 1956.

CHAPTER N + 1

The Art of Computing for Scientists and Engineers

N + 1.1 IMPORTANCE OF THE TOPIC

It is unusual in a book on computing to include a chapter on the vague, general topic of how to approach and solve problems, using computing machines. The title of the chapter is itself ambiguous; it could refer to the art as practiced by a person computing for scientists and engineers, or it could refer to the art as practiced by scientists and engineers; in fact, it is meant for both.

The subject should not be taken lightly just because at times it is opinion rather than established fact. It seems to the author to be more important than many of the specific results discussed in other parts of the book. At present, the subject is more an art than a science, but this situation is rapidly changing because the presence of computers themselves has made possible the mechanization of many processes once believed to require human thought, and active research is now going on in this important area. The more we learn about how we solve problems, the more work we shall be able to shift from ourselves to the machines. The art of solving problems, using computing machines, is of interest in its own right; it can also help the user in many situations and greatly increase the value of the machine computations now being done.

Most scientists seem to have been afraid to explore what is generally considered to be the creative process of discovery, but there have been some notable exceptions. Among the mathematicians, "The Method" by Archimedes is one of the earliest examples, and we now have the recent classic "How to Solve It" by Pólya [36]. Both, however, are concerned with the attack on well-formulated problems, whereas we are concerned with a larger framework of vaguely defined problems from which we hope to extract results to fit an equally vaguely defined situation. The motto of the book

THE PURPOSE OF COMPUTING IS INSIGHT, NOT NUMBERS

illustrates our broad frame of reference.

N + 1.2 WHAT ARE WE GOING TO DO WITH THE ANSWER?

Pólya in his book "How to Solve It" emphasizes the importance of understanding the problem. The present author, from many years' experience in computing for others, has found that usually the first question to ask is "What are we going to do with the answers?" Will the answers computed actually answer the questions that should have been asked? Do we need all the answers? Do we need more? Would something else provide a better basis for insight?

In order to answer some of these questions, typical answer sheets can be imagined and then examined for their usefulness. More times than one would expect, the requested answers will not answer the needs of the research project. The original request may have been to get the answers to a set of simultaneous equations. Sometimes this is all that the computing can give, but many times other items, such as the difficulty of solution, can add to the understanding of the situation being examined. Further, what is to be the measure of accuracy: accuracy in the unknowns, the residuals, or the change in the problem, to name but a few? Were the simultaneous equations necessary? Would an alternative formulation give more insight?

Before going on in this vein, let it also be observed that one cannot expect the problem proposer to know exactly what he wants. In research and in many stages of development, it is in the nature of the process that we do not know exactly what we are seeking. Indeed, it may be said, "In research, if you know what you are doing, then you shouldn't be doing it." In a sense, if the answer turns out to be exactly what you expected, then you have learned nothing new, although you may have had your confidence increased somewhat.[1]

Trite and obvious as the remark may be, it is important to know what you are seeking. It is less well understood that you should also plan your work to increase the chances that an unusual observation will be found. It is usually worth adding a small amount to the total machine time if in the process many side checks on the model being studied can be included. Many of the greatest discoveries were made by prepared minds making a chance observation and realizing the importance of it.

[1] P. Debye: "If a problem is clearly stated, it has no more interest to the physicist."

Thus, even though it puts a little extra burden on the output equipment, it is well to include a few wisely chosen numbers beyond the bare minimum when planning the output formats.

In summary, while there are exceptions, it is a good general rule to begin a problem in computation with a searching examination of "What are we going to do with the answers?" An active, imaginative mind can make great contributions to the whole research problem at this stage, and a dull, lazy one can prevent any real insight from emerging from all the hours of machine time used to get the obvious numbers.

One of the most common mistakes in planning a problem is to request too many answers. This is particularly true in problems having many parameters. In such situations, what is generally needed is a good statistician who understands *the design of experiments* theory. Very often he can plan the search in such a fashion that only a small fraction of the original cases need be solved. The sheer volume of answers can often stifle insight.

N + 1.3 WHAT DO WE KNOW?

Having decided, tentatively, what we expect to get out of the computation, we next ask "What do we know?" What information do we have? What is the input? Are we putting in all that we know? For example, if the answer is known to go through the origin, have we included that fact in the input?

Again we remind the reader that Pólya emphasizes the importance of understanding the problem, but again he is considering mathematically formulated problems and is supposing that there is a complete statement. In the application of mathematics this is, of course, not the situation, nor is it true even in mathematical research. Often further probing of the situation being explored will bring further information to light. In the least-squares fitting of a polynomial in Sec. 17.13 we first ignored the fact that the curve should go through the origin, and as a result we had to do the problem over again. When we used this information, we found a more satisfactory answer.

Sometimes it is difficult to include all the known information in our formulation. Thus, in the above case, we knew that the first term had a positive coefficient. We did not include this fact in the input but saved it for an output check. Nevertheless, it is well to get clearly in mind all the relevant information that is available before going on to the next stage.

Sometimes this exploration of the unknown situation will reveal alternative formulations of the problem, and these in turn may suggest new ideas to be explored. Sometimes we find that unnecessarily restrictive assumptions have been made on the model, and that, with little or no extra effort, they can be removed. In any case, their role should be understood, and checks that will throw light on the validity of some of the assumptions should be included in the computation. Thus the exploration of the input may reveal new or different output requirements.

N + 1.4 ARRANGING THE COMPUTATION

Only after we have clearly in mind where we are and where we want to be should we turn to serious work on the question "How do we get from here to there?" This is the domain of Pólya's book; all his remarks are relevant, and we assume that the reader is familiar with them.

Although a problem is posed as one suited to a computing machine, the necessity for using a computer should be questioned. An analytical answer is often very much superior to a numerical result, and sometimes, even when it is more difficult to compute than the original problem, the error estimates may be made much more accurately. Our exploration of what we know about a problem may produce one or more formulations of the problem, some of them merely simple mathematical transformations of the same problem and some, perhaps, quite different. It is expensive to explore a machine-computation plan for all of them, and some initial choices must be made. As a general rule (with many exceptions) the closer the mathematical statement to the fundamental concepts of the field the better, always assuming that scale transformations have been made to make the equations dimensionless. Fancy mathematical transformations often introduce difficult computational situations.

The plan of computation adopted should make use of as much of the initial information as can be included conveniently. The various mathematical approximations made by the formulas used should be in harmony with the model adopted. The effects of sampling should be examined.

The plan of computation should include plans for checking both the coding and the results. This point is too often overlooked; thus we recommend asking the questions "How shall I know if the answer is the one that is wanted?" and "What tests shall I apply or have the machine apply?" It is essential that some redundant information be computed or found from other sources, so that some checks can be made. It is the experience of the author that a good theoretician can account for almost anything produced, right or wrong, or at least he can waste a lot of his time worrying about whether it is right or wrong.

N + 1.5 ITERATION OF THE ABOVE STEPS

We have acted as if these three stages can be completely separated, when in fact they are often interrelated. Nevertheless, it is well to keep the three stages clearly in mind and iterate around them as new information at one stage sheds new light on another. It is the author's experience that rushing into the stage of arranging the details of the computation is the most common mistake. This is especially true of the computer expert, because there he feels at home and can show his skill. But all his skill is wasted if he works on the wrong problem or produces numbers that do not answer the real questions.

It is, of course, difficult to be expert enough in a particular field to ask suitable questions about the importance of what is being computed as compared with some other things that might be computed instead. But there is an art in asking questions. Socrates claimed not that he knew truth but that he knew how to ask the proper questions of a man and draw the truth from him. He spoke of himself as a midwife. And it is in somewhat the same way that the consulting computer expert must approach the man with a problem. In the final analysis, the man with the problem must make the choices and do the research, but well-chosen suggestions from the outsider can help to clarify the nature of the choices and aid in the decisions that must be made at many stages of the work.

N + 1.6 ESTIMATION OF THE EFFORT NEEDED TO SOLVE THE PROBLEM

In any reasonably mature science, it is necessary to make estimates as to what will happen *before* spending time and money. In a sense, the more mature the field, the more accurate the estimates. Judged by this standard, computing is in an elementary state. Often only the poorest estimates are available.

Some of the estimates that should be made are the following:

1. Will the roundoff errors be serious, and if so, to what extent?
2. Will the interval spacing be adequate?
3. If there is an iteration process, how many iterations may reasonably be expected?
4. How much time will it take to code and debug?
5. How will the answers be checked to assure that they are right?
6. How much machine time will be required?
7. When will the answers finally be available?

A glance at these questions will convince the average computing expert that the present state of the art leaves much to be desired. This is no reason for not trying to make estimates that are as realistic as possible. Furthermore, proper organization of the computer facility can greatly improve the estimates made for questions 4 and 7. The provision of monitor systems, automatic coding systems, debugging systems, and a short "turn-around time" on the computer are "musts" if these two items are to be estimated accurately.

This book has tried to approach a number of these questions in various ways. In particular, Part III is believed to provide a basis for many estimates. However, many of the approaches in the book are merely tentative starts in the direction of making reliable estimates, and much remains to be done.

A person who aspires to become a computer expert in the consulting field should not dismiss these questions as unanswerable but should recognize that reasonably accurate estimates are the mark of a capable craftsman—estimates that are neither too optimistic nor too pessimistic.

N + 1.7 LEARNING FROM CHANGES IN THE PLAN

It is almost inevitable that as the computation progresses new information becomes available and changes in the plan must be contemplated. But before adopting a change in plan, an effort should be made to understand why the wrong choice was made in the first place. Does the change shed any light on the model being used? Should we still try to get the same kind of answers? Does the change suggest new or different checks on the validity of the model? Can some new insights be obtained either from the failure or from a new plan?

A change in plan should not be hastily patched in but should receive the same careful thought and planning as went into the original plan. As remarked before, if things go as planned, not much can be learned—it is from the unexpected that the great new insights can occasionally come. Thus a situation which forces a change in plan should be regarded as an opportunity rather than a failure. Of course, if the change was due to carelessness or failure to think, then it should be taken as another example of the value of preliminary planning and should be charged to stupidity.

It is all too tempting, when involved in running a problem, to be rushed into making small changes without considering the consequences and the implications, especially if the results have been promised at a certain time. Yet haste at this time can undo much of the earlier careful work. It is well to remember that "the man should be the master, not the machine," and by proper organization of the computing facility much can be done to keep this clear to the user.

N + 1.8 THE OPEN SHOP PHILOSOPHY

If we believe that *the purpose of computing is insight, not numbers,* then it follows that the man who is to get the insight must understand the computing. If he does not understand what is being done, he is very unlikely to derive much value from the computation. The bare numbers he can see, but their real meaning may be buried in the computation.

Eddington has an illuminating story of a man who went fishing with a certain-sized net. When he found that the fish caught had a minimum size, he concluded that this was the minimum size of the fish in the sea; he made the mistake of not understanding how the fishing was done. And so it is with computing; what comes out depends on what goes in and how it is processed. Without an understanding of the processes used, it is likely that effects due to the model used in the computing will be confused with effects of the model adopted by the user when he formulated the problem.

It has further been found that frequently the process of computing sheds great light on the model being computed. Computing is a tool that supplies numerical answers, but it is also an intellectual tool for examining the world.

It is not likely that great physical insights will arise in the mind of a professional coder who routinely codes problems. If insights are to arise, and they are what we most want, then it follows that the man with the problem must comprehend and follow the computing. This does not mean that he must do all the detailed work, but without a reasonably thorough understanding of what the computer is doing it is unlikely that he can either arrange his work to get maximum benefit from the computer or achieve the insights which can and do result from properly arranged computations.

Experience indicates that it is generally easier and better to convert an expert in a given field into a partial expert in computing than it is to try to make a computing expert into an expert in the given field. But if we are to require this, then it falls on the computing experts to make every effort to reduce the burdens of learning and using the computer. Arbitrary rules, special jargon, meaningless forms, changes in the methods and form, delays in access to the machine, all should be reduced to a minimum and carefully monitored to reduce them further when the next machine offers new opportunities to lift the burden of the nonessentials of computing from the outsider.

The study of numerical methods and library routines to be used to increase insight rather than mere machine efficiency is still in its infancy, and it is one of the most important areas of future research. To work in this field requires experience in the use of computing in everyday work. It is, however, a field well worth cultivating.

N + 1.9 CLOSING REMARKS

It should not be necessary to remind the reader that most of the above remarks are personal opinions developed by the author while working in a particular laboratory, and they are not necessarily of universal applicability. But in defense of them, they seem to be grounded in common sense as well as in experience. If the reader does not like them, he should not argue the topic but should try to develop his own. It is important to the progress of machine computation that the intuitive methods that we now use become more clearly understood and reduced, when possible, to explicit formulations suitable for use with a computer.

References

1. Adams, E. P., and R. L. Hippisley: "Mathematical Formulae and Tables of Elliptic Functions," Publication 2672, Smithsonian Institution, Washington, D.C., 1947.
2. Blackman, R. B., and J. W. Tukey: "The Measurement of Power Spectra," Dover Publications, New York, 1959.
3. Boole, George: "Treatise on the Calculus of Finite Differences," republished by Stechert-Hafner, Inc., New York, 1946.
4. Bromwich, T. J. I'a: "An Introduction to the Theory of Infinite Series," 2d ed., Macmillan & Co., Ltd., London, 1947.
5. Buckingham, R. A.: "Numerical Methods," Sir Isaac Pitman & Sons, Ltd., London, 1957.
6. Campbell, G. A., and R. M. Foster: "Fourier Integrals for Practical Applications," D. Van Nostrand Company, Inc., Princeton, N.J., 1948.
7. Carslaw, H. S.: "Fourier Series and Integrals," Dover Publications, New York, 1930.
8. Erdélyi, A. (ed.): "Tables of Integral Transforms," vol. 1, McGraw-Hill Book Company, Inc., New York, 1954.
9. Forsythe, G. E., and W. R. Wasow: "Finite Difference Methods for Partial Differential Equations," John Wiley & Sons, Inc., New York, 1960.
10. Fox, L.: "Mathematical Tables," vol. I, Her Majesty's Stationery Office, London, 1956.
11. Fox, L.: "Numerical Solution of Two-point Boundary Problems," Oxford University Press, London, 1957.
12. Hartree, D. R.: "Numerical Analysis," 2d ed., Oxford University Press, London, 1958.
13. Hastings, Cecil, Jr.: "Approximations for Digital Computers," Princeton University Press, Princeton, N.J., 1955.
14. Hildebrand, F. B.: "Introduction to Numerical Analysis," McGraw-Hill Book Company, Inc., New York, 1956.
15. Hodgman, C. D. (ed.): "Handbook of Physics and Chemistry," Chemical Rubber Publishing Co., Cleveland, 1960.
16. Householder, A. S.: "Principles of Numerical Analysis," McGraw-Hill Book Company, Inc., New York, 1953.
17. Jackson, Dunham: "Fourier Series and Orthogonal Polynomials," Carus Math. Monograph No. 6, Mathematical Association of America, 1941.
18. Jolley, L. B. W.: "Summation of Series," Chapman & Hall, Ltd., London, 1925, and Dover Publications, New York, 1960.
19. Jordan, Charles: "Calculus of Finite Differences," Chelsea Publishing Company, New York, 1947.
20. Kopal, Z.: "Numerical Analysis," John Wiley & Sons, Inc., New York, 1955.

21. Kuntzmann, J.: "Méthodes numériques, Interpolation-dérivées," Dunod, Paris, 1959.
22. Kunz, K. S.: "Numerical Analysis," McGraw-Hill Book Company, Inc., New York, 1957.
23. Lanczos, Cornelius: "Applied Analysis," Prentice-Hall, Inc., Englewood Cliffs, N.J., 1956.
24. Lighthill, M. J.: "Fourier Analysis and Generalized Functions," Cambridge University Press, London, 1958.
25. Milne, W. E.: "Numerical Calculus," Princeton University Press, Princeton, N.J., 1949.
26. Milne, W. E.: "Numerical Solution of Differential Equations," John Wiley & Sons, Inc., New York, 1953.
27. Milne-Thomson, L. M.: "The Calculus of Finite Differences," Macmillan & Co., Ltd., London, 1933.
28. Mineur, H.: "Techniques de calcul numérique," Librarie Polytechnique, Paris, 1952.
29. "Modern Computing Methods," 2d ed., Philosophical Library, Inc., New York, 1961.
30. Muir, T.: "A Treatise on the Theory of Determinants," Dover Publications, New York, 1960.
31. National Bureau of Standards, "Tables of Functions and Zeros of Functions," Applied Mathematics Series No. 37, Washington, 1954.
32. National Bureau of Standards, "Tables of Lagrangian Coefficients," Washington, Columbia University Press, New York, 1944.
33. Nörlund, N. E.: "Vorlesungen über Differenzenrechnung," Springer-Verlag, Berlin, Vienna, 1924.
34. Ostrowski, A. M.: "Solution of Equations and Systems of Equations," Academic Press, Inc., New York, 1960.
35. Pairman, Eleanor: "Tracts for Computers No. 1, Tables of the Digamma and Trigamma Functions," Cambridge University Press, London, 1919.
36. Pólya, G.: "How to Solve It," Princeton University Press, Princeton, N.J., 1945.
37. Ralston, A., and H. S. Wilf: "Mathematical Methods for Digital Computers," John Wiley & Sons, Inc., New York, 1960.
38. Scarborough, J. B.: "Numerical Mathematical Analysis," Johns Hopkins Press, Baltimore, 1958.
39. Steffensen, J. F.: "Interpolation," Chelsea Publishing Company, New York, 1950.
40. Tranter, C. J.: "Integral Transforms in Mathematical Physics," Methuen & Co., Ltd., London, 1951.
41. Watson, G. N.: "Bessel Functions," 2d ed., Cambridge University Press, London, 1952.
42. Whittaker, E. T., and W. Robinson: "Calculus of Observations," Blackie & Son, Ltd., Glasgow, 1946.
43. Whittaker, E. T., and G. N. Watson: "Modern Analysis," 4th ed., Cambridge University Press, London, 1935.
44. Zygmund, A., "Trigonometric Series," 2d ed., vol. 1, Cambridge University Press, New York, 1959.

Index